Gaming Media and Social Effects

The scope of this book series is inter-disciplinary and it covers the technical aspect of gaming (software and hardware) and its social effects (sociological and psychological). This book series serves as a quick platform for publishing top-quality books on emerging or hot topics in gaming and its social effects. The series is also targeted at different levels of exposition, ranging from introductory tutorial to advanced research topics, depending on the objectives of the authors.

Yiyu Cai · Eleni Mangina · Sui Lin Goei
Editors

Mixed Reality for Education

Springer

Editors
Yiyu Cai
Nanyang Technological University
Singapore, Singapore

Eleni Mangina
School of Computer Science
University College Dublin
Belfield, Dublin, Ireland

Sui Lin Goei
Windesheim University of Applied Science
Zwolle, The Netherlands

ISSN 2197-9685 ISSN 2197-9693 (electronic)
Gaming Media and Social Effects
ISBN 978-981-99-4957-1 ISBN 978-981-99-4958-8 (eBook)
https://doi.org/10.1007/978-981-99-4958-8

This Springer imprint is published by the registered company Springer Nature Singapore Pte Ltd.
The registered company address is: 152 Beach Road, #21-01/04 Gateway East, Singapore 189721, Singapore

Foreword

Opening the Door to Education in the Metaverse Era!

It is my honor and pleasure to present to you the book Mixed Reality for Education edited by Prof. Yiyu Cai, Prof. Eleni Mangina, and Prof. Sui Lin Goei. Whether you are an educator or a VR/AR/MR practitioner, this book is worth reading if you want to stay relevant, especially when the Metaverse era is on the horizon.

Educators should have been aware of the digital transformation taking place in education, in which VR/AR/MR are bringing disruptive changes especially in terms of improving effectiveness and equity of education. Even if VR/AR/MR are already used in your practice, this book will still inspire you with tremendous examples on how these technologies can be further exploited, what improvements or impacts can be expected, and any other factors (e.g., safety, security, ethics, privacy) that you should take into consideration.

VR/AR/MR practitioners should generally know that education has been widely recognized as one of the top fields in which VR/AR/MR can be game-changing technologies. But you may still have difficulty in providing solid and convincing reference cases when being asked by your potential clients. This book includes best practices in K-12 STEAM education, tertiary/professional education, special needs education, and cultural, social and museum education. The way of thinking and methodology behind them may even trigger your new ideas of applying VR/AR/MR in other fields.

Let's also talk a little bit about the Metaverse, the latest futuristic topic capturing more and more attention. Metaverse may refer to a kind of experiences in which the outside world is perceived by the users (human or non-human) as being a universe that is actually built upon digital technologies as a different universe ("Virtual Reality"), a digital extension of our current universe ("Augmented Reality" or "Mixed Reality"), or a digital counterpart of our current universe ("Digital Twin"). In a broad sense, metaverse is the advanced stage and long-term vision of digital transformation. There are different metaverses in the eyes of different people, but the biggest consensus on the Metaverse so far is that it will have a profound impact on our daily work, play, and life, across all industries and sectors, reshaping the economy and society for all humankind. I would like to take this chance to applaud Prof. Yiyu Cai et al., who started a book series on Gaming Media and Social Effects almost a decade ago. Such a visionary effort, including this book as the latest part of the book series, helps our industry and society open the door to education in the upcoming Metaverse era.

Please let this book be one of your assistants in education, in VR/AR/MR applications, and in the journey to the Metaverse.

Yu Yuan, Ph.D.
Chair, IEEE VR/AR Standards Committee
President-Elect, IEEE Standards Association
Co-founder, VerseMaker
President & CEO, 0xSenses Corporation

Contents

Introduction

Yiyu Cai, Eleni Mangina, and Sui Lin Goei

Abstract This chapter is an introduction to the book Mixed Reality for Education which is part of the Springer series on Gaming Media and Social Effects.

1 Background

Over the past decade, there has been growing interest in Virtual Reality, Simulation and Serious Games for education application. Almost 10 years ago, since the first book 3D Immersive and Interactive Learning was published by Springer [1], a Springer book series on Gaming Media and Social Effects has been developed. The series includes (1) Simulation, Serious Games and Their Applications [2], (2) Simulation, Serious Games for Education [3], (3) VR, Simulations and Serious Games for Education [4], (4) Virtual & Augmented Reality, Simulations and Serious Games for Education [5], (5) When VR Serious Games Meet Special Needs Education [6]. In particular, the 2019 book on VR, Simulations and Serious Games for Education is among the top used publications on SpringerLink that concerns one or more of the UN Sustainable Development Goals (Fig. 1).

Y. Cai (✉)
Nanyang Technological University, Singapore, Singapore
e-mail: myycai@ntu.edu.sg

E. Mangina
University College Dublin, Dublin, Ireland
e-mail: eleni.mangina@ucd.ie

S. L. Goei
Windesheim University of Applied Sciences, Zwolle, The Netherlands

VU Amsterdam, Amsterdam, The Netherlands

S. L. Goei
e-mail: s.l.goei@vu.nl

Y. Cai et al. (eds.), *Mixed Reality for Education*, Gaming Media and Social Effects,
https://doi.org/10.1007/978-981-99-4958-8_1

Fig. 1 Helping solve the world's greatest challenges

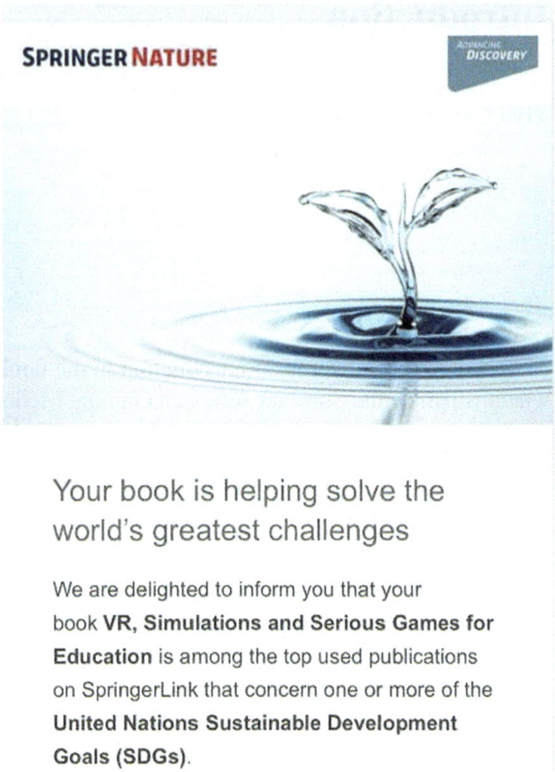

2 Overview of the Book

Mixed Reality for Education is a new book with the series edited by Cai, Mangina & Goei. This book consists of sixteen (16) chapters divided into four different sections including K-12 Science, Technology, Engineering, Art and Mathematics (STEAM) Education, Tertiary/Professional Education, Special Needs Education, and Cultural, Social & Museum Education.

2.1 Section I – K-12 STEAM Education

This section has five (5) chapters. In Chap. 2, Ulbrich, Haas, Elbedewy and Lavicza share their work on aMazing Walk-through Mathematics. In experiments they designed, positive stimulation and high participants' engagement are observed in solving mazes and labyrinths in an interdisciplinary workshop that teaches 3D modeling, AR, and 3D printing skills for pre-service mathematics teachers.

In Chap. 3, Elbedewy, Haas and Lavisca discussed Architectural Models created with Mixed Reality Technologies towards a new STEAM practice with the aid of Dynamic Geometry Software, AR, VR, CAD, and 3D Printing. Students experience inquiry principles allowing them to engage, explore, explain, elaborate, and evaluate.

In Chap. 4, Jansen, Aung, Taib, Cai and van Joolingen investigate the application of prior meta-modeling knowledge to a VR model of a biological process. The goal of this study is to find out whether students can extend prior meta-modeling knowledge, gained using 2D models, and 3D models. The pre- and post-tests in this study show that teaching students about aspects of meta-modeling knowledge and letting them work with these models is an important step in stimulating students' meta-modeling knowledge for both 2D and VR models.

In Chap. 5, Ramnarain and Penn perform a systematic review of pedagogy related to mixed reality in K-12 education. They report that pedagogies can be applied such that the affordances of these technologies are optimally exploited in supporting content understanding, long-term memory retention, improved physical task performance, and increased motivation and engagement for learners. Based on the findings of this review, it is proposed that a guiding framework for pedagogical processes is followed in integrating MR technologies for K-12 teaching and make recommendations on future research that can be pursued in this area.

In Chap. 6, Choon, and Cai share their research on transforming learning experiences through affordances of virtual and augmented reality in secondary school curriculum.

2.2 *Section II – Tertiary/Professional Education*

This section has six (6) chapters. In Chap. 7, Movania, Samad and Raza explore the Mozilla® Hubs® platform for virtual final year project exhibition. Students are asked to use their knowledge and solve a local or global problem under guidance of a faculty member as their final year project supervisor. A virtual replica of the Habib University is designed to showcase their posters and presentations. Virtual avatars are developed for the students, participants and evaluators.

In Chap. 8, Ryan, Murphy, McAuliffe and Mangina look into VRLE-based Teaching for Medical Students. Specifically, VR-Baby is discussed aiming to enhance the understanding of complex topics in Obstetrics and Gynaecology Education. The objective of this study is to describe the development of a 3D model of fetal circulation for a VRLE to teach medical students within specific pedagogical outcomes.

In Chap. 9, Yu, Huang, Ooi and Cai discuss immersive and interactive VR for training of tower crane lifting with a focus on safety and productivity improvement. VR technology is developed for fidelity modelling, immersive visualization, real-time interaction, natural user interface and their applications of tower crane lifting in a simulated construction environment.

In Chap. 10, Xu, Kilroy, Kumar, Iqbal, Mangina and Campbell conduct a systematic review of mixed reality applications in tertiary veterinary education. Intensive search is performed with Embase, PubMed, Scopus, ProQuest, Cochrane Reviews, CNKI and Xplore.

In Chap. 11, Planey and Lindgren report exploring gestural agency in collaborative and embodied rates of change simulations. They describe the theories, design, and analysis applied to a pair of collaborative mixed-reality simulations addressing different scientific content based on the crosscutting concept of "rates of change".

In Chap. 12, Ng, Azam, and Cai examine the efficacy of wearable devices for virtual reality-based aircraft maintenance tasks.

2.3 Section III – Special Needs Education

This section has two (2) chapters. In Chap. 13, Psyrra, Mangina and Treacy present a case study of AR digital literacy intervention for students diagnosed with ADHD. They evaluate AR, digital and conventional literacy programme to examine whether the interventions have a significant impact on participants' literacy skills, and to determine whether their use prevails over the standard learning methods used in school. Data collected during the intervention of the "ADHD AUGMENTED" pilot project, Learning (WWL) online literacy programme or the WWL programme are used in this study. A pre-post assessment methodology is applied for the evaluation of students' performance.

In Chap. 14, Chiazesse, Goei, Tostino, and Mangina discuss the development of AR interactive components for behaviour management within classroom settings. This study is executed within the AR Interactive Educational System (i.e., ARETE) project aimed at investigating – from a user design perspective—the development, implementation, and evaluation of embedding AR into behavioural lessons taught within the Positive Behaviour Interventions & Support (PBIS) framework. PBIS is a school-wide value-based approach for promoting prosocial behaviour by providing a framework for implementing evidence-based interventions that contribute to a safe, positive, and predictable school climate.

2.4 Section IV – Cultural, Social and Museum Education

This section has three (3) chapters. In Chap. 15, Eriksson and Sunnerstam present a new framework for learning patterns and social presence in VR enhanced learning. Existing concepts and frameworks from both research on virtual reality as a media and research on teaching methodology are combined in this study.

In Chap. 16, Iqbal, Xu, Nallur, Scanlon and Campbell examine Security, Ethics and Privacy Issues in Remote Extended Reality for Education. The study focuses on a better understanding of the human value in terms of XR for learning purposes in the remote setting with more responsible design and usage.

In Chap. 17, Gu, Li, Ye, Wang and Luo describe an Immersive Learning Environment to improve User Experience in Science Museums. The study proposes a construction framework for science museums to systematically build immersive learning environments. Force feedback technology is embedded in the work to enhance the haptic perception and increase their level of immersion.

References

1. Cai, Y. (ed.): 3D Immersive and Interactive Learning. Springer, Singapore (2013)
2. Cai, Y., Goei, S.L. (eds.): Simulation, Serious Games and Their Applications. Springer, Singapore (2014)
3. Cai, Y., Goei, S.L., Trooster, W. (eds.): Simulation, Serious Games in Education. Springer, Singapore (2016)
4. Cai, Y., van Joolingen, W.R., Walker, Z. (eds.): VR, Simulations and Serious Games for Education. Springer, Singapore (2019)
5. Cai, Y., van Joolingen, W.R., Veermans, K. (eds.): VAR, Simulations and Serious Games for Education. Springer, Singapore (2021)
6. Cai, Y., Cao, Q. (eds.): When VR Serious Games Meet Special Need Education. Springer, Singapore (2021)

Amazing Walk Through Mathematics

Eva Ulbrich, Ben Haas, Shereen ElBedewy, and Zsolt Lavicza

Abstract Labyrinths and mazes, formations with either unicursal or multicursal pathways, are based on mathematical concepts as well as design features. Both have been part of human culture for thousands of years. In recent experiments, we have observed positive stimulation and high participants' engagement in solving mazes and labyrinths in an interdisciplinary workshop that teaches 3D modelling, augmented reality and 3D printing skills. These technologies require skills in science, technology, engineering, art and mathematics (STEAM) which can be trained by this activity. The workshop was tested both online and offline at ArsElectronica at Kepler's grounds and in further workshops. It was used in courses for pre-service mathematics teachers aiming to teach them about 3D modelling and 3D printing. Participants aged five to the age of retirement experienced the geometrical attributes of mazes to develop an understanding of mathematical concepts such as mirroring, rotation, translation and algorithm basics. The created mazes from this workshop are solvable, shareable and experienceable using mobile devices.

1 Introduction

School activities that integrate the student's environment out of the classroom into teaching aim to develop skills and strategies in students which can help them solve problems in the real world later in their lives. Such activities can encourage and motivate students to explore skills in Science, Technology, Engineering, Arts, and Mathematics (STEAM). By understanding the purpose of these skills in interaction with their personal life, students are encouraged to explore these STEAM skills. New technologies such as 3D modelling and 3D printing (3DMP) or Augmented Reality (AR) can cultivate problem-solving skills, improve transitions between virtual concepts

E. Ulbrich (✉) · B. Haas · S. ElBedewy · Z. Lavicza
Johannes Kepler University, Altenbergerstraße 69, 4040 Linz, Austria
e-mail: eva.ulbrich@gmx.at

Z. Lavicza
e-mail: zsolt.lavicza@jku.at

Y. Cai et al. (eds.), *Mixed Reality for Education*, Gaming Media and Social Effects,
https://doi.org/10.1007/978-981-99-4958-8_2

and concrete physical objects, and open new possibilities to teaching these skills and transport concepts from, for example, mathematics [10, 24]. They also can help connect problem solving strategies taught in schools to students' daily lives for additional motivational boost and they help in preparing students for future challenges as having skills in such technologies will become more and more important in professional careers [18, 40]. One example of connecting skills from class to an outside-of-school environment and virtual and physical worlds can be using emerging technologies like AR or 3D printing. As some teachers struggle with using these new technologies, it can be useful to create exercises that all STEAM teachers can use, share and that might even be helpful to teach them the use of these technologies in an easy way. Kösa and Karakus [22] found that modelling with Computer Aided Software (CAD) can help develop spatial visualisation skills and thus it seems suitable to use 3DMP to develop visuospatial skills in STEAM teachers. AR can be beneficial for students to learn about complex technical interrelationships and visualise them [10] We therefore created an exercise that should combine the possibility to get familiar with these technologies in various settings as well as provide an activity for pre- and in-service teachers to be used for their STEAM education.

Combinations of representations promise to work well in particular when teaching STEAM content like virtual representations such as simulations and physical representations like hands-on activities []. Connecting not only outside and inside of schools, but also virtual and physical environments is part of studies by [12, 13]. They started creating tasks for AR and 3DMP using the GeoGebra platform offering the possibility to create geometrical mathematical objects and much more which has gained in popularity among STEAM teachers who are aiming to connect digital and physical worlds [12–14]. For teachers to adapt to new technologies, research indicates that they not only need to have access to technologies [15, 27], but they also need to be convinced that the extra effort required for adapting a new technology is beneficial for their students [21, 38, 41]. As previously stated, 3DMP is not as widely known among teachers as virtual and augmented reality is. The study of [39] investigated the familiarity of teachers in the US with new technologies, only about half of them were familiar with VR (48%), fewer knew about AR (40%) and even fewer teachers knew about 3D printing (24%) which shows that teachers could benefit from support to learn about this technology [39].

The use of technologies for teachers has to fulfil certain prerequisites. Foremost, the technology and resources for this technology need to be available and sufficient infrastructure needs to be provided and the technology needs to be usable [15, 38]. Another important influence is the perceived ease of use and perceived usefulness of a technology [4, 19]. The teacher's use of technologies also can vary from school to school in relation to teacher beliefs, school culture, self-efficacy and the knowledge of a teacher about a technology [7]. Technologies can even be perceived as too complex and to raise the workload too much, they can avoid integrating complex technologies such as robotics into their classes, even though it can be beneficial [20]. Khanlari [20] found that for some teachers, it is too cumbersome to solve technical problems and frequently having to solve them can be discouraging. Courses for a successful technology integration need to support teachers both with technology knowledge as well as pedagogical knowledge [28].

Using new technologies in lessons that are not usually a part of everyday teaching processes, such as 3DMP and AR, contains characteristics of an interactive lecture where students are invited to solve more open questions in sometimes small groups in contrast to more traditional lectures [30, 33]. As mentioned, to use this technology, teachers need to know how 3D printing machines work as well as how to get or create 3D printable models. This knowledge in 3DMP helps them support their students in developing spatial visualisation skills [11, 42]. It can bridge virtual and physical realms, can be therefore both subject-general and subject specific, can be used to create 3D models for visualisation purposes and can be combined with more techniques like painting something which helps to foster cross-disciplinarity, flipped classroom learning and hands-on-approaches [35].

Since labyrinths and mazes can be seen as puzzles with mathematical attributes and have been deeply rooted with human culture over thousands of years, we chose this exercise to motivate participants to engage with mathematics. We chose these puzzles due to their motivational and cultural aspects and used the fun of interacting with them inside and outside of classrooms to teach mathematics, other STEAM related skills and the use of certain technologies that might help future teachers.

We aimed to create an exercise for pre-and in-service teacher training that can be used as workshops and lectures in STEAM education. Those should be motivational and enable learning skills related to multiple aspects of STEAM. Moreover, it should be exciting and straightforward enough to be used as a STEAM exercise with students later. We created a teaching resource that can serve as a lesson plan for the teachers to use in their own practices. The resource was structured as a workshop or a set of activities for participants. Thus, these activities could be used by teachers as future teaching resources and as reference points.

Other goals of the are that it should connect physical and virtual worlds as well as connect learning about STEAM topics to out of classroom activities. It should also be usable as a low level introductory activity to learn about the features of technologies such as 3DMP and AR, while meeting teaching goals of the maths curriculum in Austria with younger students with a special focus on algorithmic- and spatial thinking.

The developed workshop to create mazes and labyrinths based on mathematical features uses emerging technologies such as 3DMP and AR with GeoGebra as the basis to help share and experience the created work, as a challenge or as a physical object. The workshop was created as an online and offline experience for the ArsElectronica festival 2020 during the Covid19 pandemic for an international audience. The requirements were that the activity had to be usable in a large outdoor space, in private gardens, living rooms or small office spaces. Our developed activity can be created using GeoGebra, shared as a GeoGebra resource and solved on screen as well as on mobile devices outdoors. It was used and tested as an exercise in 3D modelling for AR and 3D printing and also part of an online puzzle game including other resources such as physics or chemistry experiments.

We will first offer a brief overview of labyrinths and mazes and how they are connected to human culture and STEAM-related skills. Then, we will explain how the creation of mazes can help with specific learning goals in mathematics in Austrian schools, and will explain each step of the activity so that pre- and in-service teachers might recreate the exercise themselves. Finally, we show different settings in which we used this activity with which we investigated its uses. We specifically wanted to find out whether the activity is suitable for multiple age groups for teaching such as in teacher training, and whether technologies like AR and 3DMP can be learned by the activity to be used for training teachers.

2 Labyrinths and Mazes Connected to Human Culture

Labyrinthic decorations occasionally adorn houses in Greece and mazes are part of gardening motives such as in Schönbrunn in Austria, as a decorative element on Hindu, Buddhist, Sikh and Muslim holy sites, and can also be found as graffiti. This hints to the fact that labyrinths and mazes have permeated through human history and culture for centuries. The term "Labyrinth" is commonly used for something that is in a confusing arrangement of paths, or for structures that writers of antiquity like to describe in their manuscripts [26]. Some might call the walk-through a modern supermarket or furniture store a labyrinth or maze to lead people pass by as many things they might want to buy as possible. Others call situations with many possibilities that might not all be considerable a maze of decisions [5]. These intrinsic attributes of labyrinths and mazes render them appropriate as a basis for puzzles and challenges which is also referred to in the tale of the Minotaur of Knossos, the palace from where King Minos ruled Creta. This Cretan labyrinth can be found on old coins as seen on Fig. 1 dating back to a period when Greece used this form often and the example of it can be seen on the Knossian Coin. It also utilises concepts from mathematics and informatics resulting in entire book chapters chapters revolving around the mathematics of the setup of the Knossian labyrinth created by King Minos as well as how labyrinths can be used as an entertaining exercise for mathematics [2, 6].

Saward [31] explains that classical labyrinths can be found carved in rocks and were found to date back about 4000 years and almost always had the same form-a round shape with a crossed centre from which a path goes back and forth until one end of the cross is reached. Later, labyrinths were popular in the Roman Empire from 200 BC, where mosaics were created showing not only a pathway or ornament, but were also depicted as fortified cities. Labyrinths can be found throughout the Roman Empire in buildings and on floors as artful decoration. In mediaeval Christian Europe, they represented the curvy way to salvation [31]. In other cultures, they were used as ceremonial pathways of other religions, protections and traps, and were also used directly in cultural activities as games or by forms of cultural dancing [31].

In comparison, mazes are much younger than the classical labyrinth and date back about 500 years. They were first developed in later mediaeval periods, often as

Fig. 1 The coin of knossos with a classical labyrinthic shape (Reproduced from [32])

garden mazes [31]. These garden mazes frequently stood as puzzles for entertainment. They were mentioned in literature as a romantic hideout where lovers would have a rendezvous in the "The Passionate Elopement," by Compton Mckenzie in 1911 [26]. This younger variation of a labyrinth generally is developed to entertain, as a puzzle or game.

Mazes and labyrinths can either have an exit that needs to be found, or only an entrance where it begins or a person enters the maze, respectively. The main difference between labyrinths and mazes is that the labyrinth is always a more or less spiralic pathway from one entrance to an end of a path, while a maze usually contains two or more different pathways requiring more than just patience to solve as can be seen on Fig. 2. Mazes therefore contain much more decision-making and complexity which allows for a multitude of usages. They can be solved by following certain rules such as the Pledge algorithm where choosing an arbitrary path at each crossroads and checking by counting the edges if a person is at the same spot again, can lead to the exit or a special spot within a maze like for example the centre, while labyrinths do not require any special strategies to find the end or the exit of it.

Mazes and labyrinths are integral in architecture, in art, as decorations and gardening features, and can be motivational fun experiences. However, they are not only a simple fun experience deeply connected with human culture. They contain mathematical features and can be used as a tool to teach awareness for mathematics, to exercise computational and algorithmic thinking and problem-solving skills. Mazes in particular can be solved using algorithms that do not have to be programmed, but can simply be written and then tested [9]. This makes labyrinths and mazes valuable as they can also be a gateway to teach the usage of emerging technologies.

Labyrinths and mazes are usually created based on mathematical principles e.g. geometries and can be created using algorithmic rules. There are multiple types of maze designs, such as with more or less circuits, single or multiple groups of twists

Fig. 2 The differences between mazes **a** and labyrinths **b** a labyrinth has one pathway whereas a maze has several choosable pathways and thus more complexity

and turns, and four axes of symmetry. Previous research indicates that working with the creation of labyrinths and mazes can contain many opportunities to experience computational thinking, combinatorics, reasoning and to explore results of geometrical operations [36]. It can be educational to experience various geometrical features such as mirroring, rotations, multiplications of a feature, and scaling in two dimensions. However, the dimension of a challenge such as solving a maze can also be added to the experience [8].

3 Connections to the Austrian Mathematics Curriculum

Similar to many other mathematics curricula, the Austrian mathematics curriculum introduces spatial thinking in 2nd grade, or 7 to 8 year-old students. It is a specific goal for this age to train students in spatial orientation within their surroundings and to understand connections between two dimensional and three dimensional shapes like squares and cubes. The aim is that the students develop a feeling for sizes and for their spatial position in relation to an object.

This perception of space is also a topic later in the Austrian mathematics curriculum for older students in slightly higher classes. The curriculum asks mathematics teachers of the 3rd and 4th grade, with students' ages 9 to 11, to motivate the students to investigate first steps in geometry and quantity theory. The curriculum suggests that they should start gaining experience in these topics by inventing games, carrying out strategy games, and recognising connections between moves towards the first steps of optimising solutions. Drawing maps is one activity mentioned in the Austrian curriculum that can train these optimisations.

In geometry, students should start to interact and understand relations of spatial positions and location relations of objects. Concepts such as parallel, intersections, right angles, as well as describing geometric figures by bounding surfaces, edges and corners should be learned. The focus is especially on squares and rectangles and assembling and disassembling them.

Playful design with solids and surfaces is also an activity teachers should do with their students by various materials such as folding paper, creating mosaics, drawing symmetrical pictures on grids, and investigating edges and forms by performing tilting movements, removing boundary surfaces, and creating symmetrical pictures with for example mosaics.

Many of these actions can be found in the described exercise which makes it extra valuable for use with younger children. Fenyvesy et al. [8] describe how simple mazes, such as the so-called "Greek Key", can be created by cutting and tilting parallel lines by 90° (Fig. 3).

More complex forms like the labyrinth found on the coin of Knossos representing the Knossian labyrinth hosting the Minotaur, a monster guarding the labyrinth, can also be created by performing these cutting, tilting, and reassembling steps of geometrical figures Fenyvesy et al. [8] (Fig. 4).

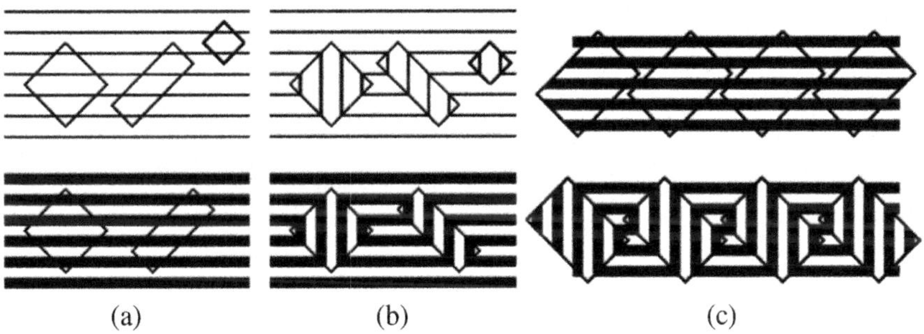

Fig. 3 Examples of **a** rectangular and **b** square shapes creating **c** labyrinthic structures by tilting 90° (Reproduced from [10])

Fig. 4 **a** Changing the classical labyrinthic form created **b** by tilting two squares by 90° (Reproduced from [10])

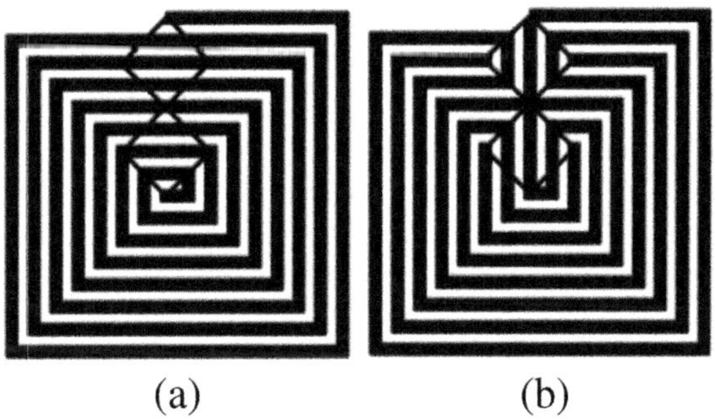

These activities contain many learning goals found in the mentioned part of the Austrian curriculum indicating that it can be a good tool for teachers to use in their classes. The mathematical learning goals previously mentioned such as drawing on grids, tilting geometrical objects, playing with symmetries and creating mosaics of geometrical objects can be trained as well by creating mazes. Creating an idea for a labyrinth or maze can be seen as an exercise of game creation with certain rules and a goal where mathematical as well as logical thinking can help find ideas for solving the maze or labyrinth in an optimised way. Various strategies might be developed by students creating and solving such a labyrinth, helping them understand connections between moves and optimization problems.

Moreover, the creation process contains many aspects of the curriculum mentioned. If tilting rectangular shapes and squares, labyrinths can be created in artistic ways. When solving other students' mazes and getting inspired for a maze by mathematics and art, geometrical forms might be not only the outside boundary of a maze but also used to create first ideas of an own maze or labyrinth. Thinking about the bases of labyrinths can be seen as highly mathematical and creative, as well as artistic process as all these skills are needed for many kinds of future problem-solving tasks, especially in mathematics.

In addition to all these concepts, not only staying in two dimensions but using cubes and cuboids can add even more depth to the exercise. While three dimensional objects usually are part of classes for students of different ages depending on complexity, the way from a square to a cube or a rectangle to a cuboid can be introduced in a playful way by modelling 3D objects out of 2D drawings. Three dimensional objects such as cuboids and cubes should be introduced to students in a simple way from 2nd grade on, so three dimensionality in itself is nothing new to young students, however, new perspectives such as looking at geometrical objects from the inside might helps to understand three dimensional objects better. It is a specific goal in training of students of this age to teach spatial orientation which can be done by letting students investigate three dimensional mazes and labyrinths. Such 3D objects can then be observed and investigated by students using for example technologies such as AR or 3D printing.

4 Benefits of Technologies for the Mathematics Curriculum

Technologies that require a more complex and abstract thinking are especially suitable for an age of upper secondary school levels and using technologies usually requires a variety of STEAM skills such as logical and computational thinking. We combine three such technologies which all contain their own aspect of benefits in STEAM lessons.

The use of mobile devices gets more and more common at that age and thus can also be integrated into lessons. Mobile devices can be used for a more mobile and out-of-classroom experience such as doing exercises with Math City Map [3] where one group of students is in the classroom solving problems with another group that walks around outside the classroom utilising and being connected by smartphones.

Another such technology that is easily accessible by mobile devices using GeoGebra is AR. It can be usable by downloading the GeoGebra 3D app that enables modelling and displaying 3D objects. It can also be used to access existing 3D models from other students or pre-prepared models by teachers through setting the maze to a publicly visible setting or even developed and shared immediately using the GeoGebra classroom function. This technology allows for overlaying visual objects in our real life mixing real and virtual environments. This mixture allows users to stay in their real worlds non immersed in a virtual environment but yet experiencing virtual objects in their surroundings. Moreover, it allows users to experience adding and analysing large virtual objects or constructions that are not possible with real physical objects [29].

AR can display 2D and 3D objects and combine a virtual world with a physical surrounding of students which can help understanding the geometrical attributes of 3D objects by looking at them from various angles and having additional information such as scales displayed. GeoGebra AR would even allow for an outdoor mathematical modelling possibility through applying symmetries, rotations, translations or any other sort of mathematical transformations which is a very useful tool for outdoor activities [1]. Game- like experiences in particular can improve student's learning attributes and learning performances in out-of-class-activities [17].

A third technology that requires even more STEAM skills to master such as engineering, informatics, and mathematics skills is 3D printing. According to [16], the technology of 3D printing needs a person to use innovative and creative skills. In addition, it trains the abilities to reflect on one's approaches and requires critical thinking. Computer Aided Design (CAD) tools such as the 3D properties of GeoGebra can create visualisations of abstract concepts like numbers, shapes and three dimensional objects while also calculating them and creating them parametrically with high probability to increase a person's mathematical and especially geometrical skills [16]. This not only means that 3D printing is highly useful for training students of older ages in STEAM skills, but it can also be very helpful in training older people including pre-service and in-service teachers.

Previous research on 3D printed mazes and games has indicated that mazes can be used to create a special kind of game type called transreality games, where virtual and physical worlds can merge into each other [34]. Such games are considered captivating and thus can be useful to teach concepts of topics that are often perceived as boring or hard as for example mathematics or engineering.

5 Creating Labyrinths and Mazes for AR and 3D Printing

Exploring mazes in artwork can help with understanding mathematical concepts like for example algorithmic thinking, basics of geometry, or the visualisation of shapes and can also inspire a participant's own ideas on labyrinths and mazes. A great source of inspiration is the collection of pictures and descriptions found throughout the internet or at labyrinthos [32].

One attribute of a maze or maze-like architecture, a geometrical idea or a repeating element, can form the basis of a maze drawn on square paper. Figure 5a shows part of a mosaic maze from Cremona in Italy [32] and a maze design that was inspired by it on Fig. 5b. Mazes can be created by principles such as mirroring, rules like always turning right, translation, or symmetries. This can train algebraic thinking as it can mean to assemble fractured parts and deductive and spatial reasoning by addressing the challenge in two as well as in three dimensions []. We developed steps for a workshop to create mazes and labyrinths with mathematical and artistic backgrounds to help develop or train these skills amongst a deeper connection to other STEAM related skills such as needed for science, engineering, or communication. The steps reach from drawing a labyrinth or maze to creating an AR or 3D printed visualisation as can be seen on Fig. 6.

Step one of the workshop is to invite participants to observe several representations of mazes that can be found on art and on architecture. Using a squared sheet of paper and a pen, the geometrical attributes of mazes and labyrinths can be investigated and played with by the workshop participants. It is useful to inspire participants and let them observe pictures of labyrinths and mazes by using a picture search in a browser or by letting them surf through labyrinthos.net [32]. The participants should observe, investigate, and formulate what they believe the mathematical basis of the mazes could be. For example, participants could see the principle ‘cross and four

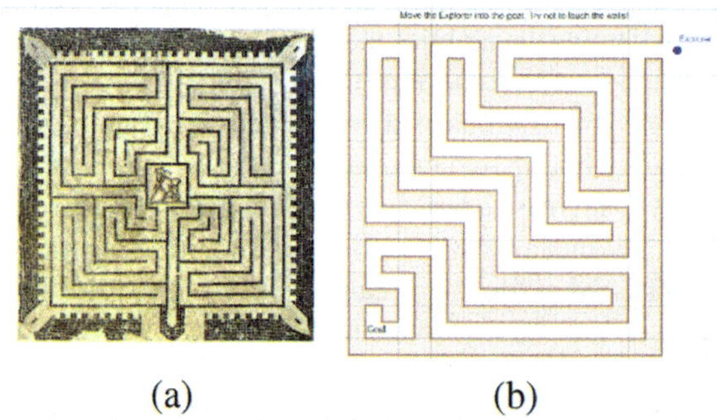

Fig. 5 **a** A mosaic from Cremona/Italy compared to **b** a GeoGebra version in the GeoGebra classics mode of a maze with similar features

Fig. 6 A drawing of a labyrinth and the AR representation of a labyrinth through the GeoGebra 3D app

Fig. 7 Prehistoric classical labyrinth from Greece (Reproduced from [32]

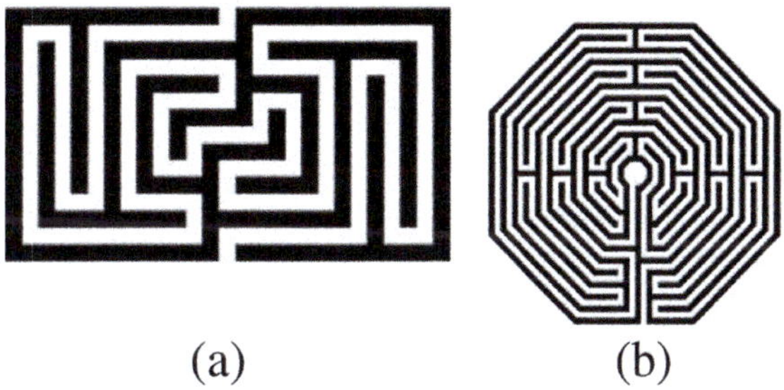

Fig. 8 A pattern on the belfast cathedral, **a** and a labyrinth on the arras cathedral **b** as inspiration-example for step one for participants creating their personal maze (Reproduced from [32]

corners' as described by [36] in Fig. 7 and could observe that the pattern in Fig. 8a is symmetrical, while the one in Fig. 8b is not entirely symmetrical but alternates between short and long corridors. Based on these observations, participants should think about their personal labyrinth or maze and which principles they would like to base it on.

Step two involves drawing a maze or labyrinth using pen and paper. Using a squared sheet of paper and a pen, the geometrical attributes of a maze or 'maze-like architecture', a geometrical idea or a repeating element can form the basis of a drawn maze a participant chooses to create. Physically drawing the maze idea can allow participants to explore their imagination and play with the impressions from art they had observed. In our example, the maze has the shape of an incomplete square by concatenating a wall by a 90° angle counter-clockwise, adding 3 units for every second wall placed and subtracting two units at the entrance wall, see Fig. 9a.

The drafts should follow the rule that a labyrinth has one entry and should consist of one single object. Mazes might have several entrances and a higher complexity and therefore might consist of more than one object. They might follow algorithmic observations, might resemble a form, can be created by tilting parts or can follow no such mathematical principles, but merely investigate squares and rectangles in a playful way. The drafts consisting out of one or more geometrical shapes will be turned into a three-dimensional version.

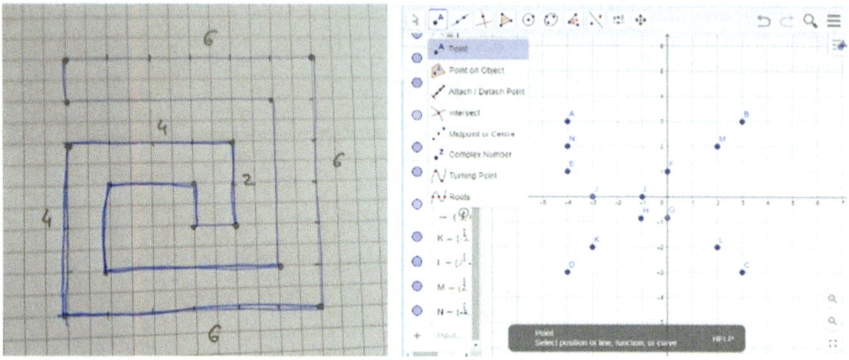

(a) (b)

Fig. 9 **a** A drawing of a labyrinth following certain rules, **b** the edges as dots transferred to GeoGebra

Physically drawing the maze idea using pen and paper serves several purposes. For one, it shows that every three dimensional object needs to have thickness and needs to be enclosed. As all 3D objects have thickness, they need an outer and inner shell and can not be just a line. While in AR this might not be necessary due to its virtual nature, drawing on a grid provides important information when modelling something for 3D printing as seen on Fig. 9. Being aware of 3D objects needing a thickness is especially needed if another CAD programm is chosen as GeoGebra has the ability to produce objects theoretically without thickness, such as lines or points which do not have three dimensions, to give them a thickness for 3D and thus, make them printable.

The second important purpose of drawing a maze idea is the prototypical nature. A participant can try out whether the idea is working and can experience geometrical ideas without any technical support. And thirdly, creating the maze in GeoGebra requires the creation of all corners or nodes and connecting them in a certain order. It is easier for participants if there is a visual aid showing the amount and position of nodes and edges of the maze as in Fig. 9b. It requires a participant to add the corners to the coordinate system as is required in 3D modelling tools.

The third step is to create the maze idea as a 3D model in GeoGebra which requires drawing all corners and then creating a polygon that connects all corners by edges, as shown in Fig. 10b and then extruding the model as seen in Fig. 10b. The participants are asked to reproduce the steps described in a GeoGebra resource to create the 3D model of their labyrinth or maze. GeoGebra has to be opened in the classical view and the points need to be drawn using the point tool. Depending on the maze's complexity, these dots need to be carefully added as seeing a participant's maze in a cloud of dots can be challenging. Next, the polygon tool is selected to connect these dots. This activity can train logical and computational thinking. It also helps to understand that every 3D model requires thickness of the object and can not only

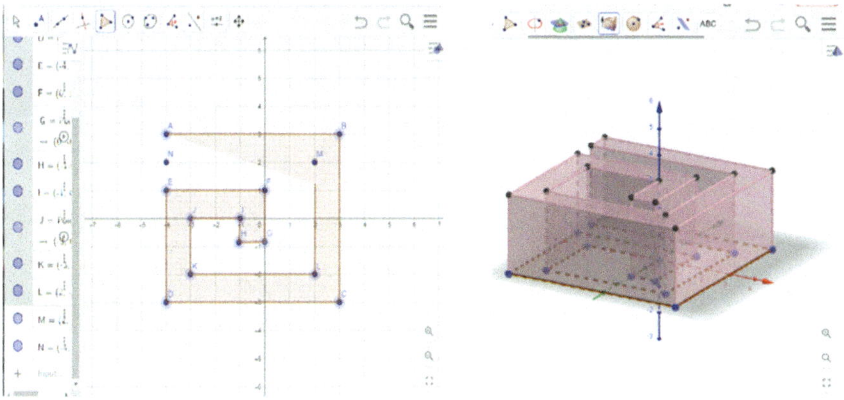

(a) (b)

Fig. 10 **a** The dots in GeoGebra connected by the prism tool and **b** the prism extruded to 3D, ready for investigation in AR or for downloading it for 3D printing

exist from drawing one line. To create an edge that can work as an outer border of a 3D model, the corners have to be clicked one by one until the first corner completes the polygon in the order that resembles the maze which is described step by step at the GeoGebra resource. After creating a 2D polygon maze, the polygon can be extruded to a three-dimensional model. After completing the 3D modelling, the model can be saved and the privacy settings can be set to "public". The resource can then be shared and used by all other GeoGebra users and, thus, the workshop participants. The maze can be experienced on most smartphones after having installed the GeoGebra 3D app.

6 Solving the Maze in AR Mode

Mobile devices are more and more common among students and using them in an educational setting teaches students and others about aspects that go beyond communication and entertainment. Students are often familiar with handling such mobile devices due to their daily use of them which means that using these devices in an educational setting does not require does not require much previous knowledge of informatics skills. Not needing much knowledge about technology is a good opportunity to introduce new technologies in classrooms as a low-level entrance point. As a further step in the workshop, participants that had installed the GeoGebra 3D app on their mobile device as described in a GeoGebra resource can open the labyrinth or maze which they had previously designed in GeoGebra, start AR mode, and walk through their virtual maze.

Tablets provide a bigger screen, but smartphones can also be used and most recent operating systems are supported by the GeoGebra 3D app as seen on Fig. 11a and b. Sometimes, teachers are sceptical about using smartphones in classrooms and

(a) (b)

Fig. 11 **a** Explanation how to navigate through a maze and **b** participants navigating a maze by holding up a tablet in front of them

moving the entire experience to a place outside of it can overcome this fear and add extra motivation for students. Using mobile devices to investigate the previously created mazes turns this exercise into an experience that can also be enjoyed out of the classroom, giving it an additional dimension after the creation and analysis on paper as well as in GeoGebra. With this step, participants can dive deeply into their own creations, observing them from the inside and investigating the labyrinth or maze symmetries and paths, where by doing so they might develop strategies to solve it using problem solving skills such as critical and algorithmic thinking.

Not only critical thinking is developed by using AR, but also informatics skills are improved and the aforementioned mathematics curriculum content is taught. Students can be motivated to train their orientation in space, learn about cubes and cuboids, can connect to historical lesson contexts if given and thus experience a true STEAM exercise.

Scaling the model is possible depending on a selected height and width which enables a participant to walk through the maze. The height and width of the walls can also be adapted by manipulating the size of the walls which can make it easier to solve if the walls are quite low or make it a lot harder if the walls are higher so a participant can not see what lies behind the corner as seen on Fig. 12. Making the mazes and labyrinths smaller and larger, higher and lower and estimating for example the volume changes can directly show the impact of manipulating vectors in geometry. Scaling can be used to enlarge a maze in a way that the path is big enough to walk through. This way, the participants can observe the influence of the choice of vector lengths and thus wall sizes on the resulting object. The size of the model is indicated in the 3D GeoGebra app to give a participant a feeling of how big or small the scaling level is compared to the initially created model.

Using the tablet and walking through the mathematical maze can serve as a testing tool of finding ideas about the physical real life impact of virtual ideas which helps identify areas that could be improved. Narrow areas or unwanted corners can be investigated up close. Participants can then alter their maze and test again whether

Fig. 12 **a** Lower and **b** higher walls of a virtual maze as **c** seen through a mobile device

they want to change it. The possibility to alter the maze can be another motivational aspect for a participant to engage directly in the mathematics behind the mazes. Moreover, it will help the participants to understand possible weaknesses of a 3D model if it is to be 3D printed in the future. Using AR and being able to scale a geometrical object really large to be able to walk through it can move the perspective of a participant directly into the maths making geometry experienceable from a new perspective.

7 Turning the 3D Model into a 3D Print

After creating a 2D polygon maze, the polygon can be extruded to a three dimensional model that can be downloaded from GeoGebra for 3D printing. This model can be created with any 3D printer. Schools often prefer a Fused Filament Fabrication (FFF) printer due to machine costs and maintenance reasons. 3D printing can be a powerful tool to visualise mathematics while requiring STEAM skills such as needed for mathematics, engineering, informatics and even art can be part of trained skills [34]. In addition, it gives students some kind of ownership feeling if they hold their personal virtual idea in their own hands and a smelly, loud and moving machine is always an attention catcher.

Using 3D printing as a visualisation tool adds additional learning perspectives to the experience, as it requires more complex software and hardware than AR on a smartphone that a student or participant owns. First, the model has to be exported from GeoGebra and might need some corrections as sometimes, parts can be under-defined. This can be done by automatic repair tools or manually if uploaded to one of the numerous CAD programs that might suit a participant's knowledge level, such as TinkerCAD. To be correctly modelled or repaired, it needs to be loaded into a so-called slicer software that defines the motor movements of a 3D printer. This can already require informatics and engineering skills, while it might even train language skills as most of these softwares use English.

Slicing a model often reveals additional optimisation opportunities of a model as it is sometimes scaled to large or to small which influences production times vastly. Sometimes, parts are modelled in a way so that a person walking the maze would not be able to pass through later, which can also be found out sometimes by slicing. Altering the model by scaling it or changing the height can contain surprises about production time and thus, intrinsically the impact on an object's volume. Changing the height of the maze invites workshop participants to investigate the differences of the maze's production time and experience the implications of volume changes towards or against the printing time. The maze on Fig. 2a and 2b for example took 6.5 min for a maze of 1 mm height and 40×35 mm bounding box. Scaling the object up by a factor of 1.5 takes over 18 min. It can be interesting and counterintuitive for younger participants that a maze that is not even twice as big can take three times the printing time.

Participants can play with the scaling, can mirror and duplicate their 3D model and can practice mathematical competencies such as scaling, duplicating, mirroring or the concept of vectors by investigating whether the faces of the 3D model all point outwards. It gives them a first connection point between their virtual mathematical maze idea and physical properties of a physical object.

After printing it, they will immediately see whether their changes solved a problem successfully, whether their idea is working and they have the possibility to manually touch and feel, turn and investigate their first virtual and now physical maze idea. The 3D print will add several more attributes for experiencing and investigating the mazes for the participants such as height, weight, texture, and more. Moreover, 3D geometry attributes can be experienced using this approach which is an addition to the geometrical and virtual mathematical attributes. Participants can thus not only use their cognitive inner representation of a mathematical concept, but use senses such as touch or smell to build up an understanding for mathematics responsible for the maze's geometry. In addition, the 3D print can be used in various more personal ways like as a map to investigate all turns of a maze as can be seen in Fig. 13b, or used as a participant's personal pendant if artistic ideas were the basis of a participant's model.

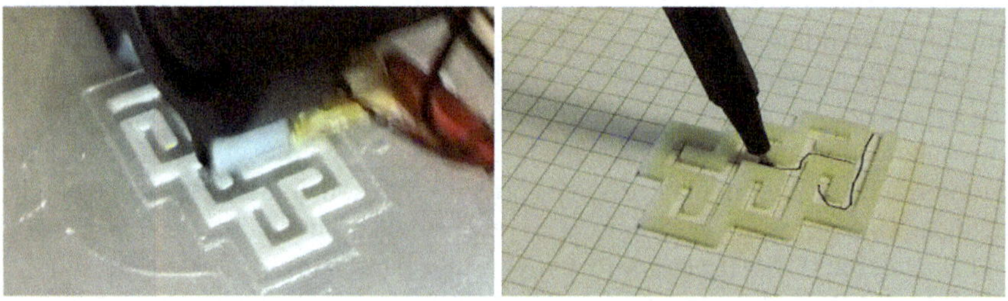

(a) (b)

Fig. 13 **a** The printing process of a maze and **b** the 3D printed maze as a tangible token

8 Testing the Exercise at the Ars Electronica Festival 2020 and in Online 3D Printing Workshop Environments

The initial idea for this workshop was to create an experience that can be used in an online as well as an offline setting due to the Covid-19 pandemic in 2020, since outside activities during summer had to switch quickly sometimes. Therefore, an offline part, an online part and a part of an activity that could even be used remotely was created.

The described exercise on creating mazes was first used as a workshop-like activity at the Ars Electronica festival and then refined with the feedback we got from the participants. We created a workshop activity with steps ranging from getting inspired by mazes to being able to share one's own maze creation with others by 3D printing it or sending an AR link, but due to the increase in infection levels, this idea had to be postponed. It was later used in online workshops and in part during online lectures about 3D printing. The offline possibility to walk through pre-prepared mazes using AR to solve them as some kind of a puzzle remained and the creation process of how to make these was shrunk down to only a theoretical introduction to participants inviting them to do the full workshop later at home.

A second variation of the activity was created as a pure online experience where partners in chemistry and physics created a trail of puzzles that had to be solved one after the other, including mazes, until the price of a gummy bear recipe was reached in case all puzzles were solved correctly. Participants of the offline part were also invited to investigate this online STEAM challenge later.

Participants at the ArsElectronica festival were invited to walk through a virtual maze with a tablet showing the 3D model of the maze, as can be seen on Fig. 14. The 3D model created in GeoGebra could serve as a printable model and also a model in an AR environment. Tablets provided a bigger screen, but Smartphones could also be used.

(a) (b)

Fig. 14 Participants of various ages walking through mazes using tablets at the Ars electronica festival 2020

Over 100 participants took part in these experiments and were given 3D printed maze pendants as a reward. The age of participants ranged between preschool and retirement age and were invited to walk through a large physical maze at the ArsElectronica festival with a tablet with a 3D maze model on it. Participants were challenged to create and solve the maze during the workshop using AR to explore geometrical attributes of mazes and labyrinths.

Younger participants in particular showed motivation and engagement indicated by asking for more mazes to solve after their first maze and they asked about the GeoGebra book to further solve and create mazes later at home.

After refining the workshop-like activity, we wanted to investigate more in-depth whether it was also useful for older students and in-training, as well as in-service teachers and in addition, find hints about the usefulness of this activity to transport information about 3DMP and AR. Thus, we asked a 5-year-old preschool student, a 16-year-old student in upper secondary, one 30-year-old masters student, and two PhD students 30 and 32 years of age respectively, to join in a 3DMP online course featuring this exercise, as seen on Fig. 15.

As the 2.5 h hour long workshop was virtually held using zoom, there was emphasis on the modelling part and the actual 3D printing was observable by a second camera pointing to a 3D printer.

First, basics of 3D printing were explained theoretically and examples were given how this technology can be used. Second, the basics of 3D models were explained in theory and the participants were asked to find good and bad examples of 3D models. Third, 3D modelling tools were presented and the maze exercise started.

One pre-prepared exercise was given to the participants to first see, what to expect, how the 3D application of GeoGebra was downloaded and how a maze could look like. Then, pictures were investigated to inspire the participants of their own personal maze idea. After first drawing it on paper and then modelling it using GeoGebra, one maze was printed out and the printing process was shown to the participants.

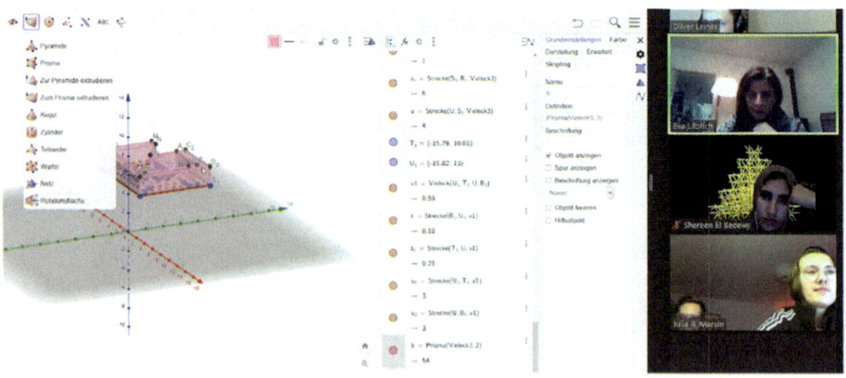

Fig. 15 An online workshop environment to model mathematical mazes or labyrinths in 3D for AR and 3D printing with 5 participants from 5 to 30 years

All of the participants engaged in the process of maze creation and reported that they were inspired to create more mazes and felt more confident with using GeoGebra as a 3D modelling tool, as well as were motivated to use these technologies in the future. The strong engagement of all participants and the motivation they afterwards showed hints to a highly motivational aspect of the workshop, especially for learning complex technologies.

This first workshop was used as the basis for developing a second online workshop focusing on the maze experiences and maze modelling and having 3D printing in the background at a conference to get more data from an older target group. To look at reactions from an older target group, we prepared a GeoGebra book containing all resources at https://www.geogebra.org/m/ekrfpfch and a presentation guiding through all the steps of the process. We then conducted a second workshop which was visited by 10 participants of mostly unknown age-most of them seemed about 40 and above, one told us that she was 29 years old and a high school teacher. They were from various countries and timezones and the workshop this time was only 90 min long. Asked about their motivations about joining the workshop, they said that they were interested in learning fun new tools.

We accentuated various examples of mazes and labyrinths and let the participants search for inspirational pictures on the labyrinthos webpage. Then, we let them try pre-prepared mazes in AR mode and showed short videoclips of the workshop presenters solving the mazes to show the participants how to open and explore such an AR object. After that, we invited the participants to draw their own mazes based on mathematical rules such as symmetries or algorithms using pen and paper. These mazes were then reproduced in a GeoGebra classroom environment. During the workshop, some of the participants joined later or had technical difficulties like slow computers or networks, so some missed out on information which might have been the reason why not all participants created mazes. Due to the shorter workshop time, the 3D printing part was left to only describing it briefly theoretically.

We asked the participants of these and another hybrid workshop to fill out non-mandatory questionnaires after the workshop. From 23 answers in total we asked if participants could use this experience to learn about 3D printing and whether they had suggestions about how this activity can be used in educational settings. The participants that filled out the questionnaire were between 16 and 48 years old with a median of 24 years of age (Table 1).

9 Case Study in Early Childhood

As this activity seems to be intriguing for especially young students as discovered during the first workshops, we highlight the work of a five year old explorer. We invited him to participate in one of the described online workshops, using the maze workshop as the basis to introduce the geometrical aspects of mazes as well as AR and 3DMP in a playful way. He participated in the workshop with four other participants

Table 1 Workshop results

Description	Data	
Age	Between 16 and 48 with a median of 24 years	
Gender	6 male, 17 female	
Country	Greece, Estonia, Egypt, Germany (2), Austria (18)	
Occupation	1 programmer, 3 students, 5 in-service teachers (4 mathematics, 1 physics), 14 pre-service mathematics teachers	
Devices	Named devices: tablet, PC, smartphone, ordinary printer, laser-engraver, UV-engraver, raspberry pi, beamer	
	One devices	3
	Two devices	9
	There devices	8
	More than 3 devices	3
Expectations of the workshop	Quote 1: "I expected to experience or learn something new. I did find out the wonders of Geogebra. My use of it was in it's early stages and it has grown so much. Wonderful!"	
	Quote 2: "Material, size of objects, creation of objects"	
	New/future skill	3
	Complex/concerns	5
	3D/mathematical objects	6
	Material	4
	Hands on/artistic exercise	5
Feeling secure with using AR and 3D printing before/after the workshop	Likert scale 1 = insecure, 5 = confident before the workshop, average 1,8	
	Security rating	Count
	1	10
	2	9
	3	3
	4	1
	5	0

(continued)

Table 1 (continued)

Description	Data	
	0	
	After the workshop, average 3,4	
	Security rating	Count
	1	1
	2	3
	3	9
	4	6
	5	4

and was invited to learn about 3D printing with them with the maze he created to be his first contact with 3D printing.

First, we took some time to investigate pictures of mazes and labyrinths for inspiration about what mazes and labyrinths can look like and which features they can have, as described. After we all took time for our personal doodle development step, he created a 2D maze in GeoGebra with a little help from his mother. Then, he extruded the 2D maze into a 3D version with some more help from his mother as this feature is a bit more hidden, the model was downloaded and his mother sent it to us as a file. Since it was an online workshop and he did not have a 3D printer at home, we created the maze remotely and he was able to observe the printing process by holding up a camera to the printer we used as well as showing him the finished object that is shown in Fig. 16.

Afterwards, he was very keen to solve the maze himself alone using AR mode on the smartphone of his mother. He was very joyful during the experience of solving it and excited when he managed to solve his personal maze himself. His mother

Fig. 16 The printed maze of the five year old explorer

observed that he also had grasped the concept of mazes and was still enthusiastic about it later since he started to create mazes out of building blocks for days after the workshop.

10 Conclusions and Future Outlook

The participants of the activity investigated geometrical and other attributes of labyrinths and mazes. After being inspired to develop their own mazes with pen and paper, they learned to use the basics of 3D modelling in GeoGebra. The created 3D model could be used in an AR setting by the GeoGebra 3D app for mobile devices and as such, could be tried out as a large version as a challenge similar to the garden mazes found in for example the Austrian castle Schönbrunn, as an out of classroom activity. The created mazes were also shared amongst other GeoGebra users or with visitors of the Ars Electronica festival in 2020 during the Covid-19 pandemic. The course was created in mind of having an exercise that can be useful both online and offline, in class and out of class depending on the situation and worked in all settings.

We experienced a massive motivational boost at a presentation in a large festival and in two workshops we held online and offline in settings amongst various age groups from preschool students to in service teachers or lecturers. One of the workshops focused more on the AR part in combination with mathematics while the other focused more on whether 3D modelling and 3D printing could be taught. The experience of the online workshops suggests that the exercise can be used as a simple first exercise to explore 3D modelling in GeoGebra and investigate 3D printing on a screen.

Not only trying out a pre-created maze or labyrinth by other people showed motivational boost effects but also the entire workshop from inspiration, model creation, and investigation using both AR and 3D printing as tools. In addition, we found that it can teach about mathematical concepts such as geometrical objects, tilting, turning and scaling them, creating and solving puzzles and learning about optimisation of paths for younger students. All these skills are part of the Austrian mathematics curriculum and many other mathematics curricula worldwide. Knowledge about technologies such as how to use computers and mobile devices as well as AR and 3DMP will also be valuable for future teachers. We therefore also looked at whether this workshop can be used to teach technologies. Further investigation will go into the 3DMP and AR exercise part for pre-service teachers in full courses. We will use this exercise in a non-mandatory course setting for pre-service mathematics teachers to observe the motivational aspects of the exercise when learning about emerging technologies. We believe that exercises such as this can provide the freedom that we need and motivate mathematics teachers to create their own 3D modelling projects. As technical and time wise constraints can be an issue to this workshop-like activity, we want to find out whether there is some optimisation potential available. Our idea is to separate the creation steps from each other and see which combination can be done differently, quicker, or should even be more prominent. We want to try different parts in different lectures to see if the activity can also be used beyond a workshop setting in small groups.

Following the activities of the 'five-year-old' explorer, being able to create mazes using building blocks adds another perspective for younger students since no technical knowledge is needed to create the maze. As the workshop seems to be a fun activity especially for early childhood, a next step could be to test the setting with this age group in a task based design environment to investigate in depth. It would be interesting to see which conclusions a student can draw from creating, investigating and solving a maze using AR and 3D printing. We are also curious about comparing drawing with pen and paper and using building blocks as the start of the workshop for pre-school and primary school children.

References

1. Ancochea, B., Cárdenas, M.I.: Exploring Real World Environments using Potential of GeoGebra AR. In: Research on Outdoor STEM Education in the digital Age. Proceedings of the ROSETA Online Conference in June 2020, pp. 41–46 (2020)
2. Budd, C. Sangwin, C.: Mathematics Galore!: Masterclasses, Workshops and Team Projects in Mathematics and its Applications. Oxford University Press (2001).
3. Cahyono, A.N., Ludwig, M.: Teaching and learning mathematics around the city supported by the use of digital technology. Eurasia J. Math., Sci. Technol. Educ. **15**(1), em1654 (2019).
4. Cuban, L.: Teachers and Machines: The Classroom Use of Technology Since 1920. Teachers College Press (1986).
5. DeWeerdt, S.: Food: the omnivore's labyrinth. Nature **471**, S22–S24 (2011). https://doi.org/10.1038/471S22a
6. Dudeney, H.E.: Amusements in Mathematics, vol. 473. Courier Corporation (1958)
7. Ertmer, P.A., Ottenbreit-Leftwich, A.T.: Teacher technology change: how knowledge, confidence, beliefs, and culture intersect. J. Res. Technol. Educ. **42**(3), 255–284 (2010). https://doi.org/10.1080/15391523.2010.10782551
8. Fenyvesi, K., Jablan, S. Radovic, L.: Following the Footsteps of Daedalus: Labyrinth Studies Meets Visual Mathematics. In: Bridges 2013 proceedings, pp 361–484. Enschede, The Netherlands (2013). http://archive.bridgesmathart.org/2013/bridges2013-361.pdf
9. Futschek, G.: Algorithmic Thinking: The Key for Understanding Computer Science. In: International Conference on Informatics in Secondary Schools-Evolution and Perspectives, pp. 159–168. Springer, Berlin, Heidelberg (2006)
10. Gunawan, P., Kwan, J., Cai, Y., Yang, R.: Augmented Reality Application for Chemical Engineering Unit Operations. In: Virtual and Augmented Reality, Simulation and Serious Games for Education, pp. 29–43 (2021). https://doi.org/10.1007/978-981-16-1361-6_4
11. Ha, O., Fang, N.: Development of Interactive 3D Tangible Models as Teaching Aids to Improve Students' Spatial Ability in STEM Education. In: Proceedings of the Frontiers in Education Conference. Oclahoma (2013)
12. Haas, B., Kreis, Y., Lavicza, Z.: Integrated STEAM approach in outdoor trails with elementary school pre-service teachers in Luxemburg. J. Educ. Technol. Soc. **24**(4) (2021a).
13. Haas, B., Kreis, Y., Lavicza, Z.: Case study on augmented reality, digital and physical modelling with mathematical learning disabilities students in an elementary school in Luxemburg. Int. J. Technol. Math. Educ (2021b).
14. Hohenwarter, M., Hohenwarter, J., Kreis, Y., Lavicza, Z.: Teaching and Learning Calculus with Free Dynamic Mathematics Software GeoGebra. In: 11th International Congress on Mathematical Education, pp.1–9. Monterrey, Nuevo Leon, Mexico (2008)

15. Holzmann, P., Schwarz, E.J., Audretsch, D.B.: Understanding the determinants of novel technology adoption among teachers: the case of 3D printing. J. Technol. Trans. 259–275 (2020).
16. Huleihil, M.: 3D Printing Technology as Innovative Tool for Math and Geometry Teaching Applications. In: IOP Conference Series: Materials Science and Engineering, vol. 164, no. 1, p. 012023. IOP Publishing (2017)
17. Hwang, G.J., Wu, P.H., Chen, C.C., Tu, N.T.: Effects of an augmented reality-based educational game on students' learning achievements and attitudes in real-world observations. Interact. Learn. Environ. **24**(8), 1895–1906 (2016)
18. ING DIBA Economic and Financial Analysis Global Economics.: 3D printing: a threat to global trade. Technology Report, THINK Economic and Financial Analysis (2017). https://think.ing.com/uploads/reports/3D_printing_DEF_270917.pdf. Accessed 5 Oct 2021
19. Joo, Y.J., Park, S., Lim, E.: Factors influencing preservice teachers' intention to use technology: TPACK, teacher self-efficacy, and technology acceptance model. Educ. Technol. Soc. **21**(3), 48–59 (2018)
20. Khanlari, A., Mansourkiaie, F.: Using robotics for STEM education in primary/elementary schools: Teachers' perceptions. In: 10th International Conference on Computer Science & Education (ICCSE) (2015). https://doi.org/10.1109/iccse.2015.7250208
21. Knill, O.: Benefits and risks of media and technology in the classroom. Talk Given ICTM (2007).
22. Kösa, T., Karakuş, F.: The effects of computer-aided design software on engineering students' spatial visualisation skills. Eur. J. Eng. Educ. **43**(2), 296–308 (2017). https://doi.org/10.1080/03043797.2017.1370578
23. Lam, Y.: Technophilia versus technophobia: a preliminary look at why second-language teachers do or do not use technology in their classrooms. Can. Mod. Lang. Rev. **56.3** (2000): 389–420 (2000)
24. Lieban, D., Barreto, M.M., Reichenberger, S., Lavicza, Z. Schneider, R.M.: Developing Mathematical and Technological Competencies of Students Through Remodeling Games and Puzzles. In: Bridges Conference Proceedings, pp. 379–382 (2018).
25. Mackenzie, C.: The Passionate Elopement. Bell and Cockburn (1911).
26. Matthews, W.H.: Mazes and Labyrinths: Their History and Development. Courier Corporation (1970).
27. Menano, L., Fidalgo, P., Santos, I.M., Thormann, J.: Integration of 3D printing in art education: a multidisciplinary approach. Comput. Sch. **36**(3), 222–236 (2019)
28. Peters, M.: Developing computer competencies for pre-service language teachers. In: Teacher Education in CALL, pp. 153–165 (2006). https://doi.org/10.1075/lllt.14.14pet
29. Piekarski, W., Thomas, B.H.: Augmented Reality User Interfaces and Techniques for Outdoor Modelling. In: Proceedings of the Symposium on Interactive 3D Graphics, C, pp. 225–226 (2003). https://doi.org/10.1145/641522.641526
30. Rodger, S.H.: An interactive lecture approach to teaching computer science. ACM SIGCSE Bull. **27**(1), 278–282 (1995)
31. Saward, J.: The story of the labyrinth (2016). https://www.labyrinthos.net/Labyrinth%20Story.pdf. Accessed 25 Sep 2021
32. Saward, J.: Mazes or Labyrinths… What's the Difference and What Types are There? (2017). http://www.labyrinthos.net. Accessed 20 Aug 2021.
33. Singleton, L., Davishahl, E. Haskell, T.: Getting Your Hands Dirty in Integral Calculus. In: 2020 ASEE Virtual Annual Conference Content Access (2020)
34. Sonali, M., Raman, S.P., Mishra, S.: 3D Printed Labyrinth Controlled By Android Device. (Doctoral dissertation, CMR Institute of Technology. Bangalore) (2020).
35. Song, M.J.: Learning to teach 3D printing in schools: how do teachers in Korea prepare to integrate 3D printing technology into classrooms? Educ. Media Int. **55**(3), 183–198 (2018)
36. Thompson, D., Cheng, D.: Square Seeds and Round Paths: Exploring Patterns within the Art of Classical Labyrinths. In: Bridges 2015 proceedings: Mathematics, Music, Art, Architecture, Culture, pp. 555–548. Phoenix, USA (2015). http://archive.bridgesmathart.org/2015/bridges2015-555.html

37. Tondeur, J.: Teachers' Pedagogical Beliefs and Technology Use. In: Encyclopedia of Teacher Education, pp. 1–5 (2019). https://doi.org/10.1007/978-981-13-1179-6_111-1
38. Top, E., Baser, D., Akkus, R., Akayoglu, S., Gurer, M.D.: Secondary school teachers' preferences in the process of individual technology mentoring. Comput. Educ. **160**, 104030 (2021). https://doi.org/10.1016/j.compedu.2020.104030
39. Trust, T., Woodruff, N., Checrallah, M., Whalen, J.: Educators' interests, prior knowledge and questions regarding augmented reality, virtual reality and 3D printing and modeling. TechTrends **65**(4), 548–561 (2021)
40. Trust, T., Maloy, R.: Why 3D print? the 21st century skills students develop while engaging in 3D printing projects. Comput. Sch. **34**(4) (2017).
41. Veermans, K., Jaakkola, T.: Bringing Simulations to The Classroom: Teachers' Perspectives. In: Virtual and Augmented Reality, Simulation and Serious Games for Education, pp. 123–135 (2021). https://doi.org/10.1007/978-981-16-1361-6_10
42. Verner, I., Merksamer, A.: Digital design and 3D printing in technology teacher education. Procedia CIRP **36**, 182–186 (2015). https://doi.org/10.1016/j.procir.2015.08.041

Architectural Models Created with Mixed Reality Technologies Towards a New STEAM Practice

Shereen El Bedewy, Ben Haas, and Zsolt Lavicza

Abstract Architecture can be used as real-world examples in the classroom by analyzing them and understanding them from many dimensions to develop STEAM (Science, Technology, Engineering, Arts and Mathematics) skills. Furthermore, these practices can allow participants to explore architecture, cultural and historical aspects and use their mathematical knowledge while modelling, thus contributing to mathematics education. The proposed architectural modelling is modelled in various ways using a broad spectrum of technologies. GeoGebra and GeoGebra AR features allow participants to explore architectural models' components from a mathematical point of view through the modelling process and support of mixed-reality technologies. This chapter will illustrate and discuss a case study using mixed-reality technologies that took place in a hybrid learning setting in Upper Austria for architectural modelling.

Keywords Architecture · Augmented reality · Virtual reality · 3D Mathematical modelling · STEAM education · Mathematics education

1 Introduction

Combining architecture with STEAM (Science, Technology, Engineering, Arts and Mathematics) learning can add a new connection to culture and history. These proposed STEAM practices that use architecture, culture and history can be of practical benefit to teachers. Teachers can use the participant's mathematical knowledge and mathematics to encapsulate architectural constructions that surround us everywhere. This allows them to collect facts and information that are related to the history and culture of such architectures. This may be a tool to foster participants' creativity, problem solving and inquiry principles in using what they already know as mathematical knowledge to build or model architecture. These practices can be achievable using a wide range of technologies to model and visualize the architectural models

S. El Bedewy (✉) · B. Haas · Z. Lavicza
Johannes Kepler University Linz, Linz, Austria
e-mail: Shereen_elbedewy@hotmail.com

Y. Cai et al. (eds.), *Mixed Reality for Education*, Gaming Media and Social Effects,
https://doi.org/10.1007/978-981-99-4958-8_3

digitally by using GeoGebra, GeoGebra augmented reality or virtual reality technologies. The architectural models can be further visualized physically by using 3D printing, Origami or 4D Frames. 4D Frames is a manipulative tool that was inspired by Korean Architecture that would reformulate the geometric forms of shapes and analyze them [19].

We refer to architectural, cultural, and historical disciplines in STEAM practices as the Trio in this chapter when participants explore and collect information (e.g.: location, date of construction, cultural background) related to the architectures they choose for modelling. By combining history, we are trying to allow teachers and students to explore the stories behind the architectures they collect, model and visualize. Mathematics is combined with the Trio while using technologies such as "dynamic geometry software" GeoGebra (https://www.geogebra.org/), Augmented Reality (AR), Virtual Reality (VR), CAD, and 3D printing. These practices allow participants to use GeoGebra and GeoGebra AR to explore and model architectural components from a mathematical point of view. Blum and Ferri [5] believe that mathematical modelling could be taught and learned on a daily basis by integrating these concepts and insights in teacher education.

Mixed-realities, AR or VR offer different supports and manipulations for participants. Therefore, we should define what is a reality and what are the different mixed realities that are lately referred to. In order for the reader to distinguish the terminologies between AR, VR, Mixed Reality (MR), and Extended Reality (XR) and understand what we are referring to, here in this chapter we provide the following clarifications. According to [31], "*Augmented reality is a system that enhances the real world by superimposing computer-generated information on top of it.*" AR is considered as a variation from VR, because VR is a technology that allows users to be fully immersed in a virtual world and settings that are unlike their actual surroundings. VR basically separates users from their existing environments and doesn't allow interactions between the real and virtual worlds. Eichenberg [12] described VR as "*advanced form of human–computer interface that allows the user to interact with and become immersed in a computer-generated environment in a naturalistic fashion*" The difference between MR and VR was noted by [1], where "MR is more commonly defined as being spatially aware, allowing real and virtual objects to interact, whereas AR does not." MR is capable of allowing physical and virtual objects to exist together in one space in the so-called XR which is extended reality. Another interesting definition is from Intel. (n.d). [16] the computer chip manufacturer on Intel's website "provides the ability to have one foot (or hand) in the real world, and the other in an imaginary place." After an overview of differences between the well-known reality terminologies, we will now explore their applications in this chapter.

Technologies that tend to immerse or impose digital overlays on real-world environments can be used in educational settings for several purposes. In this research approach, AR embedded in the GeoGebra 3D calculator is used because it doesn't require any pre-coding prerequisites and could serve the goal of providing a dynamic application that may serve any architectural modelling practices. As described by [29], the added features of the GeoGebra AR platform covers most of the problems

AR users used to face in the previous platforms. These features are converting 3D objects to AR mode or 3D objects created from scratch in AR mode. Further, there is direct access to AR objects, selecting points, lines of 3D objects in AR mode. Moreover, real-time behaviour, which changes objects in the scene, appears in real-time reflections in the AR scene, labelling objects and their mathematical functionalities as area and volume. Lastly, measurement capability can be achieved in AR mode.

We discuss the effects of these newly added features AR, VR, and dynamic geometry software on STEAM education and how they can be applied in various educational settings e.g., VR glasses in mathematical modelling. These functionalities are of high importance to our research scope as it renders a physical-digital functionality and proposes a new approach for modelling in AR, not only viewing previously modelled objects in AR mode. According to [20], as AR tools continue to evolve, their usage in education will be very significant. Moreover, this will help apply the terminology supporting "Interactive education" as AR applications can transform educational environments to be more fruitful, enjoyable, and interactive. AR can be well designed to meet each student's needs and provide each student with unique interactive content from 3D models and environments.

In this chapter, AR acts as the main technology in architectural modelling and visualization. The AR used in this research scope that is integrated in GeoGebra is based on a marker-less AR approach, where users move the device in order to define surface or ground to allow users to place virtual created GeoGebra models above it.

The AR mode allows users to do mathematical modelling in a real-world environment while not being totally immersed from their surroundings. The mathematical modelling approach followed is derived from the umbrella of "Modelling as a Vehicle" by [13]. The aim behind this modelling approach is to help in serving the mathematical syllabus, generate a motivational need for mathematical content and finally adapt mathematical modelling tasks to real-world ideas. Users can have an architectural statue (Fig. 1), or a 3D printed one in their physical world. They can start to use AR features to imitate it or overlay its virtual version above it or next to it. This concept of combining and visualizing both physical models and virtual models of the same architectural building at the same time may pave the road for 3D transformation exploration by encapsulating AR and 3D printing technologies. Figure 1, illustrates a physical real architectural model while (Fig. 2), illustrates the 3D transformation process of visualizing architectural models in a digital mode as in GeoGebra 3D and GeoGebra AR and in a physical mode as in 3D print of the architecture model.

Even though AR marker-based approach is not yet available in GeoGebra, it can be integrated into this research to allow users to scan markers and attach architectural models to them. In addition, AR marker-based approaches may allow users to explore architectural models and share them through sharing physical markers.

Compared to AR, VR technology is totally immersive because it separates users from the real world around them. Mckechnie and Wilson [22] showed that using VR technology in education can be valuable in interfacing real-world problems. Moreover, the most exciting advantage to using VR is practicing any task in a simulated safe environment. From these perspectives, VR has been adopted in this

Fig. 1 Queen hatshepsut obelisk from ancient Egyptian architecture

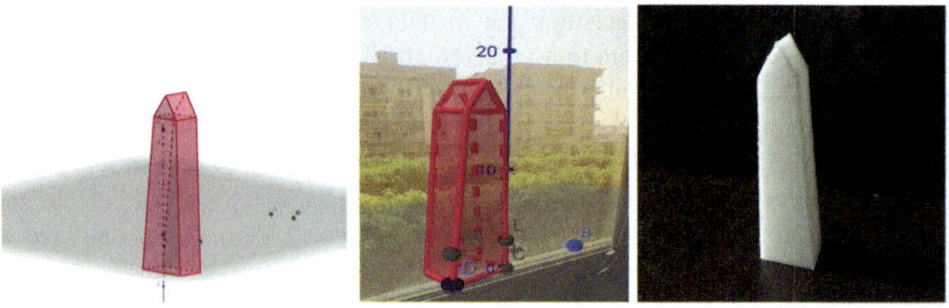

(a) (b) (c)

Fig. 2 3D transformation of obelisk in ancient Egyptian architecture, **a** 3D modelling in GeoGebra 3D, **b** GeoGebra AR, **c** 3D print of Egyptian obelisk

research as a digital visualization tool for architectural models already modelled in GeoGebra; they can be exported from GeoGebra in required formats to match the participant's headsets inputs. During the research, we were guided by the following questions. Are mixed reality technologies that use AR or VR likely to support students in the proposed STEAM practices to foster architectural mathematical modelling understanding?

In this chapter we will present a case study in Upper Austria with teachers and students implementing the mentioned practices in a hybrid form online and in-class practice. The work implemented with teachers and their students was done using the GeoGebra platform for the modelling part. GeoGebra AR was used to cover the MR part with another possibility of visualizing the architectural models in a VR environment. The chapter will further elaborate on the theoretical framework used, methodology, and description on Upper Austria intervention. We will discuss mixed realities in STEAM practices.

2 Literature Review

As mentioned earlier, teaching STEAM disciplines with reality technologies (e.g.: AR, VR or MR), involves several theoretical foundations [25]. Thus, we reviewed studies and frameworks on technologies, mathematical modelling, teacher education and STEAM teaching.

Reality technologies that immerse or impose digital overlays on our real-world environments used in our study are 3D transformation and visualization of architectural models in many forms as physical or digital. According to [15], AR can improve students' learning achievements and attitudes. They advised that anyone interested in applying the field AR game-based activities should select real-world activities, including questions that require participants to develop observations and reflect about these activities, develop content, and specify location of each AR element. According to [17], the use of AR has first to rely on objectives and goals from this application and how to apply them pedagogically to the target groups. Therefore, with his Construct 3D system, technology explorations are not the core of the proposed learnings but rather are an aid to help deliver learning to participants. That is why the content of architectural modelling and the pedagogical approaches in delivering mathematical concepts behind them are main building blocks of the offered learnings and practices which can be facilitated and then visualized using technologies such as AR or VR. Kim [18] discusses the effect of AR in education, especially in STEAM approaches and describes AR as an excellent tool to be adopted in STEAM practices because it can help participants to grasp concepts while improving creative and imaginary aspects. Moreover, on a classroom basis, AR can deliver good information that can increase learning competence. Encapsulating these findings in this chapter gives a positive reflection on adopting AR for STEAM practices.

In contrast to AR, VR is totally immersive and has its advantages in education too as raised by [26], who developed a model to show educators how and when to adopt VR and when to discard it. They argue that VR can help teach regular lesson objectives and evaluate and assess them while providing a new edge of immersion in the designed environment. From these perspectives, VR has been adopted in this research as a digital visualization tool for the architectural models already modelled in GeoGebra, and they can be exported from GeoGebra in required formats to match the participant's headsets inputs.

We referred to [1] in comparing typical traditional learning methods as paper and pen with new technologies capable of integrating and showing the learning contents of technologies like VR and MR. VR and MR showed better results than the traditional way in the following factors: student engagement, enhanced learning effectiveness, student success outcomes (academic achievements and student satisfaction), which concluded that these technologies could be adapted to such learning settings. Furthermore, while the comparison between MR and VR settings took place in the referred study, results showed that the participants enjoyed VR more than MR because of a higher level of immersion which affected their engagements. According to these findings, we believe in providing multiple visualization technologies like AR and VR for architectural modelling to positively affect the users.

Other previous studies [2] showed that VR can be more effective than the traditional and even video situations because it allows higher engagement and positive feelings. Furthermore, positive feelings or emotions and immersion concepts have been proven adequate for education and learning contents by [24]. This meta-analysis work by D'Angelo et al. [9] showed the effects of using computer-based simulation versus no computer-based simulation in mathematics learning. Moreover, their overall findings show that computer-based simulations increased students' improvements in provided learning than non-computer-based simulations.

We are using GeoGebra in these STEAM practices to architectural modelling based on mathematics goals. Also, because it is dynamic, including modelling tools such as polygons, extrusions, surface of revolutions and many others, it has AR functionality to allow participants to view their work in AR mode. Furthermore, the GeoGebra classroom allows us and teachers to view participant's work and monitor their progress, define obstacles that may meet them during their work, and assist them. Finally, GeoGebra allows model extrusions as STL format, which can be 3D printed or used in VR environments, as discussed in the following sections.

3 Theoretical Framework

The conceptual framework in this research covers all the proposed notions and various directions this research can reach, keeping into consideration the research focus and a theoretical framework's fusion of combining various theories all together. As has been highlighted above, the research direction is combining several aspects as architectural modeling, mathematical learning, culture and history connections, various technologies, and as a preparation for this fusion a lesson planning technique. All these variables formulate the research construct and in order to cover these directions, theoretical foundations play an important role in defining variables that need to be addressed in the research question and in the design phase. The theoretical foundations affect design as well as it affects practice of any research, which makes it of high importance to be well defined at the beginning of the research. The theory and designs are a connected procedure where each one affects the other as stated by [7] during experiments or practice throughout the research. The dynamic lesson plan which is proposed in this research is dynamic by its nature as it holds multiple alternatives to meet the teacher's needs as will be discussed in the lesson plan section. This makes us think of a dynamic approach from a theoretical point of view as addressed in

Theory of didactical situations (TDS). TDS was established by [6] to define didactical situations (DS) as a dynamic process that involves an interplay between the teacher, students, and the surrounding of these situations and all components that are involved in this surrounding which is referred to as the milieu. Having a closer look at the didactic milieu which has been initiated by teachers before and during lesson planning. Focuses on teachers' intentions and hypotheses they have for establishing such a milieu for their students.

The Technological pedagogical content knowledge framework (TPACK) framework is considered to help in analyzing technological usage and its relation to other didactical aspects covered in this research track. It was developed by [23]. The strength of theTPACK framework is that it combines content and pedagogy with technology best practices. It tries to use suitable technology to blend required pedagogy to cover the introduced content. This framework is focused on teacher behavior towards TPACK.

The introduced framework, which is Adaptive, Meaningful, Organic, Environmental-based architecture for Online course design (AMOEBA) design framework allows us to regulate cultural differences in our introduced designs. AMOEBA [14] came out from the study of intercultural fields such as cross-cultural psychology and computer-mediated communication. This framework is concerned with integrating learners in the co-creation of activities and delivering teaching strategies that suit teachers as well as learners. As the AMOEBA framework is adopted in this research focus and follows the components to result in developed research that may be fitting for application on a cross-cultural basis.

Connecting the presented theories in structured program development to the technology as shown in (Fig. 3), the relation of integrating technology to theoretical frameworks and program development by [10]. The theories used in this research are highlighted and connected in (Fig. 3).

4 Dynamic Lesson Plan

Our aim behind developing the dynamic lesson plan (DLP) concept is to serve this research scope and engage teachers in designing the research interventions with their students [21]. We are eager to engage teachers to form a practice that would encapsulate twenty-first century skills into their planning and how they select the resources and design the tasks. When teachers are given a tool to allow them to choose resources freely and easily that may allow them to practice various technologies in practice, this may open creative doors to them to innovate, and reflect this spirit to students. It has been found by [30] that classroom boredom and teacher enthusiasm are related and affect each other. We assume this lesson planning module may result in a different experience than the normal traditional way of designing lessons.

DLP was modeled to provide teachers the feeling of ownership and freedom of choice. This approach should allow teachers to feel motivated towards implementing various tools presented and make them curious to visit all options presented in DLP.

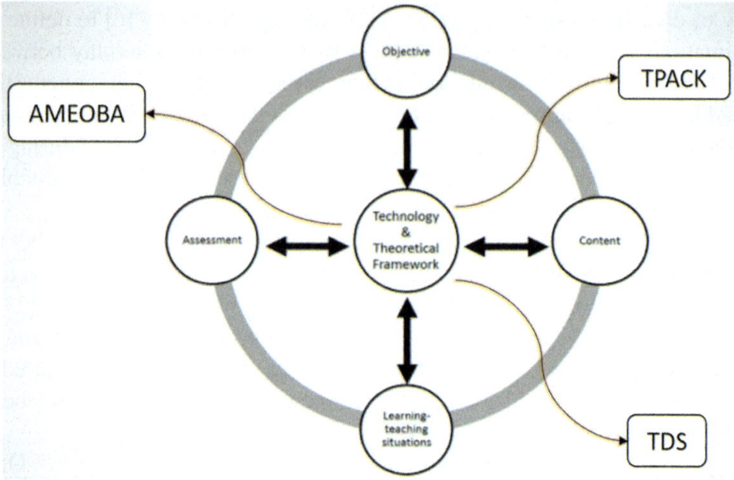

Fig. 3 Technology theoretical framework and the process of the program development

The tool was designed on Unity platform which is a game engine for interactive and real-time responsive applications. Our approach is designed to be a web portal to be accessible for all teachers and does not need any extra installation and the aim is to facilitate its usage for teachers. As teachers navigate to the provided URL, they will find a single scene with various Objects and UI elements that they can interact with as seen in (Fig. 4). In the middle, there is a large cube with 4 sides that teachers can answer each of the questions provided as who will model, what to model, where to model and how to model. The four categories in the four-sided cube specify age, architectural models, environment of the practice, and finally technology or way the teacher wishes to implement the architectural modeling. By choosing answers to these questions, the teachers will be able to complete the DLP criterion. The teacher would press the button to get a link in order for the web interface DLP to generate some GeoGebra instructional links to allow the teacher to implement the chosen criterion with students. Figure 4, shows DLP structure.

It can be seen (Fig. 5), illustrates a journey the teacher would take in order to choose all the lesson criteria from the DLP, visualizing all the options the teacher may find under each of the modules presented in DLP.

5 Upper Austria Intervention Structure

In this section, we will describe the intervention that took place in Upper Austria. The intervention started when we met teachers for the first time until the last fourth meeting /interview. We will describe what happened in each interview with the teacher. We will give an overview about the workshop that took place with teachers

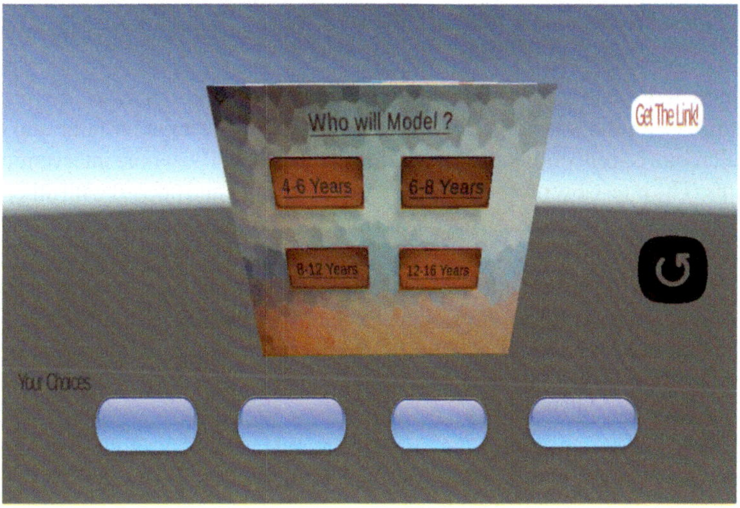

Fig. 4 Dynamic lesson plan web interface

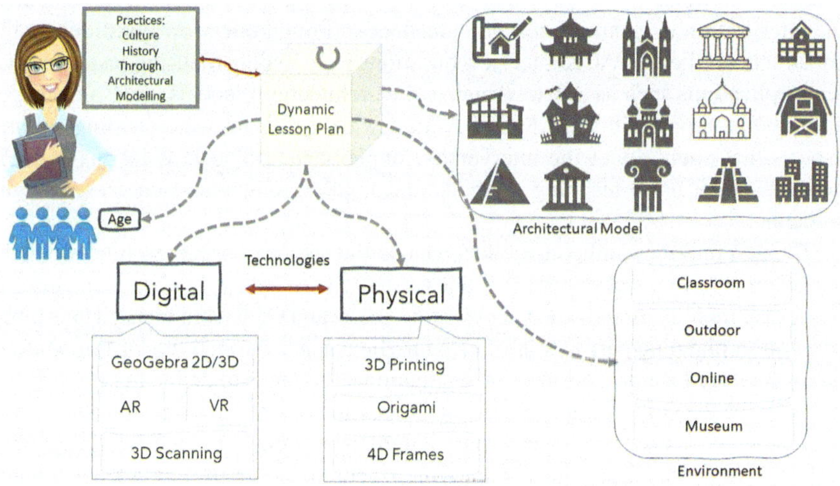

Fig. 5 The four modules explained in the map, showing the teacher journey in completing the lesson planning

and students. Following this we will discuss final projects the students presented and AR examples the students used in their final projects.

5.1 The Intervention Initialization Phase

We conducted four video-recorded interviews using the Zoom video conferencing application with the teacher. Each interview lasted on average an hour and half to share all required reflections and inputs from the teacher. The whole intervention took 5 months. The interview protocol was constructed earlier and piloted with other candidates. The protocol was a semi-structured interview with a set of questions that would take the participant into a flow first by answering questions about their backgrounds, teaching methods, technologies, STEAM learning techniques, questions related to the research scope as integrating architectural concepts and modelling, and their relation to other disciplines such as culture and history. Asking the teacher's views on integrating these principles in their learning syllabus was followed by introducing DLP and interview closing phase.

The aim of these questions in the semi-structured interview protocol was to capture teacher reflections, hypotheses about architectural mathematical modelling, cultural and historical connection, STEAM practices during each stage in the intervention. As we mentioned that theory informs design and vice versa, theory had a great impact in formulating the interview protocol. As we mentioned earlier in TDS, we are interested in the teacher's milieu which takes place during lesson preparation and after the intervention in capturing teacher's feedback through interview questions on the constructed milieu. TPACK interview questions were mainly focusing on technological applications such as AR to visualize the architectural models, their advantages and limitation possibilities. AMOEBA framework questions were tackling cultural aspects that could affect the intervention implementation such as language, technologies to be used, teaching methods and finally architectural choices with the students.

The first interview aimed to collect data on the teacher's usual teaching methods, lesson preparation, technologies' teacher usually uses, general interests the teacher has for technology used to meet her learning goals, and GeoGebra focused questions. The lesson planning part was introduced to the teacher through the DLP module in an exploratory fashion, followed by instructional guidance to define her preferences by choosing the student's age, the architectural models, the environment, and finally, technology to be used for the intervention implementation. Afterwards, we asked the teacher for her feedback. Then, we provided recommendations for the DLP module, which is a fundamental question we ask in our protocol to enhance the DLP in its future versions. Finally, we explained and showed examples of the proposed STEAM practices to the teacher inside a pre-prepared GeoGebra book.

The second interview was to define the lesson, or practice building blocks, and define criteria the teacher would like the students to follow during the workshop. These criteria were defined by the teacher using the DLP module in order to design the intervention content and structure. The teacher filled the criteria respectively as age = = 15 to 16 years old, architectural model = = ancient design, environment = hybrid online and in classroom practice, and finally technology used to be GeoGebra modelling and GeoGebra AR. The teacher in her choices put in mind cultural and historical relationships that can be practiced with students during the workshop. We captured all workshop content from the teacher during this interview as an ancient model to be used according to the teacher's criteria and introduced them to the GeoGebra classroom. We showed the teacher the basic architectures to start the workshop with to explain the basic modelling steps in GeoGebra. We asked the teacher about the assessment strategy she would like to follow for the student's work. In the second interview, the intervention duration, date, and time were defined. The teacher asked us to conduct the interview with her guidance to the students. So, the teacher invited us to join the classroom using Microsoft Teams which is a commonly used video conferencing tool in this school.

The third interview was after intervention with students, and it was during their work on projects not after they handed in their final work. This interview had two parts: the first one was to answer some technical questions the teacher and students had towards using GeoGebra modelling, and the later part of the interview was to capture reflections from the teacher's side on students' work in progress. The reflections were architectural models' choices, collaboration between the students, mathematical terminology used during the project implementation, and mathematical noticed knowledge they use in their work. Other reflections from the teacher were technology usefulness and limitations for GeoGebra modelling and GeoGebra AR and overall reflections on the conducted workshop.

The fourth and last interview that took place with the teacher was to capture all reflections from the teacher during the intervention, for example, limitations the teacher and students met, challenging and yet easily achievable experiences, teacher's impressions on student's work and their work attitude towards the research idea and the actual modelling process. This interview captured the grading strategy the teacher set for this project and assessment strategy. The teacher intended on grading this project and including it in her syllabus results. In this interview, we asked the teacher for some recommendations for future interventions.

5.2 The Intervention Setup

The teacher lacked GeoGebra knowledge especially when it comes to architectural modeling using GeoGebra. Therefore, we conducted the workshop for the students for an hour and a half with the teacher's presence and guidance to the students. The students were 35 students and due to the pandemic situation at that time of the workshop the teaching was done in a hybrid mode, half of the students attended online on teams, and the other half attended in the classroom. The Teacher had the role of organizing the workshop and giving a brief introduction to students about the workshop. We joined the online group on Microsoft Teams for half attending from their homes and for the other half attending school we were projected on a projector in the classroom. The workshop was conducted in English, but the language used during the workshop between teacher and students was German. The teacher asked the questions from students in English to our side. The workshop was recorded in

Microsoft teams and access to the recording was left available for all participants for references. The teacher introduced the topic to the students earlier one week as that they will learn to model architecture using GeoGebra. The teacher reported that the students found the topic challenging from its title. The teacher added as a first reflection that students thought architectural modelling is advanced to their knowledge and age. But after the intervention was over the teacher reported that their opinion changed after the workshop, and they were really excited to build their own architecture.

5.3 *The Intervention (Workshop) Content*

We kept in mind the teachers' criteria from the DLP module and tried to mirror them in the workshop content. We started the workshop with a broad overview by explaining the research idea and walking the participants through the GeoGebra interface. Then we introduced two simple architectural examples that show the basics of modelling with GeoGebra by creating polygons and extruding them to different forms. Then we shared with them the GeoGebra classroom link for all the students to join and to see other modelled architectural examples that were added prior to the GeoGebra classroom. The teacher chose an ancient architecture model that reflects the culture and history of Austria, the Carnuntum (Fig. 6).

Therefore, we showed the participants how to model the Carnuntum according to the teacher's earlier choice from the DLP. We presented some historical and cultural facts about this architecture as a simulation of how culture and history are connected to architectural modelling. We showed the participants how to change colors and opacity of the model to give the actual architectural look and feel. This step was very convenient for Carnuntm which was constructed using a glass shield as seen

Fig. 6 The Carnuntum Austria

in (Fig. 7). We applied the teacher's technological criteria which is GeoGebra AR. We introduced AR technology to the participants during the workshop by guiding them in installing the app on the mobile or tablet device and viewing the modelled architecture of Carnuntum in AR as visualized in (Fig. 8).

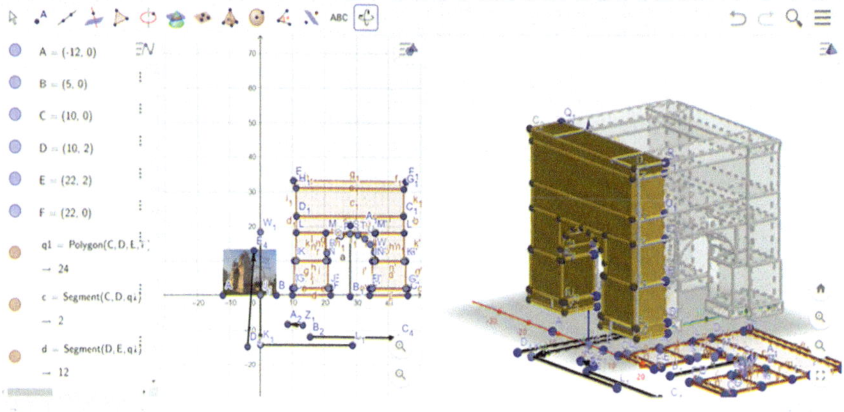

Fig. 7 The carnuntum in Austria modelling in GeoGebra

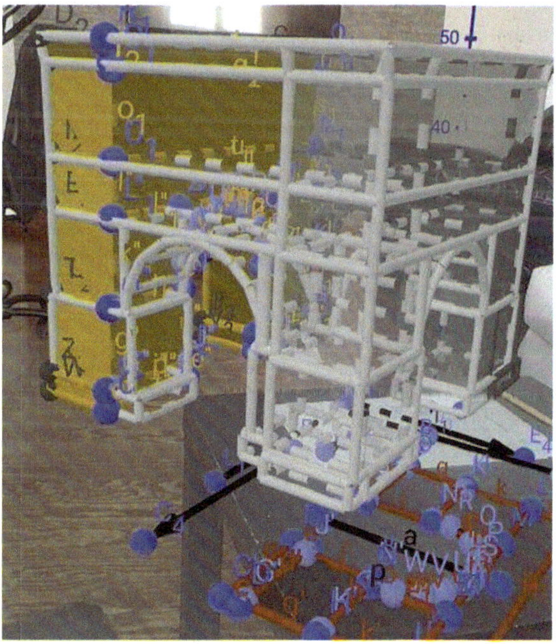

Fig. 8 The carnuntum in AR using GeoGebra AR

Fig. 9 DonaTurm, Austria

5.4 *The Student's Work*

In this section, we will show the readers some of the student's final work that was done using GeoGebra and GeoGebra AR during the intervention. The student modelled the Donau tower, which is located in Austria, the same country of their origin (Fig. 9). This reflects to the readers the cultural connection of the architectural model choice. The students connected to history by collecting some historical facts about the Donau tower which were stated in their final submission documents. The Donau tower was modelled based on mathematical tools in GeoGebra by constructing a circle followed by a prism extrusion function to give the tower a height. The student visualized the upper part of the tower by translating three circles to different heights. The first circle was extruded to a pyramid form with a negative value to look downwards and give the cone look. The middle circle was extruded using a prism extrusion to give a cylinder look and finally the last circle with the smallest circumference was extruded in a pyramid form to give the long upper part above the Donau tower. The student used the GeoGebra AR functionality to visualize the tower using AR in the middle of a large greenery space.

In (Figs. 10 and 11), two students referred to ARS Electronica Center as the chosen modelled architecture in their final projects. This center is located in Upper Austria in Linz which is the same city the students live in and therefore this visualizes a strong cultural connection that students may have visited and are familiar with. This gives the readers an idea about the motivation and connecting to places that may have an impact on the student's architectural choice. The students collected some historical facts about these architectural models that were stated in their final project's documents.

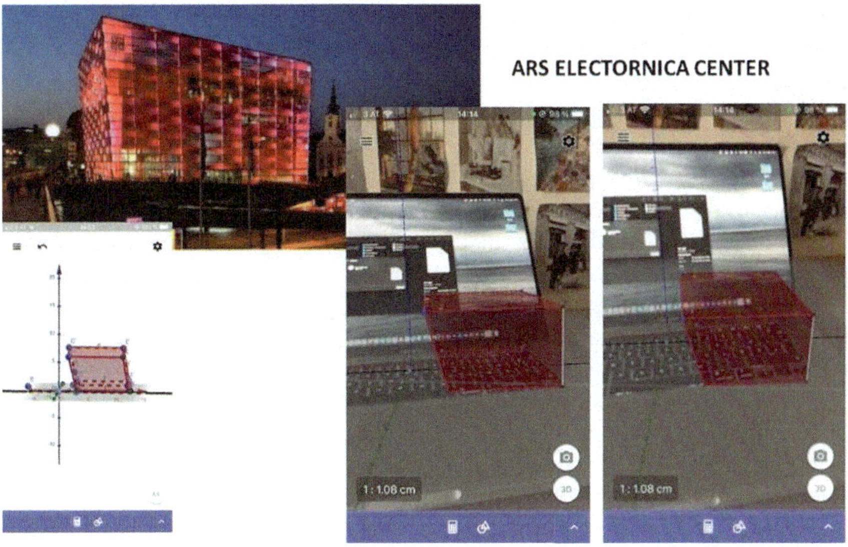

Fig. 10 Ars electronica center, Austria

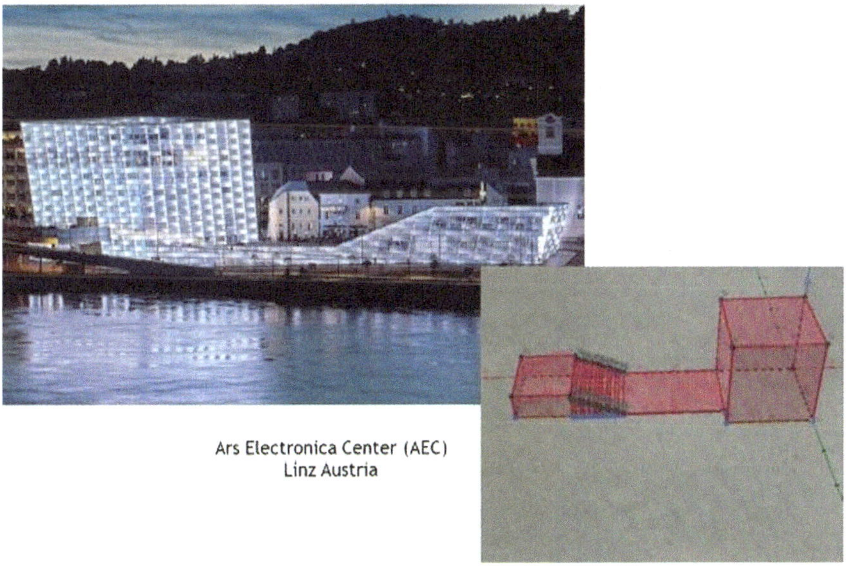

Fig. 11 Ars electronica, Austria

The two students modelled architectural models in different ways. In (Fig. 10), the students modelled the larger part of the architecture by modelling a square form polygon in the x,y,z plane and translating it to a higher height. The student then connected the two polygons giving this inclined polygon form represented in the larger part of the architecture. Followed by a GeoGebra AR representation.

In (Fig. 11), Ars Electronica center was modelled in a different way by modelling 2 square sized polygons extruded to prism form to different heights and a connecting plane. Students followed this by modelling multiple standing planes with various height extrusions to visualize this inclined form in the architecture. From Ars Electronica's example readers can understand the various modelling options that can take place in the same architectural construction. We believe this may foster creativity and imaginary skills.

Art museum architecture that is located in Austria. The student tried to imitate the architectural constructions (Fig. 12). The student constructed the museum outline as a polygon in 2D plane followed by rotation and prism extrusion in 3D plane. The student afterwards visualized this model in GeoGebra AR.

Students modelled Music Theater (Fig. 13), they stated in their documents that they used to go there and train but due to pandemic they stopped going to this place and therefore, this was a motivation to choose this theater as an architectural model for this intervention. The student modelled this architecture construction in separate parts each extruded to different heights then they translated each one to a position in 3D space to reflect the actual architectural shape.

We can see in (Fig. 14), the student represented the main plaza building located in Frankfurt Germany. The architecture was modelled as an irregular polygon to different sizes in 2D plane and followed by multiple extrusions to various heights representing the final architectural shape. The student then visualized the architectural model using GeoGebra AR on his desk.

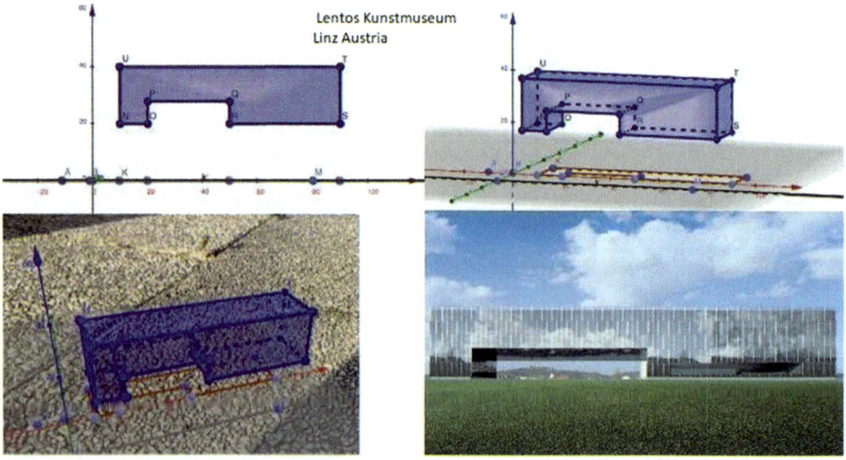

Fig. 12 Arts museum, Austria

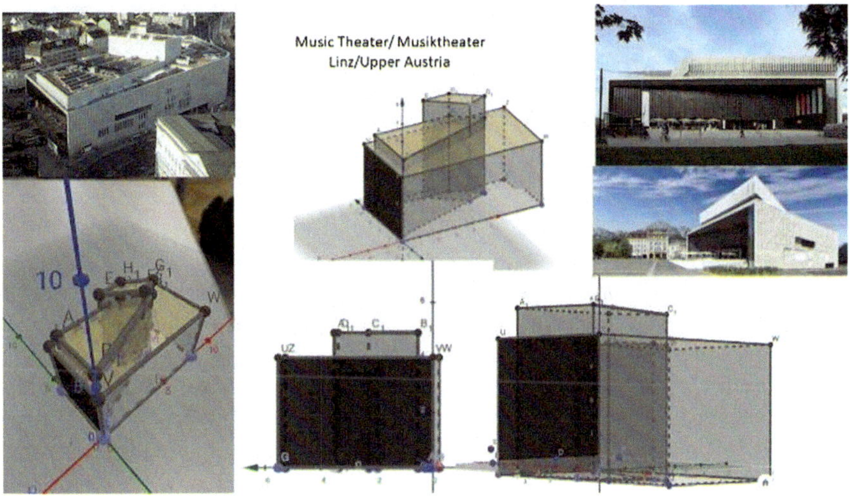

Fig. 13 Music theater, Austria

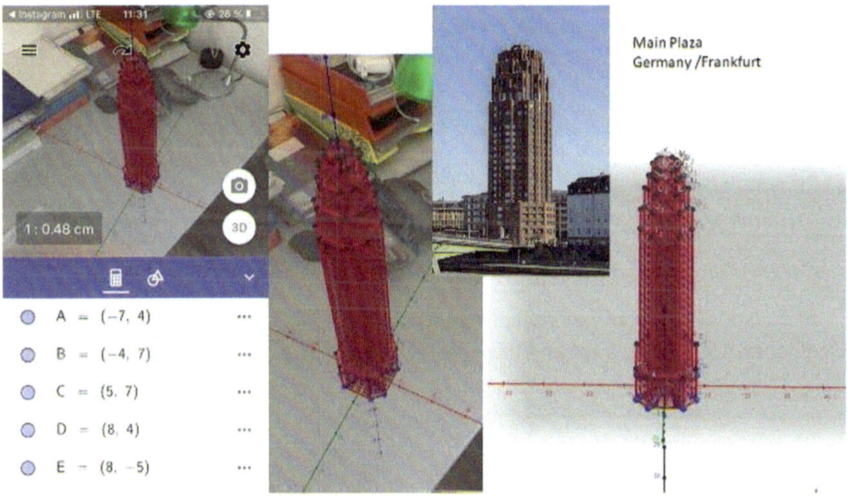

Fig. 14 Main plaza, Germany

5.5 AR Teacher Reflections

We have represented some samples from students' work of the modelled architecture using GeoGebra during the intervention. Now we will highlight teacher's reflections on their students' work using the GeoGebra AR feature. The teacher chose GeoGebra AR as a technological tool to visualize architectural models and when she was asked whether to include AR features in the workshop her reply was:

Teacher (First Interview)

00:54:42 Teacher_AU

I Think that They Would Love that.

The teacher recommended her students to use AR technology during the intervention.

Teacher (Second Interview)

Yeah, I would recommend them to do The augmented reality thing. But if someone would do something different they can. If they have those. (those referring to the other technologies presented as 3D printing, Origami, 4D Frames).

The teacher's feedback on the GeoGebra AR feature was positive, and she added that GeoGebra made it really easy with just a single click to see the constructed models in AR mode. That in other software's AR feature would require a separate development.

Teacher (Third Interview)

Especially if You Think of UM, Doing a 3D Print Afterwards or the Augmented Reality Thing then They See. OK, I Can Do Something with It and that Looks Pretty Cool. And then the Running Part Appears More.

01:12:44 Teacher_AU

OK, I Do that, but then I Have Something Completely and that's like, Uh, Motivating.

But I'm not Quite Sure if in Maya or Something like This.

01:13:00Teacher_AU

Does not Have Such an Easy Option. You Need Another Software Tool.

She also added from her reflections on the students' projects that they liked the GeoGebra AR feature a lot and they played with it in the classroom.

Teacher (Third Interview)

It's the Augmented Reality Thing. They Were like, Wow, that's Really Cool. And They Were Really Impressed.

The teacher was describing one of her students working with AR in the classroom. They constructed simple architectural models and started placing it in the classroom in AR mode as displayed in (Fig. 15).

Teacher (Fourth Interview):

And the AR Was Really Something that They Tried. You See One Did Those Twin Towers. And He Projected It in Class. And I Mean 2 Twin Towers Means Two Cubes, Right? This is Really Easy to Do, so He Did.

The Basic Thing with the Two Towers, but then He Found Out that He. Put It in the Classroom with the AR App. Yeah, that's so Cool. I Mean Even He Spent Some Time Doing Several Screenshots. That's Good to Say Out Loud.

Fig. 15 Shows the example of the student's trial of 2 twin towers with AR inside the classroom

From the teacher's reflections we may conclude that the overall GeoGebra AR experience was positive, and they may encapsulate it in future practices.

6 The GeoGebra Modelling Analysis

The architectural models the students created during the intervention were analyzed and grouped according to initial categories. Figure 16, is a map with all the architectural models that were submitted by the students and their corresponding themes and frequencies. These themes were our subjective view on the modelling quality as normal modelling: very basic modelling that didn't capture architecture details but only used basic polygon shapes with single or double extrusions. The detailed modelling used complicated figures, multiple extrusions or irregular shapes. Another criterion we found reflective of architectural modelling concept is how close did the participants try to imitate the actual architectural construction. The last theme is a non-architectural model if the participants decided to choose something else to model that is not an architectural construction. In (Fig. 16), the highest weight has the largest line width.

As mentioned earlier, teachers encouraged students to represent their work in AR. The student's work represents that 7 of overall 22 submitted projects used GeoGebra AR functionality. The students didn't use any other visualization tool although the teacher referred to some other technologies that they can use like 3D Printing, Origami or 4D frames. Figure 17, shows a pie chart of the students' opinions in the GeoGebra AR feature after the workshop.

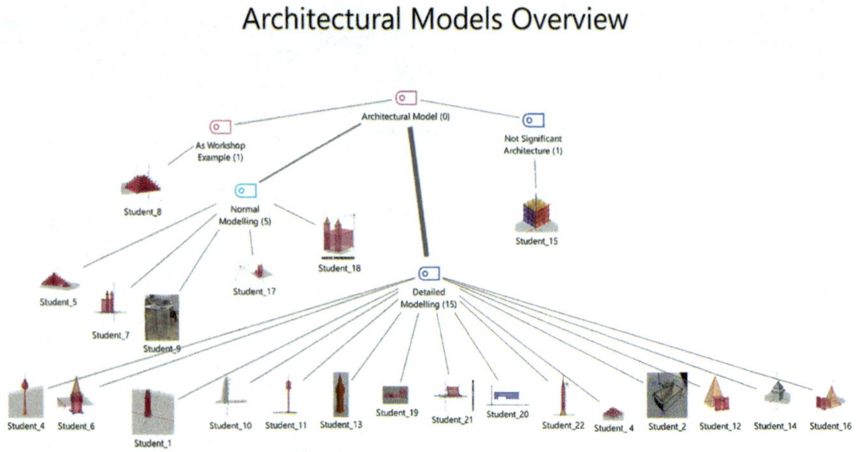

Fig. 16 A map shows all the students' architectural models' choices with corresponding frequencies to the designed themes

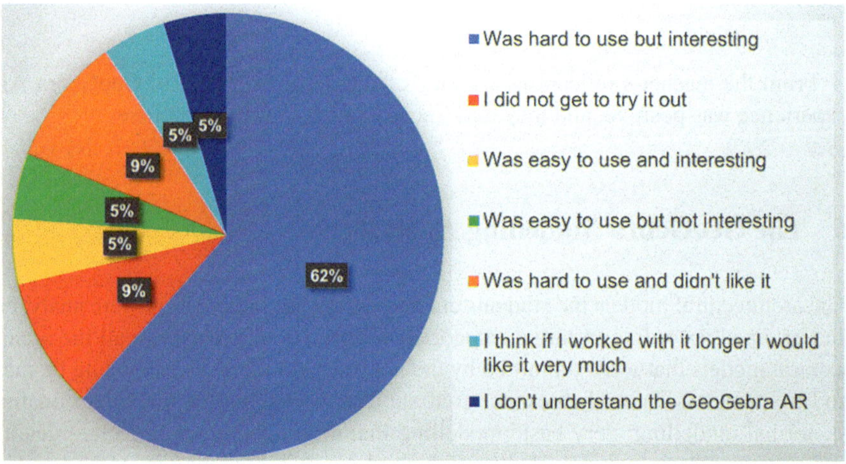

Fig. 17 A pie chart showing the students' opinions about using the GeoGebra AR feature, N = 21 students

As mentioned earlier GeoGebra AR was introduced in the workshop that took place in Upper Austria. The teacher encouraged the students to represent their final projects in AR mode. This survey was distributed after the student's submitted their final projects, so it is considered a post-test. The students' opinions summarized in the pie chart in (Fig. 17) show that most of the students liked the idea of using GeoGebra AR and representing their architectural models in AR but met some challenges that could be due to any technical problems. This is because one of the major limiting factors is that some mobile phones and tablets don't support the visualization of AR mode, and this makes AR button in GeoGebra AR invisible, this may cause some frustration for some participants who really would have wished to view their work in AR mode. For that we recommend teachers to advise students to collaborate so that they all experience the AR visualization on shared mobile devices that support that feature. This limitation may be resolved later by future mobile technology enhancement so it would support a wider range of devices specifications.

7 Architectural Models in VR

Earlier we emphasized architectural modelling in GeoGebra along with an AR visualization possibility using GeoGebra AR that is not fully immersive. We highlight in this pilot trial that VR experience may add a new learning possibility to participants from technological perspective that would allow participants to do the same architectural modelling presented in this chapter that is mathematical based but in a fully immersive environment.

Therefore, we did a pilot study with the provided Upper Austria intervention student's work using VR in order to assess the feasibility of integrating VR in future interventions. We exported the files from GeoGebra in. STL format and converted them in Blender to. GLB format. Then we imported them in the Oculus Quest virtual reality headset in the Oculus home application. We took some snapshots from the VR environment with the student's work during the pilot trial as presented in (Fig. 18).

From this pilot trial we believe that VR may add a new edge to these practices especially if they manage to find mathematical modelling applications [28] as Neotrie VR that allows modelling of 3D shapes in a VR environment or importing existing 3D models, and finally painting them in VR. Implementing these practices in VR with teachers and students using the Neotrie VR application could be a possible future publication for this research direction. Campbell [8] stated a couple of questions that could be kept in mind when implementing mathematical modelling in VR such as "how it is that learners will orient and navigate through virtual environments? How will their sense of identity evolve? What implications for collaborative learning and instructional design?" Campbell also encouraged researchers and mathematics educators to set priorities to apply mathematical modelling in these environments as VR. Virtual environments suggest that as mathematics educators, we should re-conceive our means of and priorities in teaching and learning mathematics, especially mathematical modeling and applications, and no less, as educational researchers, our ways of conducting research into these phenomena as well.

For VR applications, modelling has to be done in a way that is also suitable for 3D printing because the models are extracted from GeoGebra as a. STL format. The models have to have a thickness so as to be imported in VR and viewed in a VR environment, but single-sided planes with no thickness are not 3D printable nor VR viewable.

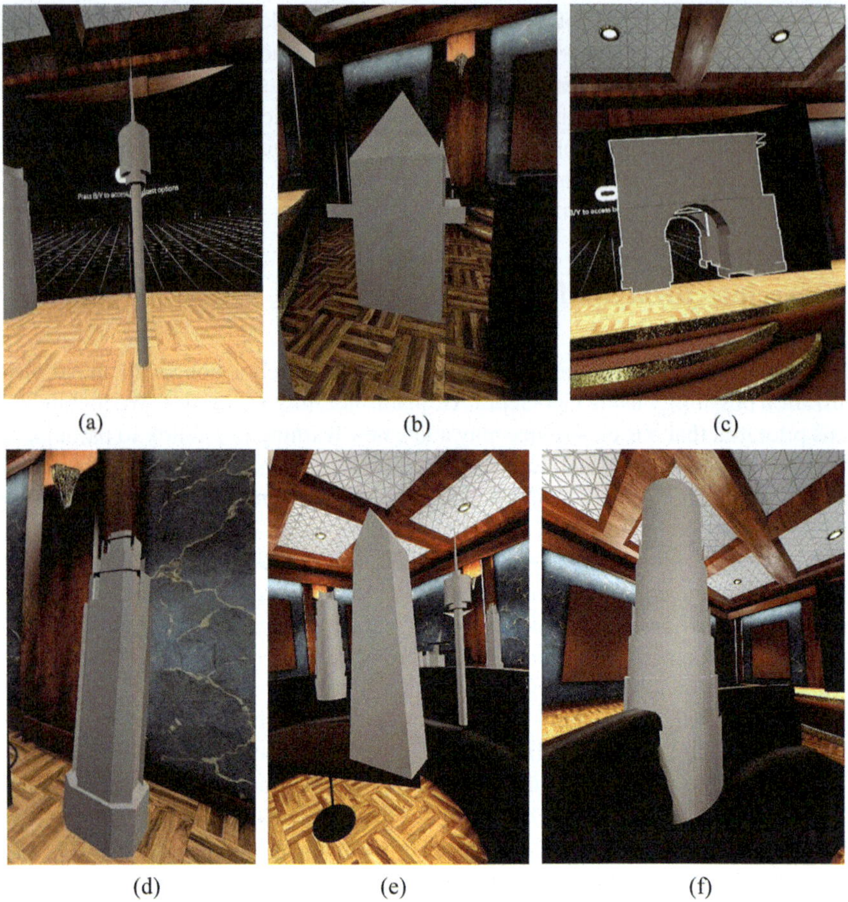

Fig. 18 Screenshots from upper Austria student's work in VR environment, **a** Donau-tower, **b** Graz clock tower, **c** Carnuntum, **d** Main plaza, **e** The Egyptian obelisk, **f** Ibiza tower

8 Discussion

The intervention in Upper Austria gave valuable insights to the authors on the possibility of integrating the Trio to STEAM learning practices and on using AR, VR technological capabilities to visualize the modelled architectures. The Austrian teacher's motivation towards research ideas was well reflected not only in her words during the interviews but well delivered to students after the workshop in her support and guidance. The whole intervention allowed application of IBL principles [4] where teachers facilitated the environment for their explorations and multiple trials. It was intentional to see the effects of applying the proposed STEAM practices in IBL teaching technique. The authors believe that the intervention demonstrated encapsulating the Trio concept and integrating it with STEAM principles, and the cultural-historical effects on the intervention and the data collected from the students during their final projects. And the technology effects especially the use of GeoGebra platform for the modelling experience. Moreover, AR, VR effects as a visualization tool to the modelled architectures.

The introduced STEAM practices encapsulated the participants mathematical knowledge in the architectural mathematical process as has been discussed in upper Austria Intervention. The teacher's role in these practices is to refer to these mathematical knowledge encapsulations during the modelling and to highlight this mathematical connection point. For example, teachers can highlight geometrical knowledge as forms, shapes, planes, polygons, relationship between the polygons. Techers can also highlight the trigonometry knowledge as the angle's calculations, the collinear and coplanar relationship between the points that constructed the modelled architectures. Other teachers may use the transformations functionality as rotation, translation reflection and scaling. These practices afford for teachers the chance to visualize these mathematical concepts using GeoGebra and GeoGebra AR. To adapt these practices to the mathematical knowledge they want to emphasize on with their students.

But the intervention also highlighted some strengthening aspects when it comes to the applications of GeoGebra AR specifically. Because GeoGebra AR is not like any other AR application, it doesn't require pre-coded contents to be responsive to the users. GeoGebra AR was built as a dynamic tool that acts as the usual GeoGebra 3D calculator but specifically in AR mode. This feature is very strong and can give users unlimited applications in encapsulation of AR technology. We believe GeoGebra AR managed to transform standard AR applications that require pre-coding prerequisites and instead introduced a new degree of freedom for users when it comes to the massive learning AR applications that can be used [3]. The GeoGebra AR applications and its integration in the mathematical walks that can practice architecture modelling through applying symmetries, rotations, translations or other transformations is an integration step to this research focus for an outdoor student's activities future possibility.

9 Conclusion

To conclude the work presented in this chapter, the Trio was integrated with STEAM practices to offer a new interactive practice that would foster mathematical learnings and help in developing several skills for the participants allowing them to use their mathematical knowledge and adopt new ones through the modelling experience. Moreover, the proposed practices foster IBL approaches that were first initiated by Dewey [11], especially in the mathematics field as stated by [4] where IBL concepts are implemented between teacher and students in exploring, designing lesson plans and various activities that can be practiced with students. Students who use these designed tasks by teachers also undergo and experience inquiry principles allowing the participants to engage, explore, explain, elaborate, and evaluate (Plan and Instrument n.d.) [27]. This chapter highlighted that technology acted as a good complementary teaching method for these practices for architectural modelling and the modelling visualization.

The limitations we can see in integrating these practices is its integrational nature the Trio, as it combines aspects from various disciplines of study as architecture, culture, history, and mathematics. This may lead teachers to find difficulty in mapping these practices to their current syllabus. Therefore, we encourage them to apply these practices as a complementary practice to their core content. Although this can be perceived as a limitation to some, but it could be considered as an opportunity to others as it may pave the way for teachers' collaboration from various disciplines that cover the various focuses of the Trio. These practices may allow teachers to look at disciplines in a more open way and to be more flexible in collaborating and integrating other educational focuses in their teaching and learning cycles.

Another limitation is the technology affordance in schools which may affect the visualization of these practices and the 3D transformational idea to take place representing the architecture models in digital and in physical forms. The AR architectural representation using GeoGebra AR would require teachers and their students to have access to mobiles or tablets that support augmented reality services while the VR architectural representation would require school to afford VR headsets. In that case we recommend teachers to know the hardware devices available at the school before introducing them to students in these practices.

For these practices application we recommend that teachers learn GeoGebra or get themselves familiar with the basic modeling tools before practicing them with students. Moreover, teachers can take a workshop that applies these practices main notions and how to integrate them together in one practice. Teachers would then have the capability of integrating these practices and utilize the student's mathematical knowledge in an efficient way. All this knowledge would qualify the teachers to guide their students later in these practices' optimum implementation.

The intervention that took place in Upper Austria provided us with data that may help and contribute to future interventions. We think that when we present these practices through other cultures this may add more valuable findings to our research, and this is what we hope to present in future papers. We would like to use GeoGebra AR functionality to apply mathematical calculations for example as calculating the height or volume of the real architectures by mapping the virtual modelled overlay on the real-life architecture construction. Moreover, using the GeoGebra AR to model architectures in AR mode. We are also keen on exploring more VR applications that would allow participants to model in the VR environment. These possibilities of future work may contribute strongly to the idea of encapsulating MR applications in mathematical education while connecting to the proposed Trio.

References

1. Allcoat, D., Hatchard, T., Azmat, F., Stansfield, K., Watson, D., von Mühlenen, A.: Education in the digital age: learning experience in virtual and mixed realities. J. Educ. Comput. Res. **59**(5), 795–816 (2021)
2. Allcoat, D., von Mühlenen, A.: Learning in virtual reality: effects on performance, emotion and engagement. Res. Learn. Technol. 1–13 (2018)
3. Ancochea, B., Cárdenas, M. I.: Exploring Real World Environments using the Potential of GeoGebra AR. In: Research on Outdoor STEM Education in the Digital Age. Proceedings of the ROSETA Online Conference in June 2020, pp. 41–46 (2020).
4. Artigue, M., Blomhøj, M.: Examples of inquiry-based activities with reference to different theoretical frameworks in mathematics education research. ZDM—Int. J. Math. Educ. **45**(6) (2013)
5. Blum, W., Ferri, R.B.: Mathematical modelling: can it be taught and learnt? J. Math. Model. Appl. **1**(1), 45–58 (2009)
6. Brousseau, G.: Theory of didactical situations in mathematics (N. Balacheff, M. Cooper, R. Sutherland & V. Warfield: Eds. and Trans) (1997)
7. Brown, A.L.: Design experiments: theoretical and methodological challenges in creating complex interventions in classroom settings. J. Learn. Sci. **2**(2), 141–178 (1992)
8. Campbell, S.R.: Mathematical Modeling and Virtual Environments. In: Modeling Students' Mathematical Modeling Competencies, pp. 583–593. Springer, Dordrecht (2013)
9. D'Angelo, C., Rutstein, D., Harris, C., Bernard, R., Borokhovski, E., Haertel, G.: Simulations for STEM learning: systematic review and meta-analysis. SRI Educ. **58** (2014) Available online: https://www.sri.com/wp-content/uploads/pdf/simulations-forstemlearning-full-report.pdf. Accessed 30 November 2021.
10. Demir, S.: Two inseparable facets of technology integration programs: technology and theoretical framework. Eurasia J. Math., Sci. Technol. Educ. **7**(2), 75–88 (2011)
11. Dewey, J.: The philosophy of the arts. John Dewey: Later Works **13**, 357–368 (1938)
12. Eichenberg, C.: Virtual Reality in Psychological, Medical and Pedagogical Applications. InTech, New York, NY (2012).
13. Galbraith, P., Stillman, G., Brown, J.: Turning Ideas into Modeling Problems. In: Modeling Students' Mathematical Modeling Competencies: ICTMA 13 (2010). https://doi.org/10.1007/978-1-4419-0561-1_11
14. Gunawardena, C.N., Wilson, P.L., Nolla, A.C.: Culture and online education. In: Handbook of distance education, pp. 753–775 (2003)
15. Hwang, G.J., Wu, P.H., Chen, C.C., Tu, N.T.: Effects of an augmented reality-based educational game on students' learning achievements and attitudes in real-world observations. Interact. Learn. Environ. **24**(8), 1895–1906 (2016)
16. Intel. (n.d).: Demystifying the Virtual Reality Landscape. https://www.intel.com/content/www/us/en/tech-tips-and-tricks/virtual-reality-vs-augmented-reality.html. Accessed 26 Sept 2021.
17. Kaufmann, H.: Collaborative Augmented Reality in Education. Vienna University of Technology, Institute of Software Technology and Interactive Systems (2003)
18. Kim, J., Kim, J.: Augmented reality tools for integrative science and arts STEAM education. Int. J. Pure Appl. Math. **118**(24) (2018)
19. Lavicza, Z., Fenyvesi, K., Lieban, D., Park, H., Hohenwarter, M., Mantecon, J.D., Prodromou, T.: Mathematics Learning Through Arts, Technology and Robotics: Multi-and Transdisciplinary STEAM Approaches. In: 8th ICMI-East Asia Regional Conference on Mathematics Education, May, pp. 110–121 (2018). https://www.researchgate.net/publication/327402165
20. Lee, K.: Augmented reality in education and training. Techtrends **56**, 13–21 (2012)
21. Maass, K., Geiger, V., Ariza, M.R., Goos, M.: The role of mathematics in interdisciplinary STEM education. ZDM-Math. Educ. **51**(6), 869–884 (2019)
22. Mckechnie, D.H.J., Wilson, E.E.C.: Immersive virtual reality as a pedagogical tool in education: a systematic literature review of quantitative learning outcomes and experimental design. J. Comput. Educ. **8**(1) (2021). Springer Berlin Heidelberg.

23. Mishra, P., Koehler, M.J.: Technological pedagogical content knowledge: a new framework for teacher knowledge. Teach. Coll. Rec. **108**(6), 1017–1054 (2006)
24. Olmos-raya, E., Ferreira-cavalcanti, J., Contero, M.: Mobile virtual reality as an educational platform: a pilot study on the impact of immersion and positive emotion induction in the learning process. **14**(6) 2045–2057 (2018)
25. Osadchyi, V.V., Valko, N.V., Kuzmich, L.V.: Using augmented reality technologies for STEM education organization. J. Phys.: Conf. Ser. **1840**(1) (2021)
26. Pantelidis, V.S.: Reasons to use virtual reality in education and training courses and a model to determine when to use virtual reality. Themes Sci. Technol. Educ. **2**(1–2), 59–70 (2010)
27. Plan, L., Instrument, S. (n.d.) (2017): A psychometric approach to the development of a 5E lesson plan scoring instrument for inquiry-based teaching. 1–32.
28. Rodríguez, J.L., Morga, G., Cangas-Moldes, D.: Geometry teaching experience in virtual reality with NeoTrie VR. Psychol. Soc. Educ. **11**(3), 355–366 (2019)
29. Trappmair, A., Hohenwarter, M.: Driving Augmented Reality: Geogebra's New AR Features in Teaching Mathematics. In: Proceedings of Conference on Technology in Mathematics Teaching–ICTMT, vol. 14 (2019)
30. Wang, M.T., Degol, J.L., Amemiya, J., Parr, A., Guo, J.: Classroom climate and children's academic and psychological wellbeing: a systematic review and meta-analysis. Dev. Rev. **57**, 100912 (2020)
31. Wellner, P.M.W, Gold, R.: Computer augmented environments: back to the real world. Spec. Issue Commun. ACM. **36**(7) (1993)

Applying Prior Meta-Modeling Knowledge to a VR Model of a Biological Process

Susanne Jansen, Siti Faatihah Binte Mohd Taib, Yiyu Cai, and Wouter R. van Joolingen

Abstract The goal of this study was to find out whether students can extend prior meta-modeling knowledge, gained using 2D models, to 3D models. Therefore, we compared a group of students who received prior instruction on models and modeling with a group of students without such preparation and looked for differences in expressed meta-modeling knowledge considering a combination of 2D models and a 3D model. For the 3D model we developed a VR application on the biological process of blood-glucose regulation. Both groups of students worked with the VR application after which they worked on an assignment where they used a combination of 2D models and the VR model to answer questions related to important aspects of meta-modeling knowledge. A pre- and post-test was used to find out whether working with the VR model influenced students' expressed meta-modeling knowledge in general. Results from the assignment showed a higher level of expressed meta-modeling knowledge considering 2D and 3D models for the group of students with prior knowledge than for the group without prior knowledge. The pre- and post-test also showed that working with the VR model lead to a higher level of expressed meta-modeling knowledge for the group of students with prior meta-modeling knowledge. For the group of students without this prior knowledge, working with the VR model lead to a lower level of expressed meta-modeling knowledge. These results suggest that teaching students about aspects of meta-modeling knowledge and letting them work with these models is an important step in stimulating students' meta-modeling knowledge for both 2D and VR models.

S. Jansen · W. R. van Joolingen
Utrecht University, Utrecht, The Netherlands
e-mail: s.jansen2@uu.nl

S. F. B. M. Taib · Y. Cai (✉)
Nanyang Technological University, Singapore, Singapore
e-mail: MYYCai@ntu.edu.sg

Y. Cai et al. (eds.), *Mixed Reality for Education*, Gaming Media and Social Effects,
https://doi.org/10.1007/978-981-99-4958-8_4

1 Introduction

Models and science are inherently intertwined. Models are a central element of scientific inquiry, research and communication, where they are helpful tools for scientists to represent ideas or describe and predict processes that occur in the natural world [1, 2]. Creating models is a human enterprise, meant to simplify phenomena and help us make sense of the complex world around us. Models can do this by highlighting certain salient features of a system while minimizing the roles of others [1].

Knowledge about the use of models in science is not only useful for scientists, but also for non-scientists. As we have seen for example during the COVID-19 outbreak in 2020, results of research often find their way into society. In the COVID-19 case, models played a very important role in predicting the growth of the pandemic and communicating to society about this growth to justify the measures taken. Being able to understand such scientific models in daily life, is considered to be a component of scientifically literacy [3–7]. Scientific literacy involves the skills that are required for understanding science in everyday life and making personal decisions on socio-scientific issues. An example of such a socio-scientific issue is the vaccination debate in society [8, 9]. Considering scientific models, scientific literacy entails knowledge about models, the creation of models and the use of models (i.e. meta-modeling knowledge) [4]. To stimulate scientific literacy, teaching students about models and the process of modeling as a scientific practice is part of the curriculum in many countries (e.g. the United States Next Generation Science Standards [10], the National science curriculum in England [11], and the science curriculum in the Netherlands for the subjects physics, chemistry and biology [12]).

In the science of biology, many different types of models are used. Biological models can range from concrete scale models, such as a model of a human skeleton, to abstract models of complex biological processes, such as the process of photosynthesis. These complex models of biological processes are considered to be abstract, because they contain dynamics such as time and movement, which are often visualized by arrows (Fig. 1) [13]. Since these models consist of several concepts that are connected by these abstract dynamics, they are often called concept-process models [14].

Even though knowledge about models and the process of modeling is part of the curriculum, students are most often not explicitly taught the required meta-modeling knowledge that is necessary to understand models as they are used in science. Most teachers only use models to illustrate biological concepts or phenomena, but neglect teaching the scientific processes of creating and evaluating a model, or using the model to formulate hypotheses [1, 16]. A solution to this problem can be to incorporate suitable model-based learning approaches into the science curriculum (e.g. [17–20]. In practice, model-based learning approaches and model-based inquiries are often reflections and extensions of the scientific method [16]. They typically consist of five steps: (1) observation and data collection, (2) construction of a preliminary model, (3) application, (4) evaluation, and (5) revision of the preliminary model [21].

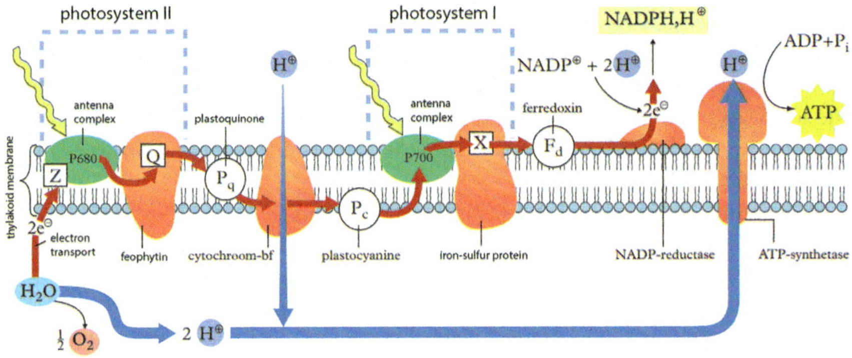

Fig. 1 A concept-process model of the light reaction of photosynthesis, as used in secondary schools in the Netherlands. The light reaction is the first step in the creation of glucose, the ultimate goal of photosynthesis. Reprinted and translated with permission from Noordhoff Uitgevers, Groningen [15]

To evaluate students' meta-modeling knowledge, various frameworks have been developed (e.g. [4, 6, 22–24]. In previous research we worked with the framework as described by Upmeier zu Belzen, [25], who combined the frameworks from [4, 22] and [23] to create a theoretical framework that can be used as an analytical framework for assessing and investigating students' understanding of models and their use in science. The five main aspects of this framework are *nature of models, multiple models, purpose of models, testing models* and *changing models*. For each of these aspects, several levels of understanding have been formulated, ranging from an initial level of understanding to an expert level of understanding (Level 3). The aspects *nature of models* and *multiple models* reflect on the way models describe phenomena. *Nature of models* focuses on the extent to which a model can be compared to the original. On the lowest level of reasoning, one can think that a model is always an exact copy of the original (Level 1). However, the creator of the model often makes choices, which can aid in emphasizing a specific part of a process. Examples are the colors that are used, the amount of detail that is present and the scale in which certain aspects are depicted (Level 2). At the highest level of reasoning, one can think of a model as a tool to highlight a certain hypothesis about how a process functions (Level 3). The aspect *multiple models* addresses the fact that multiple models can be used to represent the same original. Since no single model can show every detail of a process or explain all existing hypotheses and ideas, models are often simplified to fit the questioners' need and prior knowledge. This results in various different models showing the same process, but with a difference in focus (Level 2) or explaining different ideas (Level 3) [13, 26].

The aspects *purpose of models, testing models* and *changing models* together reflect on the use of models in science: to communicate ideas, to make predictions about future events and to test hypotheses [26]. The aspect *purpose of models* differentiates in the purpose that models serve in science. Models can be used to show what is known about a process (Level 1), to connect the depicted process to other aspects or processes (Level 2) or to formulate and test hypotheses (Level 3). The aspect *testing models* describes different ways in which models can be validated and the aspect *changing models* shows that models are per definition subject of change, due to for example falsification of a hypothesis about the original. For the aspects *testing models* and *changing models* three levels of understanding have also been described (see Table 1).

It was empirically shown by [17], that the Levels 1, 2 and 3 as mentioned in Table 1 reflect an increasing degree of difficulty. Reasoning on Level 1 is not wrong, it is just perceived as a more basic type of reasoning. In general, Level 1 focuses only on the model as a close representation of the original (*model object,* [27]. Level 2 has been reached when the model is seen as a medium that has been created based on the original (*model of something,* [27]. Level 3 shows an understanding of the use of models in science, in which models are used to test hypotheses and draw conclusions about the original (*model for something,* [27]. In contrast to Level 1–3, the initial level that has been formulated for the aspects *testing models, changing* models and *multiple models* cannot be seen as a valid way of reasoning. This level shows a lack of basic understanding of the aspects as mentioned in the framework, rejecting the fact that models can be tested or changed, or that multiple models of the same process exist [26].

Since Grünkorn et al. [26] evaluated this theoretical framework using a variety of biological contexts, this framework was the framework of our choice when working with biological models. However, considering biological concept-process models specifically, we had to extend the framework with categories that are specific for this type of models [13]. The resulting framework is shown in Table 1, where the categories that are specific for concept-process models are presented in bold. For a detailed description of the categories that are specific for concept-process models, see [13].

In previous research we combined the aspects and levels as described in Table 1 with both interventions as described by [18] and pedagogical content knowledge from teachers to develop teaching activities that focused on stimulating meta-modeling knowledge considering biological concept-process models specifically [28, 29]. In our research we mainly focused on the aspects *nature of models* and *multiple models.* These two aspects focus on models as they are presented in for example students' textbooks, without taking the modeling-process that includes testing and changing an existing model into account. Since students most often reason with existing models in class and on tests, it made sense to focus on students' reasoning considering these models. Using the framework we were able to assess the effect of these activities on students' expressed meta-modeling knowledge considering biological concept-process models for the aspects *nature of models* and *multiple models.* Results showed that the activities successfully improved students' expressed meta-modeling knowledge for these two aspects.

In our previous research we used existing educational two-dimensional models, such as the one shown in Fig. 1. However, recently 3D digital models are reaching classrooms in the form of augmented reality (AR) and virtual reality (VR) environments [30]. AR is a technology that allows virtual imagery information to be

Table 1 Framework to assess students' meta-modeling knowledge of biological concept-process models. The left column shows the five aspects that are important when reasoning with biological models. For each of these aspects up to four levels of understanding have been defined, ranging from an initial level of understanding to an expert level of understanding. Categories in bold have specifically been added to assess students' meta-modeling knowledge of biological concept-process models [13]

	Initial level	Level 1	Level 2	Level 3
Nature of models		– Model as copy – Model with great similarity – Model represents a (non-) subjective conception of the original – **Displays a process, its components and how they are related**	– Parts of the model are a copy – Model as a possible variant – Model as focused representation	– Model as hypothetical representation
Purpose of models		– Model for showing the facts – **Model for showing events**	– Model to identify relationships	– Model to examine abstract/concrete ideas
Multiple models	– All models are the same – Various models of different originals – Only one final and correct model	– Different model object properties	– Focus on different aspects	– Different assumptions
Testing models	– No testing of models – **Perceiving schoolbooks or their authors as authorities providing absolute truth**	– Testing of material – Testing of basic requirements	– Comparison between original and model – Comparison and matching of original and model	– Testing hypotheses – Testing of hypotheses with research designs

(continued)

Table 1 (continued)

	Initial level	Level 1	Level 2	Level 3
Changing models	– No reason for alterations – Alteration of how different originals are represented	– Alterations to improve the model object – Alterations when there are errors in the model object – Alterations when basic requirements are not met	– Alterations when model does not match the original – Alterations due to new findings about the original	– Alterations due to findings from model experiments – **Alterations when the focus of the model shifts to a different aspect of the process**

projected onto a live real-world environment [31]. Students can use a phone, tablet or other electronic innovations on which they see the virtual imagery projected onto their environment [32]. VR is different from AR in that it delivers an immersive experience, where a three-dimensional computer-generated virtual environment is created and the user is able to interact with this environment. In this study we will focus on the use of VR in education.

Shim et al. [33] argue that VR can be a promising addition to science teaching, since it can stimulate the multi-sensory organs of students, which motivates students and increases their interest in learning activities. Also, it is possible for real-time interaction between both the learner and a computer, and among several different users. Thirdly, the virtual environment enables experimentation to be carried out safely, in an easily controllable environment. This means students can carry out experiments that may otherwise be deemed too dangerous or expensive for the classroom. Finally, as with many other digital environments, students can participate in learning activities at their own comfort and pace. Shim et al. [33] studied the effect of VR simulations in biology education specifically and found that using VR actively engages students with the biological topic at hand and lets them immerse in the learning environment that is brought to them via the simulation. This experience is shown to be more immersive than the experience that other multimedia options offer.

In this current study we let students work with a VR model of the biological process 'blood-glucose regulation', a process that describes the regulation of the concentration of glucose in the blood involving several organs, hormones and enzymes [34]. Using both two-dimensional concept-process models covering this same process and the VR model, we focus on students' expressed meta-modeling knowledge for this combination of 2D and 3D model-types.

In our previous work [28, 29] we aimed to foster students' meta-modeling knowledge by letting them investigate various biological 2D models and by letting them construct such models themselves. In the current study our goal is to find out whether students' prior meta-modeling knowledge, obtained in these previous teaching and learning activities, was helpful in in understanding the VR model and whether this prior knowledge led to differences in learning outcomes in terms of expressing meta-modeling knowledge. Therefore, we compared the group that had received our prior instruction on models and modeling with a new group without such preparation and looked for differences in students' expressed meta-modeling knowledge after working with the VR model. This comparison was done in order to investigate the following research question:

RQ: Does prior instruction in meta-modeling result in differences in applying meta-modeling knowledge to a combination of 2D models and a 3D virtual environment?

2 Method

2.1 Participants

In total 88 Dutch eleventh-grade pre-university level students (16–18 years old) participated in our study. The group of students who already worked with important aspects of meta-modeling before consisted of 41 students (21 male, 20 female) and is referred to as the prior knowledge-group. These students had received three lessons on model-based reasoning in their previous academic year, in which both the aspects *nature of models, purpose of models* and *multiple models* were discussed and students gained experience in creating a model of a biological process themselves. The group of students who had not explicitly been introduced to these aspects before consisted of 47 students (20 male, 27 female) and is referred to as the comparison-group. All lessons were taught in the same classroom as where the students' biology lessons usually take place and informed consent was obtained from all students.

2.2 Overview of This Study

To focus on the effect of prior meta-modeling knowledge on students' expressed meta-modeling knowledge considering a combination of models, both the prior knowledge-group and the comparison-group received the same three lessons. During the first lesson students filled out a pre-test. In the second lesson the students worked with the VR model on blood-glucose regulation, after which they answered questions on paper related to the aspects *nature of models, purpose of models* and *multiple models.* The post-test was filled out during the third lesson.

2.3 The VR Application on Blood-Glucose Regulation

A VR application was developed, showing the process of blood-glucose regulation. The application can be used on mobile devices and is designed to introduce students to this biological process. This means that prior knowledge about the process of blood-glucose regulation is not necessary to work through the application. Students start in a virtual classroom where someone eats a donut. They then follow the glucose molecules that go from the small intestines, through the cell membrane of the small intestines, into the bloodstream and into a somatic cell. The scenes were chosen based on the different levels of biological organization that are commonly discussed in secondary education when this topic is discussed in class. Figure 2 shows stills from the VR application, where different environments are visible.

While working through the application, students answer questions about the process relating to either why something happens the way it does, or they predict what will happen next. All questions are multiple choice, and students receive instant feedback on the answer they choose. Figure 3 shows how questions are presented to students in the application.

After selecting an answer using the 'red dot' in the center of their vision, students hear whether the answer is correct or not. They also hear an explanation about why the answer is correct or not. When students do not answer correctly, they choose again. This cycle repeats itself until a student chooses the correct answer. The following text shows the way the example in Fig. 3 is introduced to the students and what kind of feedback they receive for each answer they choose.

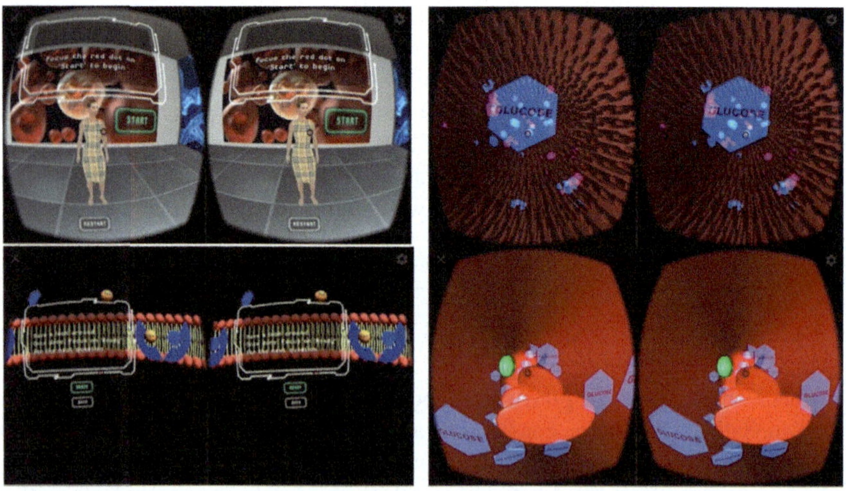

Fig. 2 Stills from the developed VR application, showing multiple levels of biological organization, from top left to bottom right: a classroom, the inside of the intestines, a cell membrane, and a blood vessel. The stills are screenshots from a mobile device. When the mobile device is inserted into VR-goggles, the two circles merge and the user experiences a 3D environment

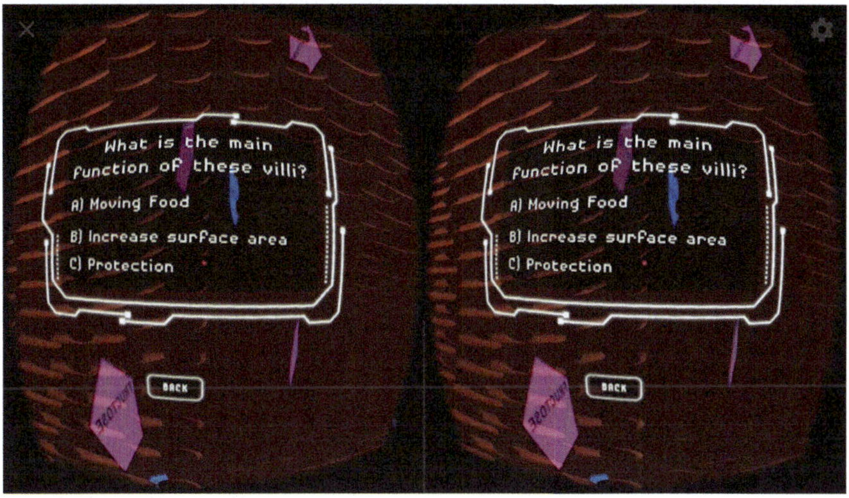

Fig. 3 A multiple choice question as presented to students in the VR application. The red dot in the center of each of the circles can be used to choose an answer

Teacher: the lumen of the small intestines is lined with villi. These are the tentacle-like structures that you see all around you. What is the main function of these villi?

A: They help digestion by moving/waving the food in the right direction (text on screen: moving food).

B: They increase the surface area for the uptake of nutrients (text on screen: increase surface area).

C: They protect the intestines from potentially damaging particles from your food (text on screen: protection).

In coherence with the given answer, the students hear a response from the teacher in the application:

A: Sorry, that is incorrect. muscle contractions of the small Intestines move the food in the right direction. let's try again!

B: You're right, awesome! a larger surface area increases the chance of nutrients being taken up by your body.

C: Sorry, that is incorrect. the shape of the villi does not contribute to this form of protection. let's try again!

To find out whether there is a difference between the prior knowledge-group and the comparison-group considering the way students work through the application, the first answer that students choose is registered for each question and for each student separately.

2.4 Pilot

The VR application was tested in a pilot session with 10 pre-university education students with a major in biology (16–17 years old) (Fig. 4).

Students worked through the application, after which they filled out an open-ended questionnaire where they could comment on the quality of the graphics, the quality of the audio, the given interaction possibilities, the way the blood-glucose regulation was explained in the application, and whether the questions in the application fitted the knowledge level of the students. Students were also given the opportunity to write down any other remarks or recommendations they had considering the use of the application. Results from the pilot study indicated that students very much enjoyed working with the application, that the audio worked well, that the feedback helped them understand the process better, that the process was easy to follow, and that the questions fitted the knowledge level of the students well. Two things did not work well according to the students: the graphics were blurry or slow on older phones, and the red dot that was implemented to interact with the virtual environment was difficult to see when the environment was red in color as well (inside the bloodstream). To make the red dot visible at all times, we placed a black circle around the red dot. Considering the blurriness, we decided not to compromise on the graphics. Instead we decided to bring a set of mobile phones on which we knew the application worked well to the classes in which we carried out our research.

Fig. 4 Testing the VR application. Students use their own mobile device, which they inserted into a Google cardboard device

2.5 Developed Student Assignment

To find out how students compared working with 2D models and the VR model, students received an assignment on paper after working with the application in which they were asked to use both types of models when answering questions related to the aspects *nature of models, purpose of models* and *multiple models.* Students were presented with four different models on blood-glucose regulation varying in level of biological organization and abstractness. Three of these models were 2D and one was the developed three-dimensional VR model. These four models were printed on paper for students to use while answering the questions during the assignment. For the VR model, a still from the application was printed and students were reminded that for this particular model they had to think of the complete VR simulation when answering the questions and not just look at the printed still.

The assignment consisted of seven tasks, which were based on the activities we developed during previous research that were used to introduce students to the aspects *nature of models, purpose of models* and *multiple models.* For a description of these activities and the theoretical background behind the tasks, see Jansen et al. [28]. The seven tasks were as follows.

1. Students had to name differences between the four models (relating to the aspect multiple models)
2. Students had to categorize the differences that they wrote down for question 1 (relating to the aspect multiple models)
3. Students had to match given aims with the four different models and explain their choices (relating to the aspect purpose of models)
4. Students had to explain whether all aims from question 3 could be met using only the VR model (relating to the aspect purpose of models)
5. Students had to explain whether all aims from question 3 could be met using only one of the 2D models (relating to the aspect purpose of models)
6. Students had to name choices that the creator of the models made in order to meet the prospected aim of these models (relating to the aspect nature of models)
7. Students had to explain whether they thought that eventually one ultimate model for the description of blood-glucose regulation would suffice (relating to the aspect nature of models).

2.6 Pre- and Post-Test

To measure the influence of having prior knowledge about the aspects *nature of models, purpose of models* and *multiple models* on students' expressed meta-modeling knowledge for these three aspects we used an online questionnaire. This questionnaire was previously developed based on work from [35] and assesses students' level of expressed meta-modeling knowledge for biological concept-process models specifically [36]. The questionnaire contains statements, relating to six different biological concept-process models. For each of these models a statement on Level 1, Level 2 and Level 3 is formulated for the aspects *nature of models* and *multiple models* (see Table 1 for a description of the levels). The test only focuses on two aspects, since including all aspects would generate too many questions for students to answer. The choice for the aspects *nature of models* and *multiple models* was made since these aspects reflect on the students' meta-modeling knowledge considering existing concept-process models, resembling both the way students are presented with models during tests in school and the way students encounter scientific information in daily life. In the questionnaire, students have to mention whether they agree or disagree with the given statements. Students' agreement with the statements is interpreted as their expressed meta-modeling knowledge for the aspects *nature of models* and *multiple models* [35, 36]. Figure 5 shows a translated example of one of the closed-question tasks in this test, containing a model showing the process of bioaccumulation. In this case the task reflects on the aspect *multiple models.*

2.7 Student Interviews

To find out how students experienced the intervention and to substantiate the results from the student material and the pre- and post-test, four students from the comparison-group and six students from the prior knowledge-group were interviewed after the completion of the lesson series (three lessons). Interviews were recorded and lasted approximately 20 min. The students were asked to reflect on both their experience with the VR application and the student material that contained questions about the aspects *nature of models, purpose of models* and *multiple models.* The interview questions were related to whether students had any experience working with VR before the intervention, whether they would recommend using this VR application in class, in what way 2D models from their biology textbook are usually discussed in class, whether their biology teacher discusses the use of models in biology in general (and if so, in what way), which question from the student material they thought was difficult and which question they thought was easy, and whether they thought it would be good to spend more time in class on the use of models in biology in general. The prior knowledge-group was also asked whether they had used the theory about important aspects of meta-modeling knowledge that they learned about in their previous year while answering the questions in the student material, and whether they thought having knowledge about these aspects helped them understand the models that are used in biology education better.

Example item

Frank is doing research on water contamination. He discovered that in areas where the water is contaminated with DDT (a chemical insecticide), many fish eating birds are dying. He creates the following model.

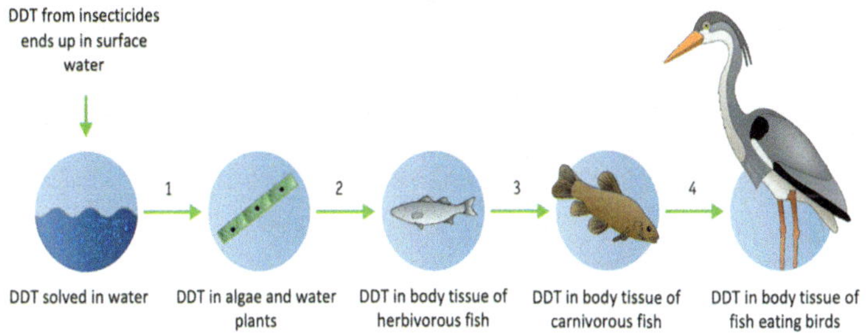

Maud is also doing research on water contamination with DDT. She creates a model of the same process. When can the model that Maud creates be defined as a different model than Franks' model?

Answer for each of the statements if you agree (yes) or disagree (no)

When Maud creates a model for an area that is contaminated with DDT, but where different species of fish are living.

[ ] Yes

[ ] No

When Maud creates a model where the focus is more on step 2 of the process.

[ ] Yes

[ ] No

When Maud has different thoughts on how DDT ends up in fish eating birds and creates a model reflecting these thoughts.

[ ] Yes

[ ] No

Fig. 5 An example of one of the tasks in the questionnaire, showing a model of bioaccumulation. The aspect of focus in this example is *multiple models* and each statement represents a level of understanding in the order: Level 1, Level 2 and Level 3 (the order of the levels differed between different tasks). The model present in this task is reprinted and translated with permission from [37]

2.8 *Data Management*

2.8.1 VR Application

We used the student answers from the VR application on the topic blood-glucose regulation to determine whether there was a difference in prior knowledge considering the process of blood-glucose regulation between the two groups on this topic. The percentage of students that answered the questions in the VR application correctly on their first try was determined for each question separately. Using an unpaired t-test we determined whether differences between the two groups could be considered significant.

2.8.2 Student Assignment

The student assignment contained questions relating to the aspects *nature of models, multiple models* and *purpose of models.* Student answers were coded using the categories, aspects and levels as described in Table 1. Ten percent of the student work was coded by a second independent coder, resulting in 84% agreement between the first and second coder. After discussing differences, 100% agreement was reached. The average agreement score (AAS) per aspect and Level was determined for both the prior knowledge-group and the comparison-group. Using an unpaired t-test we calculated whether the differences in student answers could be considered significant.

2.8.3 Pre- and Post-Test

The results from the pre- and post-test were analyzed to determine the influence of having prior knowledge about the aspects *nature of models, purpose of models* and *multiple models* on students' expressed meta-modeling knowledge for the aspects *nature of models* and *multiple models.* The results of the test showed for six different biological concept-process models whether students agreed with statements on Level 1, Level 2 and Level 3 for the aspects *nature of models* and *multiple models.* The agreement score for all levels and aspects was determined separately for each student. Since the pre- and post-test contained six tasks per aspect, and each task contained three statements (on Level 1, Level 2 and Level 3), the agreement score per aspect ranges from 0 to 6, where 0 means disagreement with all statements, and 6 means agreement with all statements. After determining the agreement score per student per question, the AAS per aspect (*nature of models* and *multiple models*) for students in the prior knowledge-group and the comparison-group was determined. Using a paired t-test we calculated whether the difference between the number of times a student agreed with statements on a certain level in the pre-test and post-test could be considered significant. Using an unpaired t-test we calculated whether differences between the prior knowledge-group and the comparison-group could be considered significant.

Student Interviews

Student interviews were transcribed verbatim and scanned for utterances related to students' experience with the VR application, their view on the use of VR in class and their view on the use of models in education. These utterances were only used to substantiate the results from the pre- and post-test and from the student assignment.

3 Results

3.1 VR Application

The VR application on blood-glucose regulation contained 11 questions. The percentage of students that answered each question correctly on their first try was determined for both the experimental group and the control group. Using an unpaired t-test we determined for each question whether there was a difference in the number of correct answers between the experimental group and the control group. Table 2 shows the results per question, where no significant differences between the prior knowledge-group and the comparison-group were found. As shown in Table 2, the number of registered student answers drops for each question in both the prior-knowledge group and the comparison-group. Even though almost all students finished working with the app, not all student answers were registered. This is because the app only registers a student answer the first time the app is started by that student. When a student exits the app and restarts the app, no new student answers are registered. Students accidentally pressing the exit button on their phone, WiFi problems causing the app to crash, or students switching from phones after a few questions because they felt their phone was 'too slow' caused the drop in registered number of student answers.

3.2 Student Assignment

Student answers on the assignment were scored using the framework as shown in Table 1. Answers to questions relating to the same aspect of meta-modeling knowledge (*nature of models, multiple models* and *purpose of models*) were combined to get an overview of students' reasoning per aspect. An unpaired t-test was carried out to compare results from the prior knowledge-group and the comparison-group. Table 3 shows these results, showing that the prior-knowledge group more often scores a Level 2 answer for the aspects *nature of models* ($t(198) = 4.24$, $p < 0.01$) and *multiple*

Table 2 Results per question for the VR application. Only those answers that were registered were taken into account (n = number of registered student answers, p-group = prior knowledge-group, c-group = comparison-group)

Question	P-group (n)	C-group (n)	Correct answers p-group (%)	Correct answers c-group (%)	t	P
1	47	64	46.8	62.5	1.64	0.10
2	40	60	82.5	78.3	0.51	0.61
3	30	59	76.7	72.9	0.39	0.7
4	22	55	86.4	92.7	0.77	0.45
5	17	54	58.8	59.3	0.03	0.98
6	15	51	93.3	96.1	0.38	0.71
7	12	49	25	51	1.74	0.09
8	13	41	92.3	80.5	1.19	0.24
9	11	35	100	100	n/a	n/a
10	10	33	80	97	1.24	0.25
11	10	28	60	60.7	0.03	0.97

Table 3 Results for the student assignment per aspect and level (n = number of scored student answers, M = mean (ranging from 0 to 1), p-group = prior knowledge-group, c-group = comparison-group)

		P-group (n)	C-group (n)	M p-group	M c-group	t	p
Nature of models	Level 1	90	109	0.71	0.86	2.59	0.01
	Level 2	90	109	0.61	0.32	4.24	<0.01
	Level 3	90	109	0.03	0.01	1.14	0.25
Multiple models	Level 1	136	193	0.90	0.97	2.70	0.01
	Level 2	136	193	0.53	0.33	3.71	<0.01
	Level 3	136	193	0.01	0.00	1.00	0.32
Purpose of models	Level 1	134	188	0.93	0.96	1.18	0.24
	Level 2	134	188	0.07	0.04	1.40	0.16
	Level 3	134	188	0.00	0.01	1.00	0.32

models ($t(328) = 4.12$, $p < 0.01$) than the comparison-group. The comparison-group more often scores a Level 1 answer for the aspects *nature of models* ($t(198) = 2.59$, $p = 0.01$) and *multiple models* ($t(328) = 2.70$, $p = 0.01$).

3.3 Pre- and Post-Test

Students' agreement with statements on the pre- and post-test can be directly related to their expressed meta-modeling knowledge for the aspects *nature of models* and *multiple models*. Students' average agreement score (AAS) for each aspect and level was determined. Since students were presented with six statements for each level and aspect, students' AAS ranges from 0 to 6. Results for the pre- and post-test were compared for both the prior knowledge-group and the comparison-group using a paired t-test. Table 4 shows these results. For the prior knowledge-group we found no significant differences in AAS between the pre-test and the post-test (see Table 4). In the control group the difference in AAS between the pre-test and the post-test for statements relating to *multiple models* Level 1 ($t(46) = 4.12$, $p < 0.01$) and *multiple models* Level 3 ($t(46) = 2.62$, $p = 0.01$) could be considered significant. Students' AAS was higher in the post-test than in the pre-test for statements relating to *multiple models* Level 1, and lower in the post-test than in the pre-test for statements relating to *multiple models* Level 3.

An unpaired t-test was used to find out whether there are differences in AAS between the prior knowledge-group and the comparison-group for both the pre-test and the post-test. Table 5 shows the results for this comparison. The difference in AAS between the prior knowledge-group and the comparison-group for the aspect *nature of models* could be considered significant for statements relating to Level 3 ($t(87) = 2.75$, $p < 0.01$) in the pre-test. The difference in AAS between the prior knowledge-group and the comparison-group for the aspect *multiple models* could be

Table 4 Comparison between the pre-test and the post-test for both the prior knowledge-group and the comparison-group (AAS = average agreement score)

		Prior knowledge-group				Comparison-group			
		AAS pre-test	AAS post-test	t	p	AAS pre-test	AAS post-test	t	p
Nature of models	Level 1	4.31	4.42	0.42	0.68	4.27	4.55	0.96	0.34
	Level 2	4.89	5.18	1.11	0.27	4.64	4.47	0.57	0.57
	Level 3	5.13	5.16	0.10	0.92	4.39	4.59	0.60	0.55
Multiple models	Level 1	1.47	1.84	1.31	0.19	1.10	2.07	4.12	<0.01
	Level 2	4.18	4.87	1.47	0.15	3.59	3.75	0.39	0.70
	Level 3	5.63	5.68	0.27	0.79	5.41	4.80	2.62	0.01

Table 5 Comparison between the prior knowledge-group and the comparison-group for both the pre-test and the post-test (AAS = average agreement score, p-group = prior knowledge-group, c-group = comparison-group)

		Pre-test				Post-test			
		AAS p-group	AAS c-group	t	p	AAS p-group	AAS c-group	t	p
Nature of models	Level 1	4.31	4.27	0.19	0.85	4.42	4.55	0.41	0.68
	Level 2	4.89	4.64	1.01	0.31	5.18	4.47	2.50	0.01
	Level 3	5.13	4.39	2.75	<0.01	5.16	4.59	1.68	0.10
Multiple models	Level 1	1.47	1.10	1.64	0.11	1.84	2.07	0.80	0.43
	Level 2	4.18	3.59	1.28	0.21	4.87	3.75	2.76	<0.01
	Level 3	5.63	5.41	1.15	0.25	5.68	4.80	3.78	<0.01

considered significant for statements relating to Level 2 ($t(87) = 2.76$, $p < 0.01$) and Level 3 ($t(87) = 3.78$, $p < 0.01$) in the post-test. For all these differences the AAS for the prior knowledge-group was higher than the AAS for the comparison-group.

4 Conclusion and Discussion

The aim of this study was to find out what the influence of prior experience with meta-modeling aspects was on students' expressed meta-modeling knowledge considering a combination of biological 2D models and the VR model on blood-glucose regulation. We compared a group of students who previously received three lessons on aspects of meta-modeling knowledge (prior knowledge-group) with a group of students who did not receive these lessons (comparison-group). Both groups worked with the same VR model on blood-glucose regulation, after which they answered questions relating to both 2D models and the VR model focusing on different aspects of meta-modeling knowledge.

Multiple questions related to the biological process blood-glucose regulation were incorporated in the developed VR application. These questions were not related to aspects of meta-modeling knowledge and were meant to see whether students' knowledge on the biological topic was comparable between the experimental group and the control group. No significant differences were found between student answers from both groups. Despite the low number of registered student answers for the second half of the questions in the VR application, we assumed that no difference in prior knowledge considering the topic of blood glucose regulation between the prior knowledge-group and the comparison-group existed.

Both groups filled out an online pre- and post-test to find out students' expressed level of meta-modeling knowledge relating to the aspects *nature of models* and *multiple models*. In this test students have to agree or disagree with statements related to the three levels of meta-modeling knowledge (Table 1) for both of these aspects. All three levels show valid ways of reasoning, but they increase in difficulty. Ideally students should be able to reason on all three levels as described in Table 1 [35].

When comparing the results from the prior knowledge-group and the comparison-group, we see that in the pre-test the prior knowledge-group significantly more often agreed with statements related to the highest level of reasoning (Level 3) for the aspect *nature of models* than the comparison-group. In the post-test the prior knowledge-group more often agreed with statements relating to Level 2 and Level 3 for the aspect *multiple models* than the comparison-group. The prior knowledge-group's higher agreement score relating to the higher levels of reasoning suggests that the lessons they received in their previous academic year influenced their expressed meta-modeling knowledge.

Looking at the difference in results between the pre- and post-test for both the prior knowledge-group and the comparison-group, we see that while there is no significant difference in student results for the prior knowledge-group, the comparison-group significantly more often agreed with statements on Level 1 for the aspect *multiple models* and significantly less often with statements on Level 3 for the aspect *multiple models* in the post-test than in the pre-test. The lower AAS for Level 3 statements in the post-test for the aspect *multiple models* shows that just answering questions that are related to the aspects of meta-modeling knowledge without receiving any theory on these aspects does not trigger a Level 3 type of reasoning. This suggests that in order to stimulate this level of reasoning, students need practice related to this aspect and level of meta-modeling knowledge. The increase in AAS for *multiple models* Level 1 statements could be due to the assignment where students were shown different models related to the same biological topic (blood-glucose regulation) and had to name differences between these models. Since a Level 1 type of reasoning for this aspect focuses on aesthetic differences, such as differences in color, having seen these differences during the lesson might have been enough to trigger this level of reasoning. The higher agreement score from the prior knowledge-group for Level 2 statements in the post-test suggests that working with the different models and/or being confronted again with questions relating to aspects of meta-modeling knowledge might have freshened up a Level 2 type of reasoning in students for this aspect. The following student quote shows that students from the prior knowledge-group were not reminded of the theory on model-based reasoning outside the three lessons they received in their previous year. This means that the current lesson, where they had to work with three of the aspects (*nature of models, purpose of models* and *multiple models*), was the first time since these three lessons in their previous year that they were reminded of these aspects.

> E2: Usually the teacher shows a picture of a model on the screen and then we just go through the model together. So the teacher says something like, you have this substance and that goes in here, then this happens, and then that. So just a bit of explanation about the process, but nothing more. Apart from those three lessons last year, nothing is said about models in general.

The assignment on paper that students received after working with the VR application consisted of open-ended questions related to the aspects *nature of models, purpose of models* and *multiple models*. Students used both 2D models and the VR model to answer these questions. Results show that the prior knowledge-group significantly more often answered with a Level 2 type of reasoning for the aspects *nature of models* and *multiple models*, and the comparison-group significantly more often answered with a Level 1 type of reasoning for these aspects. Since the questions were open-ended, students most often only wrote down one type of answer. This means that the answer they gave can be interpreted as their preferred level of reasoning, and that the students in the prior knowledge-group preferred a higher level of reasoning for the aspects *nature of models* and *multiple models* than the comparison-group. The first question, where students had to write down differences between the four models of blood-glucose regulation, can be seen as an exception. Since students had to write down at least six differences, there was room for answers on multiple levels of reasoning for the aspect *multiple models*. Results show that students from the prior knowledge-group often do not mention a Level 1 type of reasoning (aesthetic difference) among these differences. This might be caused by the lessons they received in their previous academic year on the different levels of meta-modeling knowledge. Having knowledge about the higher levels of reasoning might cause students to reject the lower levels of reasoning, even though these levels are still valid ways of reasoning. The following student quote from the prior knowledge-group illustrates this way of thinking:

> E1: I thought the first question was very difficult, because you had to name six differences between the models. And then I think, well, a difference in color is a difference, but is that difference noteworthy? That made it difficult for me, because now I looked for differences that I thought would really be noteworthy, but I couldn't find six of those.

Both the prior knowledge-group and the comparison-group mention the VR model and the 2D models when answering the questions on the student assignment. This indicates that the lessons about important aspects of meta-modeling knowledge that the prior knowledge-group received, which only contained 2D models, did not lead to a preference in type of models when answering questions that are related to these aspects. Our results also indicate that students' knowledge about important aspects of meta-modeling knowledge related to 2D models also leads to higher levels of expressed meta-modeling knowledge when working with a combination of 2D models and the VR model. However, since this study used a combination of 2D models and the VR model, future research should focus on the way students apply their meta-modeling knowledge to VR models only.

Student interviews revealed how students experienced using VR in class. Consistent with literature on this topic [33], students mentioned that they enjoyed working with the VR application. As shown by the following student quote, students found working with the app engaging.

> C1: I enjoyed the lesson, it was more fun than a traditional lesson. There was an interactive aspect in the application. I like that a lot, that's the way I like to learn something. And you're constantly busy, you cannot really get distracted. It was also relatively quiet in class, that's usually not the case. So, it really had a positive impact on the way of learning and on the ambiance in class. It motivated me more to learn, because when I just read something it feels like I'm quickly filled with information and then I get tired and bored. This interactive aspect works much better for me.

Both the comparison-group and the prior knowledge-group respond in a positive way when asked whether they would recommend using VR in biology education. They mention that using VR would bring more variety in the lessons, that the virtual environment keeps them focused on the biological topic, and that they appreciate the interactive element in the VR application because it results in a more personal experience where each student can go through the application at their own pace. In line with the research by [33], one of the students mentioned an advantage of using VR for biology lessons specifically:

> E3: Especially for biology lessons I can see the advantages. Because you can really look inside an organ. Not in the way we do in a practical lesson with a knife, but that you can also see what happens inside an organ. I think that is pretty useful. Of course there are other ways in which you can sort of learn about that, but those are usually images, or you have to use a knife to look inside an organ, or it's a static model.

4.1 Limitations

The way students have been selected for this study can be seen as a first limitation. Since we used students from a previous study as the prior knowledge-group, this can be seen as opportunistic sampling. The reason we chose this group of students was to save time. By selecting students from a similar school that is located in the same area for our comparison-group, we tried to make sure that we limited the variables in-between the comparison-group and the prior knowledge-group.

Second, even though our study showed that students are able to work with both 2D models and the VR model, it has to be noted that the VR model differs from the

2D models in more than one way. Not only provides the VR model a 3D instead of a 2D experience, it also brings the opportunity to merge multiple levels of biological organization into a single application and it creates an interactive environment where dynamics such as time and movement can be visualized without the use of arrows. This raises the question whether the VR application should be interpreted as a single model, or as multiple models in a single application, and thus whether 2D models and a VR model can both be used as ‘single models’ in studies such as the one described here.

As mentioned by [38], the use of multiple models can be beneficial for learning when a single representation would be too complicated if it presented all the information or if the information is on radically different scales. This justifies the use of multiple models, or sub-models, in the VR application. It also shows that multiple 2D models are necessary for students to understand a biological process on multiple levels of biological organization. When students worked with the VR model, they actually worked with multiple sub-models of the same biological process. In the VR model students shift between different levels of biological organization, meaning that they see the process on different scales. In similar situations where students work with 2D models, including in the current study, they usually work with multiple 2D models showing the same levels of biological organization as we used in the VR model. Since we used multiple 2D models of the same biological process, we believe the multi-model nature of the VR model did not have a major impact on our study. The following student quote shows how a student from the prior knowledge-group looked at this aspect:

> E4: I really enjoyed the application, I felt it was a good way to use multiple sub-models, so to say. As if you can zoom in from one model into another aspect [of the process] and see more details. That made things very clear to me.

4.2 *Contributions*

The VR app that was developed during this study can be used in other settings to both let students experience working in a VR environment, while teaching them about blood-glucose regulation.

Also, this study indicates that teaching students about aspects of meta-modeling knowledge and letting them work with these models is an important step in stimulating students’ meta-modeling knowledge for both 2D and VR models. Results show that students who previously learned about important meta-modeling aspects using 2D models are able to extend this knowledge to a VR model.

4.3 Future Research and Recommendations

Looking at students' view on the use of VR in biology class, we would encourage teachers to incorporate the use of VR in biology class. It can be an important motivator for students, but also brings the opportunity to show biological processes on various levels of biological organization.

Considering future research, we believe it is important to focus on the way students interpret VR models, to find out what differences they experience between 2D models and VR models. Knowing how students' interpret these models can aid in finding out what is necessary to further stimulate their meta-modeling knowledge.

References

1. Hoskinson, A.M., Couch, B.A., Zwickl, B.M., Hinko, K.A., Caballero, M.D.: Bridging physics and biology teaching through modeling. Am. J. Phys. **32**, 434 (2014). https://doi.org/10.1119/1.4870502
2. Svoboda, J., Passmore, C.: The strategies of modeling in biology education. Sci. Educ. **22**(1), 119–142 (2013). https://doi.org/10.1007/s11191-011-9425-5
3. Feinstein, N.: Salvaging science literacy. Sci. Educ. **95**(1), 168–185 (2011). https://doi.org/10.1002/sce.20414
4. Grosslight, L., Unger, C., Jay, E., Smith, C.L.: Understanding models and their use in science: conceptions of middle and high school students and experts. J. Res. Sci. Teach. **28**(9), 799–822 (1991). https://doi.org/10.1002/tea.3660280907
5. Oh, P.S., Oh, S.J.: What teachers of science need to know about models: an overview. Int. J. Sci. Educ. **33**(8), 1109–1130 (2011). https://doi.org/10.1080/09500693.2010.502191
6. Schwarz, C.V., Reiser, B.J., Davis, E.A., Kenyon, L., Achér, A., Fortus, D., Shwartz, Y., Hug, B., Krajcik, J.: Developing a learning progression for scientific modeling: making scientific modeling accessible and meaningful for learners. J. Res. Sci. Teach. **46**(6), 632–654 (2009).https://doi.org/10.1002/tea.20311
7. Sharon, A.J., Baram-Tsabari, A.: Can science literacy help individuals identify misinformation in everyday life? Sci. Educ. **104**(5), 873–894 (2020). https://doi.org/10.1002/sce.21581
8. Lundström, M., Ekborg, M., Ideland, M.: To vaccinate or not to vaccinate: how teenagers justified their decision. Cult. Sci. Edu. **7**, 193–221 (2012). https://doi.org/10.1007/s11422-012-9384-4
9. Roberts, R., Gott, R.: Questioning the evidence for a claim in a socio-scientific issue: an aspect of scientific literacy. Res. Sci. Technol. Educ. **28**(3), 203–226 (2010). https://doi.org/10.1080/02635143.2010.506413
10. NGSS Lead States : Next Generation Science Standards: For states, by states. The National Academies Press (2013). https://doi.org/10.17226/18290
11. GOV.UK: National curriculum in England: science programmes of study, Department for Education (2015). https://www.gov.uk/government/publications/national-curriculum-in-england-science-programmes-of-study/national-curriculum-in-england-science-programmes-of-study
12. CvTE: Biologie VWO syllabus centraal examen [Biology pre-university level syllabus for central examination] (2018). https://www.examenblad.nl/examenstof/syllabus-2020-biologie-vwo/2020/f=/biologie_2_versie_vwo_2020.pdf
13. Jansen, S., Knippels, M.C.P.J., van Joolingen, W.R.: Assessing students' understanding of models of biological processes: a revised framework. Int. J. Sci. Educ. **41**(8), 981–994 (2019). https://doi.org/10.1080/09500693.2019.1582821

14. Harrison, A.G., Treagust, D.F.: A typology of school science models. Int. J. Sci. Educ. **22**(9), 1011–1026 (2000). https://doi.org/10.1080/095006900416884
15. Brouwens, R., de Groot, P., Kranendonk, W.: BiNaS, 6th edn. Noordhoff Uitgevers (2013)
16. Windschitl, M., Thompson, J., Braaten, M.: Beyond the scientific method: model-based inquiry as a new paradigm of preference for school science investigations. Sci. Educ. **92**(5), 941–967 (2008). https://doi.org/10.1002/sce.20259
17. Krell, M., zu Belzen, A.U., Krüger, D.: Students ' understanding of the purpose of models in different biological contexts. Int. J. Biol. Educ. **2**(2), 1–34 (2012)
18. Quillin, K., Thomas, S.: Drawing-to-learn: a framework for using drawings to promote model-based reasoning in biology. CBE Life Sci. Educ. **14**(1) (2015). Online publication. https://doi.org/10.1187/cbe.14-08-0128
19. Schwarz, C.V., White, B.Y.: Metamodeling knowledge: developing students' understanding of scientific modeling. Cogn. Instr. **23**(2), 165–205 (2005). https://doi.org/10.1207/s1532690xci2302_1
20. Treagust, D.F., Chittleborough, G., Mamiala, T.L.: Students' understanding of the role of scientific models in learning science. Int. J. Sci. Educ. **24**(4), 357–368 (2002). https://doi.org/10.1080/09500690110066485
21. Fretz, E.B., Wu, H.K., Zhang, B.H., Davis, E.A., Krajcik, J.S., Soloway, E.: An investigation of software scaffolds supporting modeling practices. Res. Sci. Educ. **32**, 567–589 (2002). https://doi.org/10.1023/A:1022400817926
22. Crawford, B., Cullin, M.: Dynamic assessments of preservice teachers' knowledge of models and modelling. In: Boersma, K., Goedhart, M., de Jong, O., Eijkelhof, H. (eds.) Research and the Quality of Science Education, pp. 309–323. Springer (2005). https://doi.org/10.1007/1-4020-3673-6_25
23. Justi, R.S., Gilbert, J.K.: Teachers' views on the nature of models. Int. J. Sci. Educ. **25**(11), 1369–1386 (2003). https://doi.org/10.1080/0950069032000070324
24. Louca, L.T., Zacharia, Z.C., Constantinou, C.P.: In quest of productive modeling-based learning discourse in elementary school science. J. Res. Sci. Teach. **48**(8), 919–951 (2011). https://doi.org/10.1002/tea.20435
25. Upmeier zu Belzen, A., van Driel, J., Krüger, D.: Introducing a framework for modeling competence. In: zu Belzen, A.U., van Driel, J., Krüger, D. (eds.) Models and Modeling in Science Education, vol. 12, pp. 3–19. Springer (2019). https://doi.org/10.1007/978-3-030-30255-9_1
26. Grünkorn, J., zu Belzen, A.U., Krüger, D.: Assessing students' understandings of biological models and their use in science to evaluate a theoretical framework. Int. J. Sci. Educ. **36**(10), 1651–1684 (2014).https://doi.org/10.1080/09500693.2013.873155
27. Mahr, B.: Information science and the logic of models. Softw. Syst. Model. **8**(3), 365–383 (2009). https://doi.org/10.1007/s10270-009-0119-2
28. Jansen, S., Knippels, M.C.P.J., van Joolingen, W.R.: Lesson study as a research approach: a case-study. Int. J. Lesson Learn. Stud. **10**(3), 286–301 (2021). https://doi.org/10.1108/IJLLS-12-2020-0098
29. Jansen, S., Knippels, M.C.P.J., van Joolingen, W.R.: Key Activities for Supporting Students' Visual Literacy and Model-Based Reasoning With Biological Concept-Process Models [Manuscript submitted for publication]. Utrecht University, Freudenthal Institute (2021)
30. Velev, D., Zlateva, P.: Virtual reality challenges in education and training. Int. J. Learn. Teaching **3**(1), 33–37 (2017). https://doi.org/10.18178/ijlt.3.1.33-37
31. Zhou, F., Dun, H.B.L., Billinghurst, M.: Trends in augmented reality tracking, interaction and display: a review of ten years of ISMAR. In: Proceedings of the International Symposium on Mixed and Augmented Reality, UK, pp. 193–202 (2008). https://doi.org/10.1109/ISMAR.2008.4637362
32. Lee, K.: Augmented reality in education and training. TechTrends **56**, 13–21 (2012). https://doi.org/10.1007/s11528-012-0559-3
33. Shim, K.C., Park, J.S., Kim, H.S., Kim, J.H., Park, Y.C., Ryu, H.I.: Application of virtual reality technology in biology education. J. Biol. Educ. **37**(2), 71–74 (2003). https://doi.org/10.1080/00219266.2003.9655854

34. Ackerman, E., Gatewood, L.C., Rosevear, J.W., Molnar, G.D.: Model studies of blood-glucose regulation. Bull. Math. Biophys. **27**, 21–37 (1965). https://doi.org/10.1007/BF02477259
35. Krell, M.: Assessment of Meta-modeling Knowledge: Learning from Triadic Concepts of Models in the Philosophy of Science. Online publication, Science Education Review Letters (2019)
36. Jansen, S., Knippels, M.C.P.J., van Montfort, M., van Joolingen, W.R.: Secondary School Students' Meta-modeling Knowledge of Biological Concept-Process Models [Manuscript submitted for publication]. Utrecht University, Freudenthal Institute (2021)
37. Bos, A., Marianne, G., Jansen, A., Kalverda, O., de Rouw, T., Smits, G., Westra, R., et al.:. Biologie voor jou 5a VWO [Biology for you 5a pre-university level], 5th edn. Malmberg (2012)
38. Ainsworth, S.: DeFT: A conceptual framework for considering learning with multiple representations. Learn. Instr. **16**(3), 183–198 (2006). https://doi.org/10.1016/j.learninstruc.2006.03.001

A Systematic Review of Pedagogy Related to Mixed Reality in K-12 Education

Mafor Penn and Umesh Ramnarain

Abstract The recent development of applications for virtual reality (VR), augmented reality (AR) and mixed reality (MR) in education has stimulated interest in how these technologies can support teaching and learning at the school level. Of particular interest, is what pedagogies could be applied such that the affordances of these technologies are optimally exploited in supporting content understanding, long-term memory retention, improved task performance, and increased motivation and engagement for learners. This chapter is a systematic review of 18 peer-reviewed journal articles from the last decade starting 2011–2021 September related to MR pedagogy in K-12 education. The main aim of the review is to establish pedagogical approaches, theories, and the impact of integrating MR in K-12 education. Data from selected articles analysed by means of content analysis revealed that though most studies adopt interactive learner-centered pedagogies like inquiry-based, activity-based, discovery, and collaborative learning for MR interventions, they are quite inexplicit about the pedagogical dimensions of the MR-enhanced learning interventions. Based on the findings of this review, we propose a guiding framework for pedagogical processes to be followed in integrating MR technologies for K-12 teaching and make recommendations on future research that can be pursued in this area.

Keyword Augmented reality · K-12 education · Mixed reality · Pedagogy · Virtual reality

M. Penn · U. Ramnarain (✉)
University of Johannesburg, Johannesburg, South Africa
e-mail: uramnarain@uj.ac.za

M. Penn
e-mail: mpenn@uj.ac.za

Y. Cai et al. (eds.), *Mixed Reality for Education*, Gaming Media and Social Effects,
https://doi.org/10.1007/978-981-99-4958-8_5

1 Introduction

The "Reality-Virtuality continuum", which was proposed by Milgram and Kishino [42], has served for more than twenty years as the reference framework for classifying the different realities [20]. However, there is still a lack of consensus on the boundaries between these different realities (virtual, augmented, mixed) and so it is worth clarifying the meaning of these realities. Mixed reality (MR) is regarded as a combination of both augmented reality (AR) and virtual reality (VR) that offers the ability to physically interact with virtual objects in the real world [42]. MR differs from AR in that while AR enables the overlay of a virtual/digital object into the real world, MR offers a means for one to interact seamlessly with real and virtual elements across the reality continuum [29, 35]. MR is generally seen as an extension of AR, whereby it is difficult to separate what is virtual from what is real. Within an MR environment, the user is not only able to 'see' the virtual/digital overlay or object, but they can also physically or mentally interact with and/or manipulate it [42]. It is this ability to physically interact with digital/virtual overlays in the real world that differentiates AR and MR [38]. Virtual reality (VR), on the other hand, uses 3-dimensional (3D) computer-generated simulations that enables a person to interact with an artificial environment whereby the mind is tricked into the feeling of realism [21, 48]. VR applications are usually immersive and stimulate a false sense of reality in a user's mind as they use interactive devices (head-mounted displays [HMD], sensors, gloves, eye trackers) to explore the virtual world completely or partially shut out from the real environment. For this review, therefore, studies that focus on VR are not included as the element of virtuality alone does cover the scope of mixed reality. Our focus on the review is situated within AR/MR education in K-12.

As educational applications emerge, research on the educational uses of VR, AR and MR has been conducted [6, 35, 60]. The benefits of learning through VR, AR and MR include increased conceptual understanding of spatial structures, learning of language associations, long-term memory retention, improved physical task performance, and increased motivation and engagement [6, 33, 35, 49].

With these recent developments in recognising the affordances of VR, AR and MR in education, there is a renewed interest in how these technologies can support teaching and learning at school and tertiary education levels. However, with a focus on K-12 education, a total of eight reviews related to the augmented reality/virtuality continuum derived from our search have been conducted in the decade starting 2011–2021. These studies include [2, 4, 6, 23, 31, 38, 41, 49].

The review by Akçayır and Akçayır [2] examined 68 studies and concentrated on the advantages, and challenges, associated with using AR in education. From this review, the most reported advantage of AR is that it enhances learning achievement. Some noted challenges imposed by AR from this review include usability issues, cognitive overload, and frequent technical problems. One of the limiting factors of this review was the way in which the review articles were selected and the fact that the focus of the review was not streamlined to a particular educational setting.

In a systematic mapping study on collaborative educational environments by Ali et al. [4], the researchers reviewed scholarly articles from 2007 to 2017 on mixed reality technologies and the associated subcategories applied to collaborative education environments, including K-12 learning environments. Findings from this review highlighted the scarcity of research in mixed reality, with particular attention to the augmented virtuality category. The authors also mentioned technical limitations in the development of mixed reality technology for collaborative environments as well as the limitations teachers face with replicating mixed reality experiments in the real environment. Recommendations from this review were focused on the development of MR artifacts rather than approaches for integration in educational settings.

Bacca et al. [6] reviewed articles that focused on the advantages, limitations, effectiveness, and personalisation processes in AR educational applications as well as the use of AR for addressing the special needs of students in diverse contexts from six indexed publications in the decade starting 2003–2013. Key findings from this review showed that AR-enhanced learning for most participants fostered affective constructs like motivation, collaboration, and interactivity [6]. Some of the key findings of the study included the prevalent use of marker-based, location-based and marker-less AR applications. From this review, prominent challenges associated with the integration of AR at the school level included the difficulty related to students' focus on the technology rather than the learning objectives. With Bacca et al. [6] focusing both on K-12 and higher education, the scope becomes too wide to assess the pedagogical approaches infused while integrating the technology.

In the review by Garzón et al. [23], the researchers conducted a meta-analysis of 46 empirical studies with the goal of analysing the pedagogical strategies employed with AR technology in education set-ups. They also looked at the moderating variables that affect learner outcomes in AR interventions. Findings from this meta-analysis revealed that collaborative pedagogical approaches like inquiry-based learning, situated learning and project-based interventions were the most suitable for AR interventions. The limitations of the study related to the fact that the findings were not definitive due to the wide range of educational settings that were considered in the review.

The review by Koutromanos et al. [31] examined the use of AR game-based apps using mobile devices in primary and secondary school environments. The researchers selected seven peer-reviewed journal articles from 2000 to 2014 for two main databases, Science Direct and Eric. The findings revealed that AR games in informal learning environments could enhance active learning and positively affect learning outcomes through increased student involvement and participation.

A recent review by Maas and Hughes [38] explored studies that involved the use of virtual reality (VR), augmented reality (AR) or mixed reality (MR) technologies in the instruction of students in elementary, middle, or high school. The review revealed common themes including collaboration, communication, critical thinking, attitude, engagement, learning, motivation, performance or achievement, and technology. However, not much was revealed on the pedagogical approaches and theories considered for MR integration.

The review by Merchant et al. [41] focused on the use of desktop-based VR games, virtual worlds, and simulations. The researchers for this review looked at 67 empirical studies from 1987 to 2011. Findings from the study revealed that the VR-based learning applications had a positive influence on learning and that learners were able to retain more of what they had learned and tended to spend more time in the virtual environment. Limitations of the review were the focus on non-immersive VR and the broad timeframe selected.

Radu [49] examined 26 studies, including conference and journal articles in a meta-analysis that compared the effect of AR versus non-AR applications on learning. The benefits of AR that were revealed included increased content understanding, increased physical task performance and collaboration, while the detriments to using AR included problems with paying attention to the virtual object and the real environment simultaneously. Difficulty in classroom integration and some difficulties in using the technology were also reported as part of the challenges. The primary limitation of this review was the scope of article selection which was quite broad.

While VR, AR, and MR are well positioned to have an impact on K-12 education, research on the successful integration of these learning technologies in K-12 settings are at an infancy stage [7, 35]. Despite the numerous reviews reported, the study by Garzón et al. [23] is the only review that presented a meta-analysis with a focus on pedagogical approaches in general educational settings, including tertiary education. This pattern suggested the need for more reviews which report findings regarding the pedagogical strategies implemented in VR, AR and MR interventions. Pedagogical approach refers to the method that teachers use to deliver knowledge so that students engage in the learning process [53]. In addressing this gap, the chapter provides a systematic review of research on pedagogies associated with the use of MR in K-12 education. In this way, the chapter contributes toward knowledge on understanding pedagogical approaches that have efficacy and promote learning gains in mixed reality-enhanced teaching and learning. The chapter fits well with the theme for this book on "Mixed Reality for Education". The following research questions are addressed in this review:

- What research methods, subject areas, and education levels dominate studies integrating MR in K-12 education?
- What are the pedagogical approaches used in incorporating MR in K-12 education?
- Which pedagogical theories/frameworks are shown to be most effective when using MR in K-12?
- What pedagogical approaches in the use of MR in K-12 lead to high learning gains?

To answer the research questions, the following objectives were set:

- To establish the research methods, subjects, and level of education (primary or secondary) where MR is integrated.

- To present the pedagogical approaches used in incorporating MR in K-12 education.
- To examine the pedagogical theories that underpin the integration of MR in K-12 classrooms.
- And finally, to examine the pedagogical approaches that led to higher learning gains for K-12 learners.

Based on this current review, we propose a framework for examining pedagogical practices in the use of MR. In this framework, we highlight pedagogies that result in greater learning gains for students compared to traditional learning instructions.

In addressing MR in K-12 education settings, several studies have been conducted across primary and secondary school levels encompassing different subjects. In primary schools, MR has been integrated in subjects including general sciences, social sciences, and natural sciences [25, 28, 47, 59] that consist of formal [55, 63] and informal [1, 24] learning environments. Studies like [11, 34, 36, 39, 40] also looked at secondary school level integration. Natural sciences in secondary school studies could also be cited in studies like [12, 27].

In all these studies, the target is to ensure that students can enhance embodiment and spatial visualisation of concepts as they learn, creating adequate mental schemas that will facilitate deep learning and, subsequently, long-term memory [36, 51]. These learning gains attributed to mixed reality technologies are mainly because of the fundamental affordances of interactivity and immersion [47]. The MR environment becomes a useful tool for teaching and learning as students are not only able to immerse but learn interactively with technological devices while achieving lesson objectives [25].

2 Situating Pedagogy

For the purpose of this chapter, it is essential that we situate and discuss pedagogical approaches, theories and their relevance in any learning environment. In a nutshell, Pedagogy refers to the broader methods, teaching styles and strategies used by teachers to facilitate learning [53]. As expressly stated by Alexander [3], pedagogy is "the act and discourse of teaching" [3, p. 8]. It considers all the interactions that take place in a learning environment, including teaching, learning, assessments and the different socio-political contexts in which schools are situated [8, 17]. For teachers and researchers, this implies that when one examines the tenets of pedagogy a broader discourse beyond just teaching approaches should be considered. Hence, for this review, we consider a wider definition of pedagogy, including teaching approaches, theories, assessments, cognitive intention and all other contextual factors, including the curriculum and social context. The excerpt below emphasises a critical narrative that situates the relevance of this review in the words of Anthea Millet as captured in Alexander [3].

> I am always struck by how difficult teachers find it to talk about teaching … They prefer to talk about learning. By contrast, they can talk with great clarity about … curriculum, assessment … [and] classroom organisation … almost anything except teaching itself', an agenda which she said should cover 'competence, excellence and failure in teaching methods [3, p.9].

True to this observation and with the fast-changing design of new educational technology, a strong misconception continues to dominate education platforms where specialists assume that integrating the latest technologies will automatically enhance learner attitudes and learning gains in school subjects [37, 58]. This assumption typically leads to teachers focusing on illustrating technology use rather than carefully crafting the pedagogical approaches that best situate technological tools within the learning objectives and goals [9, 17, 58].

Pedagogical theories are closely associated with pedagogical approaches and learning theories in that they postulate the way teaching should be enacted to enhance students' learning and the acquisition of skills [52]. Three different perspectives of pedagogical theories are accepted in modern education including herbatianism, new London group and learning theory perspectives. Popularly regarded as the father of pedagogy and initiator of herbatianism, Johann Friederich Herbart (1776–1841) proposed five components of pedagogy, *including preparation, presentation, association, generalisation and applications. Preparation* involves getting ready for the teaching; *presentation*, the actual teaching; *association* focusing on cumulating the learnt ideas in one's memory; *generalisation* entails reasoning from minute facts to broader principles while the last component of pedagogy in herbatianism focuses on *application* which entails practicalising what has been learnt.

The second school of thought on pedagogy was proposed by the new London group [10], an international group of academics who placed their focus on literacy pedagogies. This group identified four components of pedagogy which are neither sequential nor hierarchical but can be used independently. These components include:

- *Situated practice* which takes learners through relevant processes in the form of practical/project-based learning that will help them acquire specific knowledge.
- *Overt instruction* which capitalises on what learners already know and the relevance of what needs to be learned and acquired.
- *Critical framing* a pedagogical style that develops the learner as a critique of learnt concepts whereby knowledge is critically analysed in relation to its context.
- *Transformed practice* whereby authentic learning experiences are promoted and learners are both products and transmitters of literacy learning, with the swapping of the student–teacher role being a key characteristic. This component sees learners working in new situations where their knowledge can be applied.

The last but not the least set of pedagogical theories stem from learning theories. Learning theories provide conceptual frameworks through which teaching and learning are attained [53]. Major learning theories include.

- *Behaviorism*: Centred on the conditioning and modifying of learners' behaviour.
- *Cognitivism*: Centred on the complexities of the human memory and its relationship to learning. The typical pedagogical approaches considered with this theoretical lens are strategies that promote long-term memory and retention of learned concepts. Prior knowledge is also highly esteemed with cognitive theories as well as the developmental stage in which learners are.
- *Constructivism*: Broadly divided into three categories, *cognitive constructivism, social constructivism,* and *radical constructivism* advocates the active involvement of learners in the knowledge construction process. This implies that pedagogical approaches framed by these theories must incorporate active rather than passive learning environments, a social and interactive environment and a learning environment that considers learners' prior knowledge in the construction of new knowledge.

With an outline of these pedagogical theories, evidence from research suggest that constructivism is the most used learning theory for designing interventions with educational technology [5, 23].

While MR and related technologies promise a wide range of educational solutions for K-12 education settings [38], not many studies have reported the details of the pedagogical approaches and theories that have underpinned the use of MR technology in K-12 learning interventions. In fact, two studies by Turan et al. [57] and Garzón and Acevedo [22] suggest that future research be conducted to address the impact of integrating different pedagogical approaches, especially for AR interventions. This gap highlights the need for education researchers to situate the use of specific technological tools within the principal aspect of education which is pedagogy.

3 Methods

In crafting this chapter, a systematic review of literature in the mixed reality spectrum was considered as it provided the opportunity to objectively analyse findings from several studies. A three-step process was followed for the review as suggested in Kitchenham [30], including the planning phase, the review phase, and the reporting phase.

In the planning phase, the Preferred Reporting Items for Systematic Reviews and Metaanalyses (PRISMA) procedure was considered [44]. The PRISMA provides a suitable frame for the methodology followed as it helps the researchers to describe the criteria and process of article selection. Using the PRISMA procedure, one is able to provide criteria for the selection of articles/sources, eliminating duplicates, the data collection processes, and synthesis of the results of the selected studies.

Following the PRISMA, the current review focuses on looking at the pedagogical approaches and frameworks that are used for the integration of mixed reality in K-12 education and how these affect learning gains. The researchers took advantage of the research gap that exists by analysing existing pedagogical approaches to mixed reality education. The sections below focus on the steps followed in the review.

3.1 Planning

3.1.1 Why Systematic Review of MR for K-12 Learning Environments?

In the last decade, there has been several review studies in tertiary education carried out in fields such as medical education, engineering and mining education on the affordances of MR. For example, in a study by Stretton et al. [56] the researchers looked at the use of mobile MR technology in healthcare education with the main aim of analysing how users used the technology in the acquisition of skills, including surgical and clinical assessment skills and the understanding of mostly human anatomy concepts. Findings from this review showed that students recorded better performance and skills towards concepts with the use of MR applications.

Another review by Chen and Duh [14] in tertiary education focused on 20 years of using MR in learning mining methods. One of the findings from the study broadly indicated that research in MR education was targeted especially at improving student performance and the usability of the technology.

From these mentioned tertiary education studies, several recommendations including, the need to clarify the theoretical underpinnings considered when designing and developing MR applications in alignment with pedagogical approaches was proposed. Both previous reviews suggested that MR seems to promote better learning achievements and outcomes for students.

Based on these recommendations and the limited number of reviews focused on MR in K-12 educational settings, it became imperative to target a review of the pedagogical strategies and theories considered when integrating MR applications in primary and secondary schools. At different school levels and ages, the technological expectations, learning demands, personal interaction skills, and attitudes are expected to be different and will continue to evolve as students develop in their schooling years. Only few studies like Garzón et al. [23] so far have considered providing a systematic review of relevant studies aimed at pedagogical approaches in relation to MR integration and how students' learning gains are projected in K-12 learning subjects. Furthermore, there is the need for a review that focuses on peer-review scholarly journal articles, excluding conferences and book chapters.

With the scarcity of reviews on MR pedagogy for K-12, this current review leverages on previous research to provide evidence on MR pedagogical approaches that work in K-12 educational settings. The review is significant in proposing a framework to guide teachers in seamlessly integrating MR technology in teaching different subjects across K-12, considering contextual factors like technological devices, applications, educational settings, curriculum and policies.

Criteria for Literature Selection

Based on the research questions, the following criteria were considered for the selection of literature for this review.

Broad Inclusion Criteria

- Studies from January 2011 to September of 2021.
- Studies targeted at the use of MR and covering the greater AR applications in K-12 educational settings.
- Studies that provided results of educational potential that were particularly based on a research design (or on robust) method using MR environments (prototypes).
- Peer-reviewed journal articles with an abstract and written in English.

Narrow Inclusion Criteria

- Peer-reviewed journal articles which reported the pedagogical approaches, strategies, and learning gains related to MR/AR-enhanced teaching and learning in K-12 across different school subjects
- Studies that described the MR/AR applications used for learning specific K-12 subjects
- Studies that presented pedagogical approaches/theories and the possible learning gains from assuming these approaches in various educational settings for K-12 education.

Exclusion Criteria

- Non-peer-reviewed journal articles were not considered (e.g., books, chapters, colloquiums, conferences, etc.).
- Studies that had no mention of mixed reality in the title or keywords and did not allude to the mixed reality continuum were excluded.
- Studies where findings were not clear and targeted to MR/AR learning for K-12 students and related to aspects of pedagogy.
- Lastly studies that focused mainly on VR as these excluded the real environment typically included in MR and AR technology.

3.2 Conducting the Review

3.2.1 Search Strategy

Using the researchers' institutional library pages and institutional access, databases related to educational studies were considered, including ERIC, E-journals (EBSCO-host), and Scopus. ScienceDirect, Google scholar, Web of Science, and Wiley were also included in a later advanced search just to ensure that a broader coverage of scholarly articles was attained using a forward and backward approach [22, 56].

The search process was conducted manually with pre-set parameters covering peer-reviewed journal articles for the last decade starting 2011–2021 September. Keywords were based on the research questions that the review intended to address, including *"mixed reality" and "pedagogy," "mixed reality in* K-12 education," "mixed reality instruction," and "mixed reality and pedagogical approaches."

Articles that also had the combination of AR and MR were considered based on the suggestion by Milgram et al. [43], who indicated that MR was more aligned to AR in the reality-virtuality continuum. With the focus on K-12, the starting point for the literature search was on MR/XR/AR in education and related publications. Journals related to the use of educational technology are indexed in the Journal Citation Report Social Science Citation Index (JCR SSCI) with other journals on general education research. The list of identified journals was validated by the authors to ensure that articles covered the relevant integration of technology and were peer-reviewed. The guidelines suggested by Kitchenham [30] were followed in selecting studies based on the key elements of K-12 education and MR/XR/AR. All authors were involved in extracting and discussing the articles to exclude all disagreements. Figure 1 below shows the PRISMA procedure for the articles selected.

From Fig. 1, a total of 232 articles were obtained from primary and advanced search engines. After duplicates were screened, 58 peer-reviewed articles related to MR/AR for broader educational purposes were considered for abstract screening. The researchers read all abstracts and based on the broad focus on educational settings

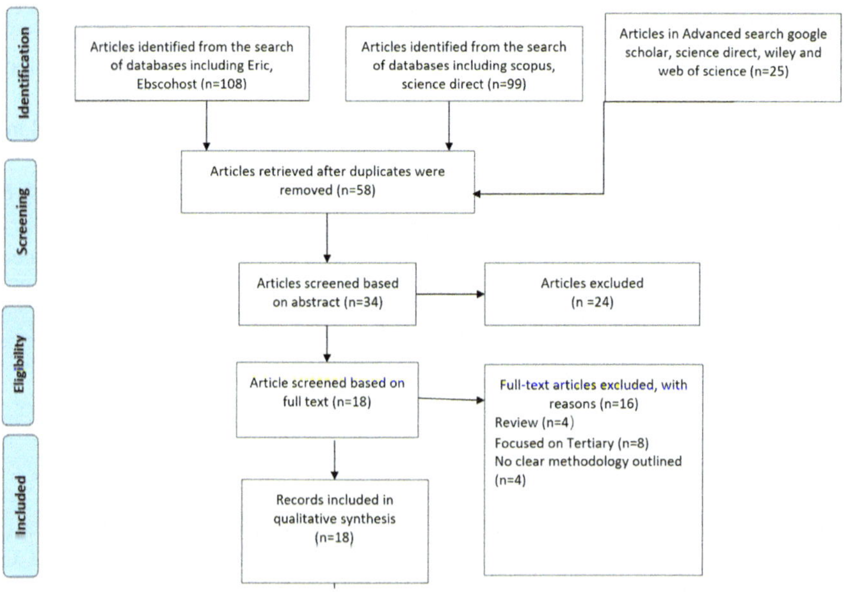

Fig. 1 PRISMA procedure for selection of literature

24 articles related to industry and vocational training were further excluded. The remaining 34 articles were then fully engaged and a further 16 articles were excluded with four (n = 4) being review articles, eight (n = 8) falling within the tertiary education category and four (n = 4) where the methodology for data collection was completely unclear, leaving only 18 articles for the actual review covered in this chapter. Once a paper was selected, the relevant data were extracted and coded in accordance with other review recommendations and the direct relationship to the research questions posed in the current study. Themes generated from coding through the data included school/informal subject, level of education within K-12, MR/AR application used, methods for data collection, pedagogical approaches, pedagogical theories, and possible learning gains noted from each study. Two iterations of code validation were done by the authors to ensure that all there was a 100% inter-coder reliability.

3.2.2 Data Analysis

From the peer-review of indexed journal articles, a total of 18 articles that met the criteria for the review were selected. Articles that considered the educational impact of MR technologies for teaching and learning in the K-12 environments and had markers related to pedagogy made the list. All articles which primarily focused on MR in tertiary education were excluded from the review process. In terms of studying the educational potential of MR environments as related to pedagogy, Table1 provides an overview of the review data analysis.

From Table 1 above, each article was analysed under different headings of interest, including the selected article, the school subject into which the MR technology was integrated, the grade level, MR applications used, the method considered for the research conducted, the pedagogical approaches mentioned in the article, pedagogical theories and the learning gains recorded post-interventions.

4 Results

In this section, the results are presented to address the objectives and answer the research questions anchoring the review.

4.1 Research Methodologies, Subject Areas, and Education Level K-12 MR

The first objective of the review was to establish the different research methods used for data collection in the literature, the subject, and the level of K-12 education where

Table 1 Data analysis

Article	Subject	Education level	Apps used	Method	Pedagogical approaches	Pedagogical theories	Learning gains
Palaigeorgiou et al. [47]	Geography	Primary grade 4	Fingertrip	Mixed methods	Interactive learning	Indirectly anchored on embodied learning	Enhanced spatial learning of Geography concepts
Yannier et al. [61]	STEM	Primary grade 1 and 2	mixed-reality AI vision technology	Experiment-al	Guided discovery Active learning instruction	Constructivi-sm and deliberate practice	Guided instruction had four times more learning gains than just discovery learning and hands-on learning Better learning of concepts
Weng et al. [59]	Science	Primary grade 5	Digital book	Mixed methods	Not explicit	Spatial visualisation	Enhanced learning outcomes for experimental group
Mateu et al. [40]	Inclusive education	High school age 12–16	Virtual touch	Qualitative case study	Tangible user interfaces	Not explicit	Learning gains, more interactive learning, and engagement
Leonard and Fitzgerald [34]	Technology-enhanced learning	Secondary school grade 7–12	Micrsoft Hololens	Design- Based Research	Not explicit	Not indicated	Engaging though difficult to use
Hung et al. [25]	Biology	Elementary school grade 5	ARToolKit	Experiment-al	Direct instruction	Spatial cognition	Practical, hands-on learning enhanced
Johnson-Glenberg et al. [27]	Chemistry and Biology	High school	SMALLab/ EMRELE	Experiment-al	Inquiry-based teaching	Collaborative learning/ Embodiment	Enhanced learning and interactivity

(continued)

Table 1 (continued)

Article	Subject	Education level	Apps used	Method	Pedagogical approaches	Pedagogical theories	Learning gains
Lindgren et al. [36]	Physics/Space science	Middle school	MEteor	Experiment-al	Activity based learning	Embodied learning	Significant learning gains, higher levels of engagement
Mateu and Alaman [39]	Computer science	Secondary school	CUBICA	Descriptive	Direct instruction/ collaborative learning	Not explicit	Increased motivation for learning and better understanding of abstract concepts
Rowe [50]	Projection mapping	Primary	Mixed reality bugs	Qualitative	Interactive learning, immersion, and engagement	Theory of exhibit attractors and sustainers	Enhanced immersion and engagement of participants
Chao et al. [12]	Chemistry	High school	Augmented virtual labs	Descriptive	Guided inquiry	Not indicated	Experimental group outperformed the control in post-test
Kalpakis et al. [28]	History	Primary grade 6	Virtual Museum	Implied mixed methods	Direct guidance student centered learning	Indirectly experiential	Learning environment was innovative, easy to use and enjoyable
Aguayo et al. [1]	Ecological literacy	Primary/ el-ementary	Pipi's world	Case study/DBR	Active learning	Activity theory Constructivism, place-based learning	
Yoon et al. [62]	Science	Primary and junior secondary	Not clear about AR system used	Quasi-experimental mixed methods	Knowledge building scaffolding	Not explicit	Conceptual development in science knowledge Learning in museum systems can be enhanced through scaffolding and group work

(continued)

Table 1 (continued)

Article	Subject	Education level	Apps used	Method	Pedagogical approaches	Pedagogical theories	Learning gains
Yoon et al. [63]	Informal learning	Primary and junior secondary	Magnetic map	Qualitative data analysis	Scaffolding based on constructivism	Text based, and collaborative scaffolds	Student learning increased when all three Scaffolding techniques were employed
Enyedy et al [19]	Physics	Primary grade 2	AR simulation	Qualitative	Interactive learning	Cognitive ethnography, laminal blends, distributed cognition	Enhance the way in which second graders get introduced to physics
Han et al. [24]	Dramatic play	Kindergarten	AR-mediated software	Experiment	Drama	Embodiment indirectly implied	Enhanced sensory engagement
Sugimoto [55]	Robotics and storytelling/dramatic play	Elementary grade 4–6	GENTORO	Mixed methods	Creative hands-on learning	Embodied interactive learning	Enhance children's embodied participation and manipulation skills

MR was integrated. From the analysis of data extracted and coded from 18 articles selected, seven studies used a purely quantitative and experimental design [12, 24, 25, 27, 36, 39, 61], while five studies employed a mixed method research design [28, 47, 55, 59, 62]. A design-based research (DBR) approach was followed for two studies Aguayo et al. [1] and Leonard and Fitzgerald [34], while the remaining four qualitative studies included Enyedy et al. [19], Mateu et al. [40], Yoon et al. [63] and Rowe [50].

As seen on Table 1, school subjects from the review included eleven studies on the natural sciences and technology [1, 12, 19, 25, 27, 34, 36, 39, 59, 61, 62], three on the social sciences [28, 47, 50], two on art [24, 55], one study on inclusive education [40] and one informal learning based on virtual museums [63].

Out of the 18 reviewed articles, 13 articles were situated within the context of primary schools (K-7), while the remaining five studies are from the high/secondary (8–12) school spectrum. However, two overlaps that involved participants from both primary and secondary schools were reported in the studies Yoon et al. [62, 63], as can be seen in Table 1.

4.2 Pedagogical Approaches Used in Incorporating MR in K-12 Education

The second objective of the review was to present pedagogical approaches used in incorporating MR in K-12 education. This objective was the primary motivation for the review as studies seldom provide explicit information about the pedagogical or teaching strategies infused in mixed reality-enhanced interventions, especially in K-12 settings. From the analysis of data, it is evident that the pedagogical approaches are still, for the most part, shallow or inexplicit, as could be seen in studies like Leonard and Fitzgerald [34] and Weng et al. [59]. Other pedagogical approaches that came to the fore included interactive learner-centered pedagogy [19, 47, 50, 55], where participants in the study were expected to work in an interactive manner with technology and their peers in order to attain the intended learning goals. Other critical pedagogical approaches include inquiry-based teaching and learning [12, 27], activity-based learning [1, 36], guided inquiry and discovery [61], Scaffolding [63] and direct instruction [39]. The general observation after reading through articles in this review is that these approaches are explained in terms of the models considered for learning interventions. The relevant theoretical lens and research which backs the efficacy of the selected pedagogical approach, and the context suitability of the pedagogical approach are not discussed extensively. Furthermore, most of the articles seldom indicate the relevant curriculum criteria that motivates the integration of the MR applications selected.

4.3 *Effective Pedagogical Theories/Frameworks for Integrating MR*

The third objective looked at the underpinning theoretical frameworks for MR pedagogy in K-12 learning environments. The pedagogical theories that underpin the learning environments in these different MR studies were similar in many ways. It was observed that while some authors were not explicit about the theoretical underpinnings of the learning interventions [12, 34, 39, 40, 62], the majority of the literature was underpinned directly or indirectly by three fundamental theories, including embodiment [24, 27, 36, 47, 55], cognitive sciences [19, 25] and constructivism [1, 28, 61]. Three sub-theories related to the aforementioned theories also emerged, including spatial visualisation [59], collaborative scaffolding [63] and the theory of exhibit attractors and sustainers [50].

4.4 *Pedagogical Approaches Associated to High Learning Gains*

Except for the study by Leonard and Fitzgerald [34], who reported that though the MR technology was engaging, it was difficult to use, all the other studies underpinned by different pedagogical theories reported learning gains, including enhanced spatial learning of Geography concepts [47], four times more learning gains for experimental groups with guided inquiry [61] as opposed to just discovery learning. Of high mention in 16 of the reviewed studies were learning gains related to enhanced interactivity, hands-on practical learning, enhanced conceptual understandings, improvement in attitude and motivation [1, 12, 27, 40, 50]. Furthermore, MR interventions were seen to enhance embodied participation [55], enhanced sensory engagement [24] and many other learning gains like procedural understandings across almost all the studies. What seem to be the limitation in these studies is that the integration of the MR technology was always reported as short-term interventions with a scarcity of longitudinal studies or non-experimental studies that focus on the actual scholarship of teaching and learning (SOTL).

5 Discussions and Conclusion

The main aim of this review was to provide a cross-examination of the pedagogical approaches, theories, and associated learning gains that underpin MR-enhanced interventions in K-12. From the findings, we propose a framework for planning teaching with MR technology as our contribution to the chapter. It was also important to look at the methodological approaches employed for MR-related learning interventions. In this section, we discuss and conclude on the different findings, followed by suggesting a pedagogical framework for integrating MR.

5.1 Methodology, Subject and Level

The review of literature indicated that most researchers preferred a quantitative or mixed methods study whereby learning gains could be assessed through surveys, post-tests and semi-structured interviews. However, there is need to enforce and track the efficacy of MR integration in K-12 as it relates to the pedagogical practices for longer periods of time. It is suggested from some studies that more longitudinal and qualitative research over long periods of time be engaged in K-12 education settings [13, 22]. Despite this suggestion, the meta-analysis by Garzón et al. [22] indicated that there is a "significant advantage of designing AR interventions that last between one week and four weeks and that any other intervention duration category has no significant differences in learning" [22, p. 9]. This notion seems to have been supported by an older study by Cheung and Slavin [15], who reported that the impact of technological tools becomes more positive on short-term basis. With these arguments, we conclude that introducing MR or any other learning technology should be outcome situated so that maximum learning gains can be realized.

Looking at the different subjects and levels, it is evident that primary school (k-7) was the most prevalent level for MR-related interventions within the natural sciences than other domains, including social sciences, art, and informal settings like museums. The rationale for this choice could be related to flexibility with MR applications and curricula at the primary school level. The rationale for more interventions in science could be the abstract nature of scientific concepts and the need to enhance visualisation [26] when teaching concepts at the microscopic level.

5.2 Pedagogical Approaches for MR in K-12 Education

Based on the findings of this review, it is evident that communication on pedagogical approaches and theoretical frameworks for MR learning interventions within K-12 learning environments are not clear [22, 54]. The pedagogical approaches that emerged from this review included inquiry-based learning, activity learning, scaffolding, embodied interactive and collaborative learning [1, 12, 27, 47, 55, 61, 63]. These findings were congruent with literature that suggests that constructivist-based pedagogical approaches are the most applied theoretical underpinnings in technology-enhanced education. Following on these are approaches that have their underpinnings in cognitive sciences, direct instruction, spatial cognition strategies, tangible user interphase and enhanced sensory strategies [24, 25, 39, 40]. Some researchers also allude to the fact that not having formal pedagogical approaches may be frustrating to the students [37], as students are usually not sure what the end goal of the intervention is. With this, it then becomes crucial that teachers/researchers pre-plan and document the pedagogical steps that will be enacted in MR-enhanced learning interventions. Using MR technology would require those design elements which support different activities based on preferred instructional approaches in both formal and informal K12 settings to be assessed [23].

5.3 Pedagogical Frameworks

Findings also indicated that the pedagogical frameworks that underpin the integration of MR in K-12 are anchored on learning theories, including constructivism, cognitivism, embodiment and collaborative teaching/learning theories. Other sub-learning and teaching theories anchored on constructivism like interactive learning, activity-based learning, and scaffolding are also seen in the different studies [12, 34, 39, 40, 62]. It is therefore important that the underpinning frameworks be explicitly explained especially when integrating MR in K-12 learning environments. Chang et al. [11] and Leonard and Fitzgerald [34], also make similar suggestions in advocating that teaching and learning theoretical underpinnings like cognitivism and constructionism be considered when crafting MR related teaching strategies during interventions. This will ensure that there is maximum engagement and interactivity in the learning experience. From this review, it is suggested that theoretical underpinnings and the associated pedagogical approaches that emerge from these theories be considered and elaborated upon when facilitating MR-enhanced learning in K-12.

5.4 Learning Gains

Review of literature in this domain over the decade has also revealed that over 95% of technology-enhanced learning interventions always lead to learning gains. What remains questionable are the different debates about the impact of technology on learning. Some researchers like Clark [16] argue that technology has no effect on learning per se but instead on the factors like attitude, motivation and interest that will make students want to learn. On the other hand, proponents like Kozma [32] emphasise that technology provides a rich learning environment and affords learning opportunities that will otherwise not be feasible. The findings from this review strongly concur with Kozma [32] that in combination with all other factors, mixed reality technology, when scaffolded adequately with the right pedagogical approaches, has the ability to enhance learning, visualisation, interactive, collaborative and inquiry-based learning. Other constructs that emerged from the review include self-directed learning [39, 61], the development of higher order skills like problem-solving, creativity and critical thinking [28, 47], and increased student participation as seen in some of the literature [36, 59]. For the natural sciences, in particular, mixed reality technology provides K-12 learners the platform to visualise, engage and interact with concepts that may not be accessible to them [27]. All these affordances are feasible because of the way in which MR learning environments are created to include real and digital context leveraged on embodiment and immersion.

In essence MR learning environments provide relevant additions to teaching and learning in K-12 and will have positive impacts on students' learning experiences [22]. From all dimensions when the right digital tools like robots [11, 55], HMD [34], high end computers [18], MR books with sensors [12, 40], tangible interfaces [28, 47] and augmented virtual labs [12] are used in the right context and learning environment, learning gains will be made.

5.5 *Proposed Pedagogical Framework*

From the analysis of articles in this study (K-12) and other studies that were conducted in tertiary learning environments, there is a scarcity of pedagogical frameworks which could aid direct teaching practices when integrating MR technology in K-12 learning environments. We, therefore, propose a pedagogical framework for mixed reality integration, which provides a trajectory/steps for integrating AR/VR/MR technologies in lessons within K-12 educational settings. This framework also provides a yardstick for MR application developers and researchers for appropriate educational MR applications that will enhance pedagogy and at the same time, foster the acquisition of learning goals for K-12 learners. Figure 2 below shows the proposed content, prior knowledge, pedagogy, learning outcomes and assessment (COPPLA) framework for sound pedagogical integration of mixed reality within specific teaching scenarios in K-12.

The COPPLA pedagogical framework in Fig. 2 proposes due processes to be followed in integrating MR technologies in K-12 learning environments. The framework proposes a Content, Prior knowledge, Pedagogy, Learning outcomes and Assessment (COPPLA) trajectory that should be considered in identifying and integrating MR technologies that will foster learning and meet the cognitive intentions of teachers and students. Based on the context of a K-12 lesson and the educational setting of a given learning experience, teachers following this COPPLA framework should:

- Plan the target content to be covered for a specific timeframe.
- Establish the relevant prior knowledge that learners need to have about this content before engaging with the technology.
- Select a pedagogical approach or a cocktail of approaches best suited to attain the learning outcomes.
- Choose the appropriate MR technology to be integrated with the pedagogical approach and lesson outcomes while ensuring that cognitive intentions are mapped out.

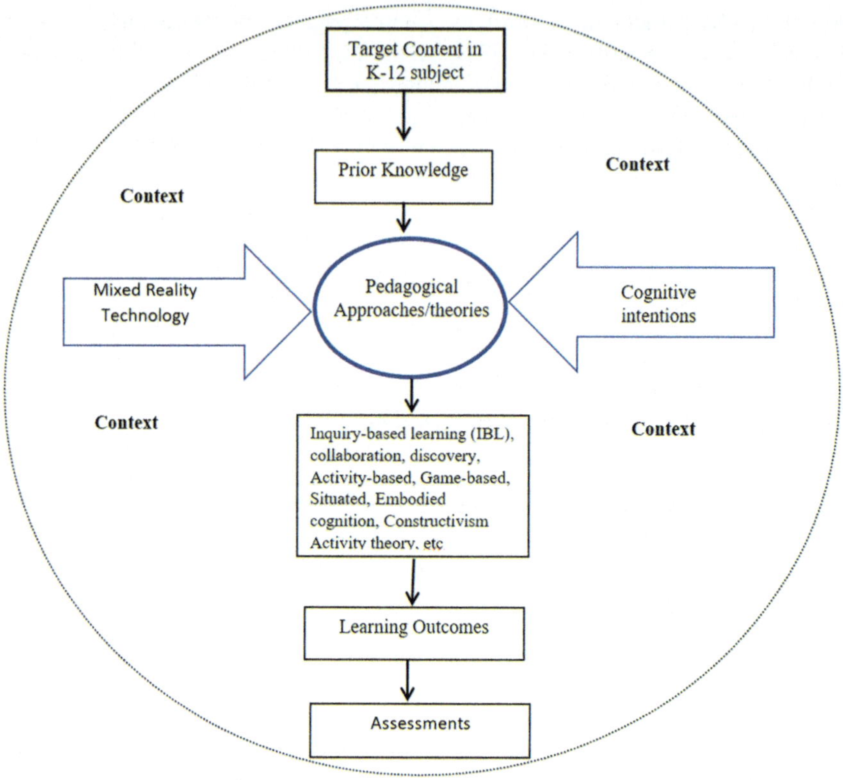

Fig. 2 COPPLA pedagogical framework for mixed reality integration in K-12 environments

- Moderators like context, including classroom set-up, technological resources available, and connectivity issues, among other contextual factors, must always be considered when planning for MR integration.
- Finally, ensure that quality assessments are designed to assess the attainment of learning outcomes at the end of the MR-enhanced lesson, all things being equal.

It is worth noting that one pedagogical approach may not be enough for a given learning intervention but rather a 'cocktail' of approaches could be used in planning action when integrating MR technology in view of cognitive intentions to be attained by students. This proposed framework is only suggestive and is currently being enacted in the researchers' own practice for supporting MR-enhanced teaching and learning.

5.6 Contribution of the Review

The fundamental contribution of this review is the proposition of a COPPLA framework that could help educators and researchers plan for an MR-enhanced pedagogy in K-12 or other educational settings. The framework guides researchers and instructors to adequately and explicitly situate planned pedagogical approaches supported by theories when mixed reality interventions are planned for K-12 teaching and learning across different subjects. Global trends show that MR, AR and VR -enhanced learning interventions with carefully selected pedagogical strategies will significantly impact teaching and learning in school settings [29, 45, 46]. This is because teachers will know exactly what lesson outcomes will be enhanced by selected technology and learners will know the expected outcomes of learning interventions.

5.7 Limitations

The review, though addressed to the best of the researchers' ability, still has certain limitations that might have risen with the search words used or the exclusion of other academic sources. While numerous bibliographic databases were used to obtain a wide range of publications for K-12 education, it is still possible that there are articles within the selection criterion that were excluded from the review. The search terms could also be limiting in that the focus was on *"Mixed Reality" and "pedagogy"* combined with keywords like *"K-12", "Primary", "elementary," and "secondary" schools.*

5.8 Conclusion and Recommendations

The main purpose of this review was to analyse aspects related to pedagogy for MR in K-12 educational settings. Based on the analysis of 18 studies, it is concluded that mixed-reality learning has a positive impact on learning and the affective domain, including motivation, interest, and achievement. However, all factors that contribute to these gains, including pedagogical approaches and practices, contexts, the nature of assessments, the timeframe for MR integration and the learning goals, should always be succinctly and explicitly discussed in studies. The results of the current review are consistent with the claims that minimal guidance during MR-enhanced instruction is not enough. Instead, an intentional plan of the pedagogy and its outcomes should be created to ensure that MR learning interventions are enriching for both teachers and students in each unique educational context.

For future reviews and research, we recommend an analysis of the pedagogical strategies that are employed in enacting MR-enhanced learning in specific school subjects and in tertiary education courses. It will also be worth looking at different frameworks that can guide teachers on best practices when integrating MR in their classrooms.

References

1. Aguayo, C., Eames, C., Cochrane, T.: A framework for mixed reality free-choice, self-determined learning. Res. Learn. Technol. **28**, 23–47 (2020)
2. Akçayır, M., Akçayır, G.: Advantages and challenges associated with augmented reality for education: a systematic review of the literature. Educ. Res. Rev. **20**, 1–11 (2017)
3. Alexander, R.: Still no pedagogy? Principle, pragmatism and compliance in primary education. Camb. J. Educ. **34**(1), 7–33 (2004). https://doi.org/10.1080/0305764042000183106
4. Ali, A.A., Dafoulas, G.A., Augusto, J.C.: Collaborative educational environments incorporating mixed reality technologies: a systematic mapping study. IEEE Trans. Learn. Technol. **12**(3), 321–332 (2019)
5. Anderson, T.: Theories for learning with emerging technologies. In: Veletsianos, G. (ed.). Emergence and innovation in digital learning, pp. 35–50. Edmonton: AU Press (2016)
6. Bacca, J., Baldiris, S., Fabregat, R., Graf, S.: Augmented reality trends in education: a systematic review of research and applications. J. Educ. Technol. Soc. **17**(4), 133–149 (2014)
7. Birchfield, D., Megowan-Romanowicz, C.: Earth science learning in SMALLab: a design experiment for mixed-reality. J. Comput. Supported Collaborative Learn. **4**, 403–421 (2009). https://doi.org/10.1007/s11412-009-9074-8
8. Black, P., Wiliam, D.: Classroom assessment and pedagogy. Assess. Educ.: Principles, Policy Pract. **25**(6), 551–575 (2018)
9. Bliuc, A.M., Goodyear, P., Ellis, R.A.: Research focus and methodological choices in studies into students' experiences of blended learning in higher education. The Internet Higher Educ. **10**(4), 231–244 (2007)
10. Cazden, C., Cope, B., Fairclough, N., Gee, J., Kalantzis, M., Kress, G., Luke, A., Luke, C., Michaels, S., Nakata, M.: A pedagogy of multiliteracies: designing social futures. Harvard Educ. Rev. **66**(1), 60–92 (1996)
11. Chang, C., Lee, J., Wang, C., Chen, G.: Improving the authentic learning experience by integrating robots into the mixed-reality environment. Comput. Educ. **55**(4), 1572–1578 (2010). https://doi.org/10.1016/j.compedu.2010.6.23
12. Chao, J., Chiu, J.L., DeJaegher, C.J., Pan, E.A.: Sensor-augmented virtual labs: using physical interactions with science simulations to promote understanding of gas behavior. J. Sci. Educ. Technol. **25**(1), 16–33 (2016)
13. Chauhan, S.: A meta-analysis of the impact of technology on learning effectiveness of elementary students. Comput. Educ. **105**, 14–30 (2016)
14. Chen, S., Duh, H.: The interface of mixed reality: from the past to the future. CCF Trans. Pervasive Comput. Interact. (2018). https://doi.org/10.1007/s42486-018-0002-8
15. Cheung, A.C., Slavin, R.E.: The effectiveness of educational technology applications for enhancing mathematics achievement in K-12 classrooms: a meta-analysis. Educ. Res. Rev. **9**, 88–113 (2013)
16. Clark, R.E.: Media will never influence learning. Educ. Tecnol. Res. Dev. **42**(2), 21–29 (1994)
17. Cowling, M., Birt, J.: Pedagogy before technology: a design-based research approach to enhancing skills development in paramedic science using mixed reality. Information **9**(2), 29 (2018)
18. De Lima, E., Feijo, E., Barbosa, S., Furtado, S., Ciarlini, A., Pozzer, C.: Draw your own story: paper and pencil interactive storytelling. Entertainment Comput. **5**(1), 33–41 (2014)
19. Enyedy, N., Danish, J.A., DeLiema, D.: Constructing liminal blends in a collaborative augmented-reality learning environment. Int. J. Comput. Supported Collaborative Learn. **10**(1), 7–34 (2015)

20. Flavián, C., Ibáñez-Sánchez, S., Orús, C.: Impacts of technological embodiment through virtual reality on potential guests' emotions and engagement. J. Hosp. Market. Manag. (2020). https://doi.org/10.1080/19368623.2020.1770146
21. Freina, L., Ott, M.:.A literature review on immersive virtual reality in education: state of the art and perspectives. In: The International Scientific Conference eLearning and Software for Education, vol. 1, no. 133, pp. 10–17 (2015)
22. Garzón, J., Acevedo, J.: Meta-analysis of the impact of augmented reality on students' learning effectiveness. Educ. Res. Rev. **27**, 244–260 (2019)
23. Garzón, J., Baldiris, S., Gutiérrez, J., Pavón, J.: How do pedagogical approaches affect the impact of augmented reality on education? A meta-analysis and research synthesis. Educ. Res. Rev. **31**, 100334 (2020)
24. Han, J., Jo, M., Hyun, E., So, H.J.: Examining young children's perception toward augmented reality-infused dramatic play. Education Tech. Res. Dev. **63**(3), 455–474 (2015)
25. Hung, Y.H., Chen, C.H., Huang, S.W.: Applying augmented reality to enhance learning: a study of different teaching materials. J. Comput. Assist. Learn. **33**(3), 252–266 (2017)
26. Hwang, G.J., Chen, C.H.: Influences of an inquiry-based ubiquitous gaming design on students' learning achievements, motivation, behavioral patterns, and tendency towards critical thinking and problem solving. Br. J. Edu. Technol. **48**(4), 950–971 (2017)
27. Johnson-Glenberg, M.C., Birchfield, D.A., Tolentino, L., Koziupa, T.: Collaborative embodied learning in mixed reality motion-capture environments: two science studies. J. Educ. Psychol. **106**(1), 86–104 (2014)
28. Kalpakis, S., Palaigeorgiou, G.K., K,: Promoting historical thinking in schools through low fidelity, low-cost, easily reproduceable, tangible and embodied interactions. Int. J. Emerg. Technol. Learn. **13**(12), 67–82 (2018)
29. Kerawalla, L., Luckin, R., Seljeflot, S., Woolard, A.: Making it real': exploring the potential of augmented reality for teaching primary school science. Virtual Real. **10**(3–4), 163–174 (2006). https://doi.org/10.1007/s10055-6-36-4
30. Kitchenham, B.A.: Guidelines for performing systematic literature reviews in software engineering (Version 2.3). EBSE Technical Report, Keele University and University of Durham (2007)
31. Koutromanos, G., Sofos, A., Avraamidou, L.: The use of augmented reality games in education: a review of the literature. Educ. Media Int. **52**(4), 253–271 (2015)
32. Kozma, R.B.: Will media influence learning? Reframing the debate. Educ. Tech. Res. Dev. **42**(2), 7–19 (1994)
33. Lee, K.: Augmented reality in education and training. TechTrends **56**(2), 13–21 (2012)
34. Leonard, S.N., Fitzgerald, R.N.: Holographic learning: a mixed reality trial of Microsoft HoloLens in an Australian secondary school. Res. Learn. Technol. **26** (2018)
35. Lindgren, R., Johnson-Glenberg, M.: Emboldened by embodiment: six precepts for research on embodied learning and mixed reality. Educ. Res. **42**(8), 445–452 (2013)
36. Lindgren, R., Tscholl, M., Wang, S., Johnson, E.: Enhancing learning and engagement through embodied interaction within a mixed reality simulation. Comput. Educ. **95**, 174–187 (2016)
37. Liu, T.Y., Chu, Y.L.: Using ubiquitous games in an English listening and speaking course: impact on learning outcomes and motivation. Comput. Educ. **55**(2), 630–643 (2010)
38. Maas, M.J., Hughes, J.M.: Virtual, augmented and mixed reality in K–12 education: a review of the literature. Technol. Pedagog. Educ. **29**(2), 231–249 (2020)
39. Mateu, J., Alaman, X.: CUBICA: an example of mixed reality. J. Univ. Comput. Sci. **19**(17), 2598–2616 (2013)
40. Mateu, J., Lasala, M., Alamán, X.: VirtualTouch: a tool for developing mixed reality educational applications and an example of use for inclusive education. Int. J. Human-Comput. Interact. **30**(10), 815–828 (2014)
41. Merchant, Z., Goetz, E.T., Cifuentes, L., Keeney-Kennicutt, W., Davis, T.J.: Effectiveness of virtual reality-based instruction on students' learning outcomes in K-12 and higher education: a meta-analysis. Comput. Educ. **70**, 29–40 (2014)

42. Milgram, P., Kishino, F.: A taxonomy of mixed reality visual displays. IEEE Trans. Inf. Syst. **77**(12), 1321–1329 (1994)
43. Milgram, P., Takemura, H., Utsumi, A., Kishino, F.: Augmented reality: a class of displays on the reality-virtuality continuum. In: Telemanipulator and telepresence technologies. International Society for Optics and Photonics, vol. 2351, pp. 282–293 (1995)
44. Moher, D., Liberati, A., Tetzlaff, J., Altman, D.G., PRISMA Group: Preferred reporting items for systematic reviews and meta-analyses: the PRISMA statement annals of internal medicine **151**(4), 264–269 (2009)
45. Murphy, K.M., Cook, A.L., Fallon, L.M.: Mixed reality simulations for social-emotional learning. Phi Delta Kappan **102**(6), 30–37 (2021)
46. Önal, N.T., Önal, N.: The effect of augmented reality on the astronomy achievement and interest level of gifted students. Educ. Inf. Technol. **26**(4), 4573–4599 (2021)
47. Palaigeorgiou, G., Karakostas, A., Skenteridou, K.: Touching and traveling on 3D augmented tangible maps for learning geography: the finger trips approach. Interact. Technol. Smart Educ. **15**(3), 279–290 (2018). https://doi.org/10.1108/ITSE-12-2017-0066
48. Radianti, J., Majchrzak, T.A., Fromm, J., Wohlgenannt, I.: A systematic review of immersive virtual reality applications for higher education: design elements, lessons learned, and research agenda. Comput. Educ. **147**, 103778 (2020)
49. Radu, I.: Augmented reality in education: a meta-review and cross-media analysis. Pers. Ubiquit. Comput. **18**(6), 1533–1543 (2014)
50. Rowe, A.: Designing for engagement in mixed reality experiences that combine projection mapping and camera-based interaction. Digital Creativity **25**(2), 155–168 (2014)
51. Rutten, N., van Joolingen, W.R., van der Veen, J.T.: The learning effects of computer simulations in science education. Comput. Educ. **58**(1), 136–153 (2012)
52. Rutto, D.K.: Pedagogical theories. Int. J. Sci. Eng. Res. **8**(6), 2025–2030 (2017)
53. Schunk, D.H.: Learning Theories an Educational Perspective, 6th edn. Pearson, Boston (2012)
54. Sickel, J.L.: The great media debate and TPACK: a multidisciplinary examination of the role of technology in teaching and learning. J. Res. Technol. Educ. **51**(2), 152–165 (2019)
55. Sugimoto, M.: A mobile mixed-reality environment for children's storytelling using a handheld projector and a robot. IEEE Trans. Learn. Technol. **4**(3), 249–260 (2011)
56. Stretton, T., Cochrane, T., Narayan, V.: Exploring mobile mixed reality in healthcare higher education: a systematic review. Res. Learn. Technol. **26**, 21–31 (2018)
57. Turan, Z., Meral, E., Sahin, I.F.: The impact of mobile augmented reality in geography education: achievements, cognitive loads and views of university students. J. Geogr. High. Educ. **42**(3), 427–441 (2018)
58. Watson, D.M.: Pedagogy before technology: re-thinking the relationship between ICT and teaching. Educ. Inf. Technol. **6**(4), 251–266 (2001)
59. Weng, C., Rathinasabapathi, A., Weng, A., Zagita, C.: Mixed reality in science education as a learning support: a revitalized science book. J. Educ. Comput. Res. **57**(3), 777–807 (2019)
60. Wu, H., Lee, S.W., Chang, H., Liang, J.: Current status, opportunities and challenges of augmented reality in education. Comput. Educ. **62**, 41–49 (2013)
61. Yannier, N., Hudson, S.E., Koedinger, K.R.: Active learning is about more than hands-on: a mixed-reality AI system to support STEM education. Int. J. Artif. Intell. Educ. **30**(1), 74–96 (2020)
62. Yoon, S., Elinich, K., Wang, J., Steinmeier, C., Tucker, S.: Using augmented reality and knowledge-building scaffolds to improve learning in a science museum. Int. J. Comput.-Support. Collab. Learn. **7**(4), 519–554 (2012)
63. Yoon, S.A., Anderson, E., Park, M., Elinich, K., Lin, J.: How augmented reality, textual, and collaborative scaffolds work synergistically to improve learning in a science museum. Res. Sci. Technol. Educ. **36**(3), 261–281 (2018). https://doi.org/10.1080/02635143.2017.1386645

Transforming Learning Experiences Through Affordances of Virtual and Augmented Reality

Choon Guan Pang and Yiyu Cai

Abstract Educational reforms across the globe in recent years have focused on increasing the quality of learning through innovation. Immersive Virtual and Augmented Reality (VR/AR) technologies have enabled schools and universities to raise the quality of learning by transforming learning experiences from passive to active, and empowering educators to leverage on four key affordances—Multimedia Augmentation, Enhanced Visualization, Experiential Learning and Learning Motivation. A review of studies has shown that using VR/AR can positively affect students' learning experiences and students are generally favorable towards the use of these technologies. Experiments show that there is a significant increase in students' engagement in learning and improvement in conceptual understanding through enhanced visualization. These findings have implications on the pedagogical design of lessons that use VR/AR.

1 Introduction

The 2020 Horizon Educause Report [1] identified XR technologies, primarily VR/AR, as one of six emerging technologies and practices that will have a significant impact on the future of teaching and learning. Technology affordances change learning experiences, affect instructional design and influence teaching pedagogy [2, 3]. Nearly a century ago, Dewey [4] asked the question "how many (students) lost the impetus to learn because of the way in which learning was experienced by them?" [5] points out that emerging technologies such as VR/AR have affordances that transform learning experiences in three ways—allowing multiple perspectives, situated learning, and transfer of learning. According to [6], learning is "the process whereby knowledge is created through the transformation of experience.". The central tenet

C. G. Pang (✉)
National Institute of Education, Nanyang Technological University, Singapore, Singapore
e-mail: choonguan.pang@nie.edu.sg

Y. Cai
School of Mechanical and Aerospace Engineering, Nanyang Technological University, Singapore, Singapore

Y. Cai et al. (eds.), *Mixed Reality for Education*, Gaming Media and Social Effects,
https://doi.org/10.1007/978-981-99-4958-8_6

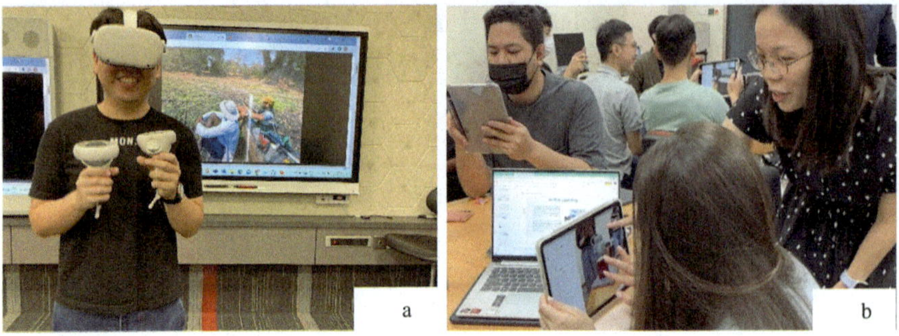

Fig. 1 **a** Student teacher watching a 360° video of "Elephant Encounter in 360: The Okavango Experience" by National Geographic; **b** Student teachers using Geogebra AR app

of Experiential Learning Theory (ELT) is that learning is best conceived more as a process rather than an outcome. The focus on process requires a shift in the focus of design of a lesson or training from one that is focused on cognitive-based outcomes to a more holistic assessment of the experience of learning. This is necessary to capture the essence of classroom learning that occurs through a dynamic process of forming and re-forming ideas and applying them to create new knowledge. New knowledge is formed and tested through a continuous process of experience, reflection, abstract and application. For example, Singapore's National Institute of Education (NIE) have been transforming teacher education through the incorporation of emerging technologies to enhance reflective skills and thinking dispositions. In the author's pre-service training classes, student teachers are encouraged to apply AR/VR to enhance learning such as increasing spatial visualization in mathematics and promoting conservation in values education (see Fig. 1).

The use of multisensory approaches in VR/AR enables pedagogical strategies to train and engage learners in twenty-first century core skills such as problem solving, critical thinking, creativity, collaboration, and digital literacy [7]. The merging of 3D graphics with the real world enhances human perception by augmenting or creating experiences. Yilmaz et al. [8] in their teaser "Are augmented reality picture books magic or real for preschool children aged five to six?", described the results of their study which showed that the animations in the AR picture books were perceived as magic, made the children happy and they found it more enjoyable than conventional books. They found a positive and meaningful relationship between the children's learning with use of AR picture books and happiness. By transferring positive emotions to them, educational technologies can lead to positive learning experiences [9].

There are some high quality off-the-shelf educational software mobile AR apps that teachers can use in their classrooms to arouse the attention of learners and give them opportunities to visualize and interact with the artefacts or objects in 3D. Examples of free apps include the Mission to Mars AR by Smithsonian Channel [10], Civilizations AR by BBC and WWF Free Rivers. AR environments can help make learning active by assisting the learner to commit information to long-term memory

[11]. The elements of fun, challenge and curiosity in AR learning environments enhance active learning [12].

The use of immersive VR technologies using head-mounted displays (HMDs) manufactured by companies such as Oculus and HTC will enable learning through a new experiential paradigm [13]. The latest VR HMDs allow users to experience a high degree of immersion [14]. Immersion can be defined as "a perception of being physically present in a non-physical world by surrounding the user of the VR system created with images, sound or other stimuli" [15]. Jaron Lanier [16], who first coined the term "virtual reality", said that VR gives people "this sense of being able to be who we are without limitation, for our imagination to become objective and shared with other people." Emphasizing that VR is "reality" that is "virtual", Slater [17] explained that the real power of VR is to go beyond what is real, not only to simulate but to create. This allows people to step out of the bounds of reality and experience paradigms that are otherwise impossible.

Emerging technologies of VR/AR transforms the learning experience by helping learners not only to recall knowledge but to critically evaluate assumptions, beliefs, values, and perspectives. This leads to meaningful change [18]. This is referred to as transformative learning, a concept first proposed by [19, 20]. One powerful example of transformative learning is in using VR to create empathy for other individuals, cultures, or even environmental issues [21–23]. VR-mediated virtual trips can offer a powerful medium to educate, engage and inspire students to action about climate issues [22, 24, 25].

In 2020, a South Korea mother had a tearful VR reunion with her daughter who had passed on three years ago from blood-related diseases [26]. The mother Jang Ji-sung burst into tears as an avatar of her seven-year-old daughter, Na-Yeon, emerged from a neighborhood park which was her usual playground. The VR system gave the mother the experience of talking to her daughter and touching her with a pair of virtual hands. The presence, immersion, and sense of embodiment are affordances of the VR system that made the experience so memorable and emotional.

While immersive VR completely replaces reality, AR supplements reality with virtual objects that appear to coexist in the same page as the real world [27]. Put it in another way, VR entirely immerses a user in a virtual environment, but AR allows the user to see a real world with overlapping virtual elements in real-time [28]. As a form of mediated immersion, AR provides experiences within a digitally enhanced context [29] with the "unique ability to create immersive hybrid learning environments that combine digital and physical objects, thereby facilitating the development of process skills such as critical thinking, problem solving and communicating through collaborative exercises".

The term 'Augmented Reality' was first coined by [30], who envisaged AR as a system to support workers with dynamic and useful information by augmenting his view with a computer-produced diagram superimposed on a real-world object (see Fig. 2b). Azuma [31] saw AR as a variation of virtual environments with the following features: (a) combines real and virtual elements, (b) is interactive in real time; (c) is registered in three dimensions. VR technology has its origins in 1965 when

Sutherland [32] published a paper detailing the very first three-dimensional (3D) head mounted display (HMD) developed at MIT Lincoln Laboratory (see Fig. 2a).

Milgram and Kishino's [33] conceptualization of Virtuality Continuum provided a taxonomy of the ways in which real and virtual elements may be combined, in varying degrees from AR to Augmented Virtuality (AV) (see Fig. 3). The Virtuality Continuum ranges from a completely real environment to a completely virtual environment, where real world and virtual world objects are presented together within a single display. Based on the Virtuality Continuum, Mixed Reality (MR) includes systems in which the virtual aspects are dominant as well as those in which the physical reality is dominant [34]. MR has the added affordance of being able to give their users the illusion that digital objects are in the same space as physical ones [34]. To create this illusion, the digital objects need to be precisely positioned and aligned with the real objects in real time [27].

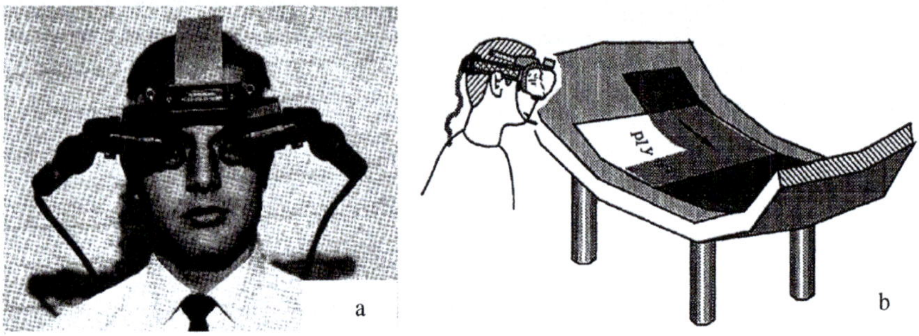

Fig. 2 **a** HMD of Sutherland (1968); **b** [30]'s AR device

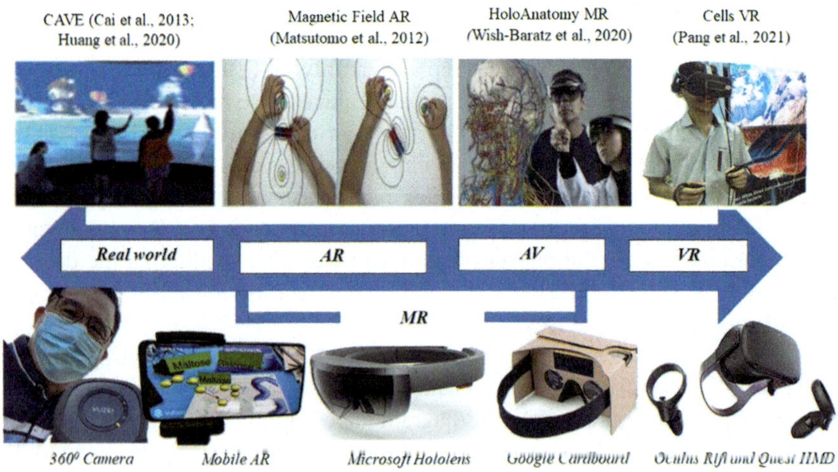

Fig. 3 Virtuality continuum [33]

Microsoft HoloLens is a MR HMD device that allows the user to engage and interact with both real-world and virtual objects in the form of holographic-like images [35]. For example, [36] carried out a study of the use of an MR application in remote online medical education. MR HoloLens headsets were shipped to 185 students who then participated in a 50-min online anatomy lesson. 84% believed that students can effectively learn human anatomy using the MR application. The design of an AR learning environment by [37] to help students visualize virtual magnetic fields emerging from physical magnet blocks is another example of the use of AR as a tool for visualization for conceptual change. Students can move the magnets around to see changes in the shape and direction of the magnetic fields. Another option for users to experience immersive VR besides wearing HMDs is through the Cave Automatic Virtual Environment (CAVE). It is a room where the walls, the floor, and even the ceiling, are made up of rear-projection screens. The user wears 3D glasses to enter the immersive world and he can move around freely in the room to interact with the system with their hands or by using a "wand" controller [38, 39]. An example is the virtual dolphinarium designed for children with autism for them to interact with virtual pink dolphins to improve their communication skills [38, 40].

2 Ecological Approach to Affordances

The concept of affordance was first proposed [41] to highlight the inherent possibilities of action of an environment, object, or tool. This was later expanded by [42] into the concept of 'perceived affordance'. While [41] focusses on the fundamental characteristics of the object in relation to the user (utility), [42, 43] emphasizes how the object is perceived (usability). As a cognitive psychologist, Norman [42, 43] stressed the importance of visual cues in the environment without which users might ultimately struggle to use a device. Norman's concept of "perceived affordances", since its introduction in 1988 has led to its widespread use across disciplines [44].

Since the introduction of the concept of affordances by [41], the approach of studying the organism and its environment has laid the foundation for the field of ecological psychology. Gibson's ecological approach was shaped largely by the historical context as he joined the U.S. Army Air Force during World War II and was tasked to investigate perception of pilots and service members. He formulated an ecological approach to visual perception that is independent of mental representations of the user which may be based on past experiences or knowledge [45]. The ecological approach focuses on the way in which organisms detect and respond to information in different environments [46].

The ecological approach that is used by [47] views learning as a unique combination of technological, social, and educational contexts. He proposed the Pedagogical-Social-Technological framework of affordances. While pedagogical or educational affordances refer to the structure and types of learning tasks supported by a particular technology [47], social affordances refer to the properties of a computer-supported collaborative learning environment that promote peer-to-peer social interactions [48].

Technological refers to the features of the technology to help students learn with ease and efficiency.

In the context of using technology for learning, [49] argued that technological affordances are functional and relational aspects which frame the possibilities for action. Affordances are functional in the sense that they are enable or constraint a given organism's attempt to engage in some activities, but it is also relational in terms of the interpretations that are made of them by social actors such as designers or users. This concept gives rise to the notion of affordances as an analytic tool [49].

Taking the ecological approach can generate fresh insights on how the affordances be used to maximize learning opportunities. For example, the concept of presence in mediated virtual environments implies that the reality of an experience must be defined by functionality [50] and usability [41, 42]. A study by [51] found that using HMD-based immersive VR game for educating passengers about aviation safety was not only superior to the safety card in increasing knowledge retention, it was also more engaging and fear-arousing. Taking an ecological approach gives a more holistic perspective that can help to explain the better learning outcomes due to the affordances of immersion and presence.

The concept of presence enables VR/AR to be defined in terms of human experience rather than hardware [52]. Presence refers to the reported sensation by users of being there in the virtual world (Schumie et al. 2001) and this is the fundamental experience of all VR environments [17]. The experience of presence is linked to Csikszentmihalyi's [53] concept of flow [54]. This is due to the sensorimotor nature of immersive technologies that can provide learners with direct experience, without the need to navigate abstract symbolic expression. In a study by [55], participants went through an MR experience that enabled them to see and hear their own hands burning while looking through a Video See-Through HMD. Despite no change in perceived anxiety, they found a significant increase in skin conductance for all participants. 6 out of 12 participants experienced an involuntary heat sensation on their hands, and these participants also had higher skin conductance.

Shin [23] outlined the affordances of immersive VR in terms of technological and affective affordances—Presence, Immersion, Embodiment, and Empathy and how they can bring about learning benefits (see Fig. 4). This signifies a shift from popular definitions of VR that usually refer to a particular technological system with devices such as computer, position trackers, gloves, and head-mounted stereoscopic display [52]. It is the combination of immersive and interactive features of VR that enables experiential learning [56].

Dede et al. [57] defined psychological immersion as the mental state of being completely absorbed or engaged with something. The participant can turn and move as they do in the real world, and this maintains the illusion of presence of one's body in a simulated setting. Embodiment is the process by which the person's body is substituted by a virtual one, by tracking the person's real movements and map these to movements of the virtual body [58].

Dalgarno and Lee (2010) identified five affordances of 3D virtual learning environments: (a) Learning tasks that support enhanced spatial knowledge representation of

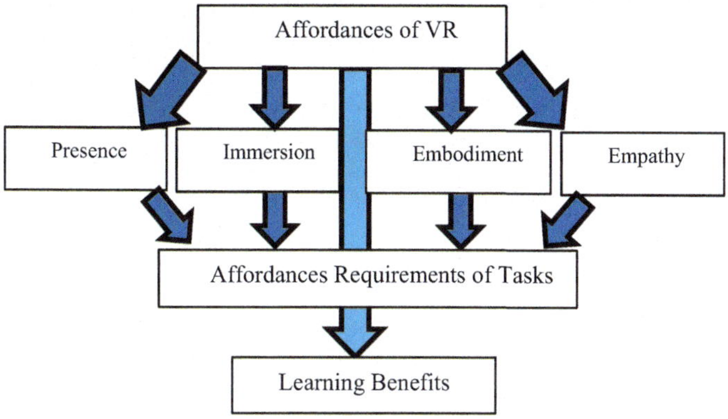

Fig. 4 Technological and affective affordances of VR

the object, (b) Experiential learning tasks that are impractical or impossible to undertake in the real world, (c) Learning tasks that lead to engagement and motivation, (d) Contextual learning tasks that lead to improved transfer or application of skills, (e) Collaborative learning tasks. For example, [59] developed an interactive VR training to teach participants the techniques of proper drilling through a digital replica of the actual drilling machine. Students that used the VR simulation achieved better results than participants in the traditional approach group. They also performed tasks at a faster speed and showed more confidence.

Bertrand et al. [21] theoretical framework of different mechanisms of practices for the design of VR experiences for empathic training in individuals represents a novel ecological approach to studying the affordances of VR/AR. Empathy can be defined as feeling the same emotion as another individual without having experienced it directly [60–63]. Empathy enables us to learn from others' pain and to know when to offer support [21]. For example, Clouds Over Sidra is an 8-min immersive 360° VR film created in partnership with the UN Millennium Campaign and UNICEF Jordan to highlight the plight of 80,000 refugees who escaped war and famine in Syria to the Za'atari camp in Jordan [64]. Schutte and Stilinović [64] carried out a study of 24 university students in Australia randomly assigned to either watch the film Clouds Over Sidra in immersive 360° VR using Samsung VR HMD (experimental group) versus the control group participants who also donned the VR HMD but watched the film in two-dimensional (2D) format. Results show that the 360° VR learning experience resulted in greater engagement and a higher level of empathy for the refugee girl compared to the control condition. This shows that VR has the potential to increase empathy.

VR/AR are the next generation interfaces that affords a different way of interaction with information which can be used to design better learning experiences [3, 5]. In terms of comparing the use of AR with more traditional forms of instruction, [65] review found that a high degree of satisfaction and enthusiasm, even when the AR tools may seem more difficult to use.

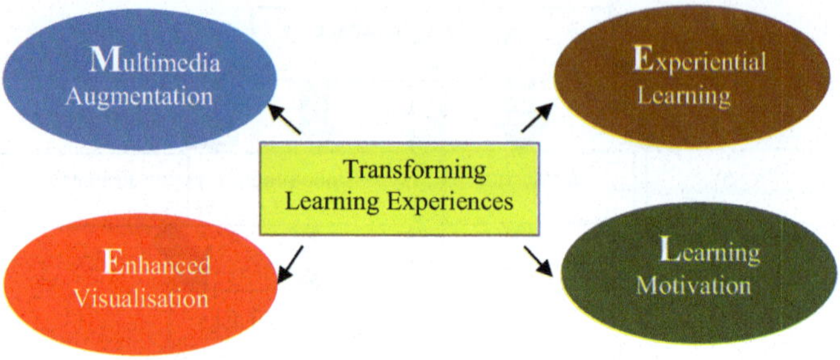

Fig. 5 Educational affordances of VR/AR (**MEEL**)

It is only recently that scholars began to discuss the design of virtual learning environments that are supported by learning theories such as [6] ELT (Fromm 2020). This chapter seeks to fill the gap in the literature by proposing a categorization of the educational affordances of VR/AR that are based on findings from significant peer reviewed studies for the past two decades with the relevant supporting learning theories. The authors grouped the educational affordances of VR/AR into four categories—**M**ultimedia Augmentation, **E**nhanced Visualization, **E**xperiential Learning, **L**earning Motivation (**MEEL**) (see Fig. 5).

3 Multimedia Augmentation

3.1 Cognitive Theory of Multimedia Learning

Three metaphors or paradigms of learning were developed during the twentieth century, behaviorism, cognitivism, and constructivism [66–68] (Mayer 2011). Behavioristic theories of learning rests on the assumption that knowledge is a repertoire of behavioral responses to environmental stimuli [69, 70]. Cognitivism is based largely on information acquisition through direct instruction with a focus on retention of knowledge [71]. Cognitivism views learning in terms of cognitive processes such as thinking, problem solving, concept formation and information processing [69]. Constructivism is based on knowledge construction, in which the learner actively builds a mental representation in working memory [71].

Wittrock's ([72], p. 349) generative learning theory posits that learners construct meaning by actively building relationships between stimuli and their stored information such as knowledge and experiences. What is learned depends on assimilating what is presented with what the learner already knows. Wittrock used the [72, 73] generative learning theory and [74] cognitive load theory led [66] to propose the Cognitive Theory of Multimedia Learning (CTML). There are three assumptions in CTML (see Fig. 6 [75] (Mayer 2003):

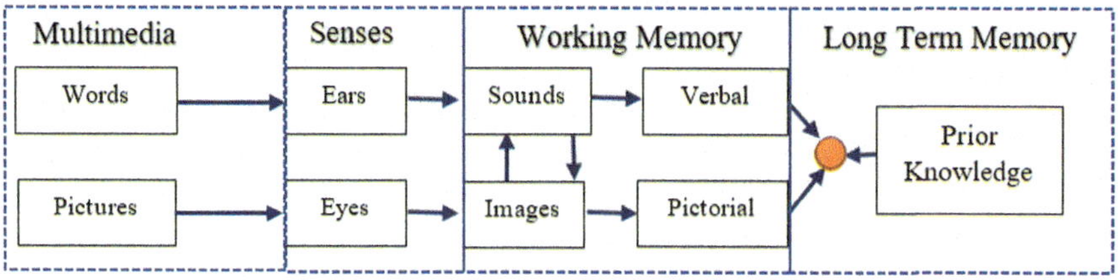

Fig. 6 Flow Chart for CTML, adapted from [66, 76, 77]

- Dual channel—There are two separate channels, auditory and visual, for processing information.
- Limited capacity—Each channel has a finite capacity that can be overloaded when there is too information presented in one mode.
- Active processing—Learning comprises of active processes such as filtering, selecting, organizing, and integrating information based on prior knowledge.

The CTML offers a useful guide on how to implement effective cognitive strategies into multimedia to help learners learn efficiently [78, 79]. The move towards the constructivism learning paradigm involves the recognition that meaningful learning rests less on the learner's behavioral activity and more on the learner's cognitive processing. This has led to a shift in focus on how to facilitate meaningful learning through active cognitive processing [80]. Meaningful learning enables deep understanding of the material [75]. The learner identifies key concepts, mentally organizes the information, and integrates this information with prior knowledge [81].

For example, the use of AR can scaffold the writing process in language learning. This taps the principle of multimedia learning that people learn more deeply from words and pictures than from words alone [76, 77]. Wang [82] found that an AR-based writing support system improves essay writing performance by twelfth-grade students. The AR system provided the augmented prompts as a form of scaffolding that enhanced students' writing motivation and creativity. The first AR mode involves learners walking around the school. They used tablet PCs with AR capabilities to read the pre-embedded text information with video, or audio that gave writing stimulation and guidance. The second mode involves using AR app to scan the paper-based worksheets in classrooms. The multimedia prompts assisted the learners in recalling and reflecting.

Santos et al. [83] applied Mayer's multimedia learning theory as a framework for developing a mobile AR system to learn Filipino and German words. They carried out a within-subjects study to evaluate the effects of a mobile AR application to learn vocabulary compared to the non-AR treatment of flashcard type application on a tablet. By presenting digital information within the context of a real environment, it allows users to form memory retrieval cues based on the real environment which aids helps memorization. This has led to better retention of words and improved student attention and satisfaction. Situating the AR application in the environment increased relevance.

A study by Estapa and Nadolny (2015) on the use of an AR application to teach algebra and geometry compared the experimental group with the control group which only had website interaction. They created an inquiry cum problem-based activity based on a spring break road trip. Along the way, students encounter challenges and issues with mathematics problems which must be solved. The AR group students could directly interact with the page to explore the multimedia content such as video, slideshow and weblinks, quiz and feedback while control group participants could only access digital resources through their mobile devices and had to toggle between the print media and mobile devices. Both types of conditions led to improvement in mathematical learning and both groups were motivated by the activities. This confirms the value of multimedia learning to support active cognitive learning processes.

3.2 Cognitive Load Theory

According to [74, 84, 85], the working memory is severely limited in both capacity and duration when dealing with new information. Humans are only able to manage two or three items of information simultaneously [86] and hold about seven items, or chunks of information, at a time [87]. When the amount of information exceeds the limitations of processing information, cognitive load results. To reduce cognitive load, both visual and auditory channels should be used [85]. Instructional materials should consider the limits of students' working memory [88]. This is known as the Cognitive Load Theory (CLT). The CLT can help guide the design of instructional materials [89] and content for AR [56].

In CLT, three types of cognitive load are recognized—intrinsic, extraneous, and germane [85] (see Fig. 7). While intrinsic cognitive load is attributed to the difficulty of the learning materials, extraneous cognitive load is attributed to the inappropriate presentation of the instructional materials and germane cognitive load is the amount of invested mental effort by learners [90, 91]. Extraneous cognitive load may be caused by cognitive activities that are not directly related to the learning goals [90].

In areas such as medical training and manufacturing operations, studies have shown that AR can reduce cognitive load. Kirschner et al. [92] compared visualizations with conventional ultrasound and an AR-guided system on the learning of needle insertion in surgery. Whereas the usual approach requires significant cognitive processes for localizing targets, the AR system reduced germane load by providing

Fig. 7 Types of cognitive load in CLT

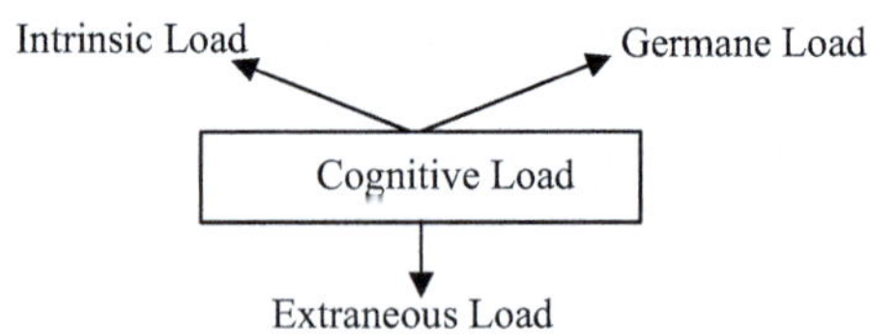

sight of the patient's body, together with components such as slice image, target, needle, and hands that are all superimposed on each other. In three experiments, the AR-guided system resulted in higher accuracy and lower variability in aiming and endpoint placements than did the conventional method.

According to CLT, traditional instructional formats can impose extraneous cognitive load when they require novice learners to expend cognitive resources to understand instructions [88]. Tang et al. [93] carried out a study on find out the effectiveness of AR instructions in an assembly task. The AR system was compared with three other conditions: a printed manual, computer assisted instruction (CAI) using a monitor-based display, and CAI utilizing a head-mounted display. Results indicate that AR which affords overlaying 3D instructions on the actual work pieces reduced the error rate for an assembly task by 82%. Measurement of mental effort indicated decreased mental effort in the AR condition. In their review of AR systems, Jeffri and Awang Rambli [56] found a positive correlation between effects of AR on mental workload and on task performance. Effectiveness of AR systems were influenced by the type of display device used, relevance and timeliness of content, information presentation, user characteristics and task characteristics.

Meyer et al. [94] investigated the relationship between media and method on learning science. The learning intervention consisted of either learning with immersive VR and HMD using the simulation The Body VR: Journey Inside a Cell or watching a video recording of the simulation in a laptop. Both conditions took place after a 'barebones' pre-training lesson to familiarize participants with parts of the cell. The VR application provides an immersive, first-person perspective as the student travels around the body as a player shrunk down to cellular size. The results showed an interaction between media and method only for the immersive VR group, indicating that pre-training had a positive effect on knowledge, transfer and self-efficacy directly following the intervention. This suggests that cognitive load may hinder the efficacy of VR learning, and the introduction of a pre-training lesson can mitigate the negative effects of extraneous load. No such interaction was found for the video condition.

In [94] study, the VR group learners may have been overwhelmed with sensory information and did not have resources available to effectively select, organize, and integrate information into long term memory. The pre-training material could have helped the immersive VR group learners to recognize and organize important concepts as prior knowledge which would help integrate the VR experience into a sort of 'spatial mental map'. Previous studies also found that learners experience cognitive overload more easily when a multimedia lesson is presented in immersive VR compared to a less immersive format. This is due to increased amount of sensory information in the form of extraneous load [95, 96]. This could explain why an earlier study by Moreno and Mayer (2002) found no significant differences in outcomes between immersive VR learning and 2D desktop in terms of knowledge retention and problem-solving in science. However, the students gave higher ratings of presence when learning with immersive VR.

3.3 Principles of Spatial Contiguity and Signaling

The ecological approach [41] to visual perception, and subsequently the work by [42, 43] reinforced the importance of visual cues in the environment. Multimedia instruction enhanced with AR provides affordances for teaching content. When words and pictures are presented contiguously, or side-by-side each other, learning can be increased by reducing germane load. This is called the spatial contiguity principle by [97]. The premise is that given the limits on the capacity working memory, students may learn more easily when words and pictures are presented contiguously.

AR can reduce cognitive load through the spatial contiguity effect. The unique affordance of AR lies in situating digital information within the context of a real environment which can aid memorization. Fujimoto et al. [98] carried out a within-subjects study involving users memorizing the number of blocks" and the location of drawers from which the blocks are picked. Users were presented with either type 1 where the information is displayed near the location of each drawer (with visual and textual instruction) or type 2 where the information is displayed at a fixed location (with only textual instruction). Results show that type 1 condition aided memorization better than type 2, confirming the spatial-contiguity effect.

The authors developed a simple mobile AR app that augmented the science textbook to show molecules in 3D with text labels to help students understand the concept of digestion of nutrients (see Fig. 8a). Students reported that it supported their understanding. Previously, they could not understand the concept by looking at the 2D textbook diagrams which lacked labels.

AR technology affords the signaling principle of multimedia design that helps learners recognize and learn information easier when call outs, arrows, and highlighting are used [99]. Patzer et al. [100] found that undergraduates who used an AR system to learn how to play the guitar were able to perform more of the melody and scale after 2 weeks compared to the control group. The AR group used the Optek Fretlight ® guitar with an interface of precise lights spots on the fret board (Fig. 8b).

Patzer et al. [100] discussed a study to find out if spatial contiguity principle affects learning effectiveness in heart anatomy. Students learn in either one of the three conditions—digital AR models with directly labelled anatomical labels, AR

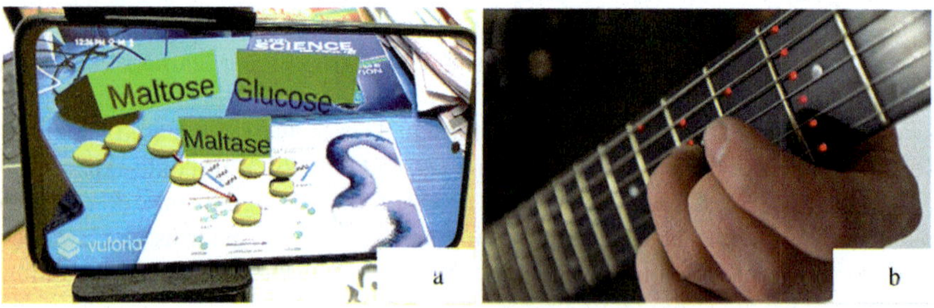

Fig. 8 **a** AR science textbook; **b** Fretlight guitar with AR signaling cues

models without labels, and physical fiberglass models without direct labels. It was found that the addition of labels directly onto the AR model significantly boosted learning when compared to an unlabeled AR model. The mobile AR group students found the learning more enjoyable, curiosity inducing and easy to use compared to the fiberglass model.

4 Experiential Learning

4.1 Kolb's Experiential Learning Theory

The origins of Kolb's experiential learning theory (ELT) reside in the writings of Dewey in his books such as Education and Experience [4], group dynamics theory by Lewin, and Piaget [6]. Kolb's experiential learning theory is based on the premise that knowledge results from the interaction between theory and experience [101]. Klippel et al. [6] proposed four components in the ELT (Fig. 9), each of which requires learners to invoke specific abilities [102]:

(a) Concrete experience—learner's willingness to experience new things directly
(b) Reflective observation—finding meaning from a variety of perspectives
(c) Abstract conceptualization—analyze and integrate new ideas and concepts, and draw conclusions
(d) Active experimentation—apply new learning to practice, problem solving, and decision making, which leads to new concrete experiences

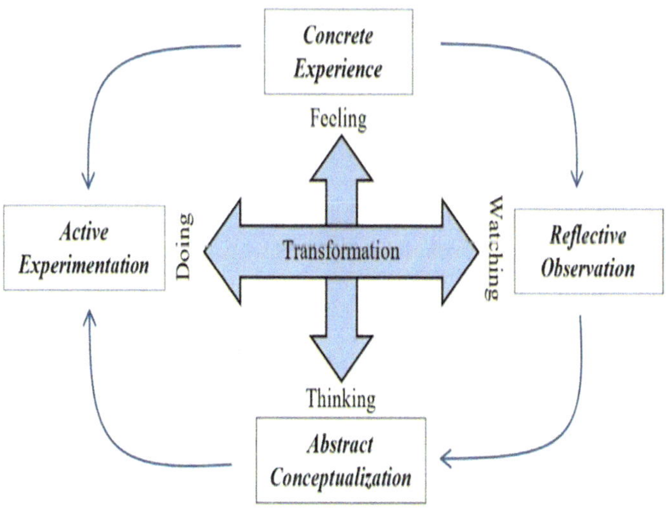

Fig. 9 Experiential learning theory, adapted from Kolb (2004)

ELT provides a framework for designing active, collaborative, and interactive learning experiences that support the process of transformation [6]. Experiential learning is active learning as it engages students to do meaningful learning and reflect on the experience [103]. Efstathiou et al. [104] studied the use of AR for experiential learning in learning history for 3rd grade students. The study showed a significant increase in historical empathy and conceptual understanding for the AR group compared to the control group, which used the traditional approach of commentary by history teachers. The experimental group used a "Young Archaeologists" location-based AR mobile application. A virtual archaeologist presented the information in the form of text, videos, images, and graphs at five hotspots in the archaeological site (Concrete Experience) and prompted students to reflect to bridge the distance between the present and the past (Reflective Observation). The AR learning environment served as a collaborative scaffolding mechanism, as the students worked in groups to view the AR multimedia content and engaged in discussions (Abstract Conceptualization).

As direct experience is the starting point of ELT, the degree by which a virtual experience resembles a real experience must first be ascertained [13]. A useful scale of experientiality is developed by [105] which classifies how much a "student's learning experience is between indirect and direct experience", and the depth of the student's involvement in ten levels of experience (see Fig. 10). It is a useful guide in assisting designers to distinguish between the stage of direct experience afforded by VR experiences. The lowest level is a form of indirect simulated experience comprising of slides, pictures, videos, and other simulations of reality [105]. At higher levels, VR offers a form of direct experience characterized by vividness and interactivity.

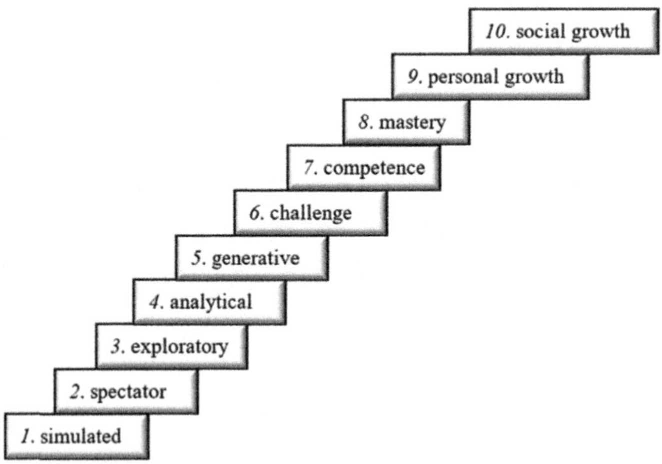

Fig. 10 Scale of experientiality (adapted from [105]

4.2 *Direct and Concrete Experience*

The first step in Kolb's experiential learning cycle is direct and concrete experience. The distinctive feature of ELT is that it considers direct experience as the foundation of learning [106, 107].

AR is ideal to enhance experiential learning [108] because it:

(a) Allows real-time interactivity in an ecological setting improving concentration and motivation [109]
(b) Provides multimedia augmentation helping users to develop skills and knowledge in a more effective way [12]
(c) Situates learning in the real environment facilitating the transfer of the acquired skills to the real world (example [110]).

Dale's Cone of Experience [111] is a theory that shows a hierarchy of learning experiences. Baukal et al. [112] updated the theory, calling it the Multimedia Cone of Abstraction to include VR technologies (see Fig. 11). VR/AR technologies can provide a more authentic and realistic experience than desktop simulations by providing a stronger presence and immersion. This leads the user to recognize a virtual experience as real [13].

The concept of presence in VR gives users the illusion of non-mediation, meaning that a person perceives and responds to the content of a particular medium as if the medium were not there [54]. In history education, [113] developed a novel mobile technology enhanced pedagogical approach called REENACT to engage students in immersive experiences from the points of views of reenactors and historians. It reveals the potential value of the approach to provide new edutainment collective experiences in history learning.

Hsu et al. [114] developed an experiential learning game situated in a virtual bathroom in which participants (n = 165) were asked to repeatedly use a 600-ml bottle to fill a specific water tank to flush a toilet and take a one-minute shower.

Fig.11 Multimedia cone of abstraction adapted from [112]

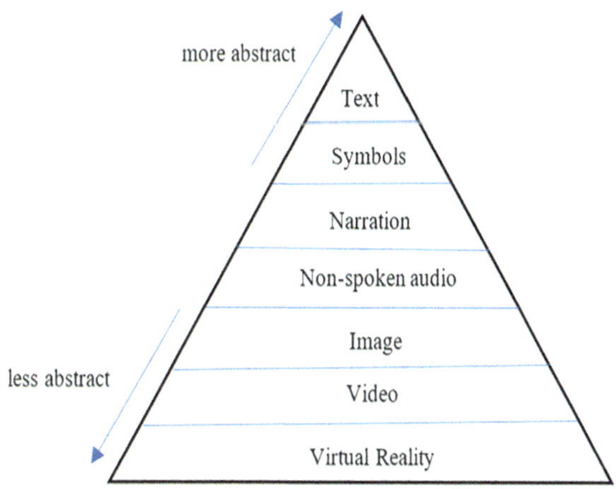

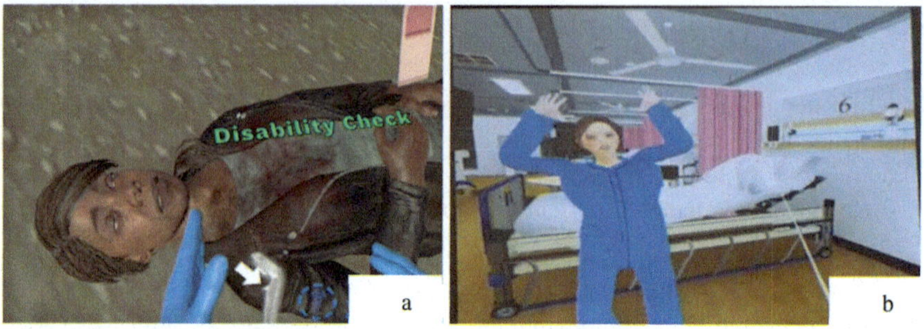

Fig. 12 **a** VISE VR app; **b** VR app to train workers to handle psychotic patients

Participants received exaggerated feedback on the negative consequences of water consumption and/or environmental damage. The concrete experience aims to increase awareness of the amount of water used, for example in flushing the toilet and the impact on the environment. This game caused significant changes in cognition and behavior intention.

An immersive VR software called VISE that affords the experience of dealing with trauma emergency situations is used by the National University of Singapore (NUS) Yong Loo Lin School of Medicine students to practice the skills of responding and handling patients injured in these scenarios. The medical students use the VR controllers to intuitively interact with the patient and assess the conditions such as excessive bleeding, wound or even death. The highly realistic simulation allows medical students to familiarize with the triage methodology process by going through the actual motions of assessing casualties (Fig. 12a) (https://cutecenter.nus.edu.sg/projects/vise.html).

At the time of writing, NUS has developed another software (see Fig. 12b) to train healthcare workers to handle agitated patients [115]. In a particular scenario, users had only eight seconds to make each decision to manage the situation of calming a psychotic patient. One medical student commented that the experiential learning experience was "very real, and there are sounds from the TV remote and sounds from other patients' beds, it really paints what is like in the hospital and this really gave me a lot of confidence on what to say and what decisions to make". In a trial with 65 students, more than 90% reported gaining more confidence in dealing with agitated patients.

4.3 Embodied Learning Experience

Embodiment theory proposes that knowledge is grounded in sensorimotor systems and learning is facilitated when it activates those systems [116]. Physics concepts are ideal for embodied learning because the body experiences the world through physical interactions [116]. Enyedy et al. [117] developed an AR modelling environment called "Learning Physics through Play Project" (LPP) that makes use of both the affordances of embodied learning and AR simulations. The study engaged 6–8-year-old students in a series of scientific investigations of Newtonian force and motion using AR. The program uses first-person play (*Concrete Experience*), for example, one student pretended to be the ball and used his physical motion to predict and represent the motion of the ball (*Reflective Observation*). Students then reflect using their own embodied understanding to form a concept (*Abstract Conceptualization*) followed by experimenting with own representational system (*Active Experimentation*) in the microworld. Results show that young students were able to develop a conceptual understanding of forces, friction, and two-dimensional motion after participating in the embodied LPP curriculum.

Madden et al. [118] studied the use of VR as an embodied learning teaching tool to teach the phases of the Moon. The students in the VR condition used the Oculus Rift HMD with hand controllers that enabled them to grab the Moon to move it around in its orbit. The participants in the desktop simulation and VR conditions could transport to Earth's surface to observe rise and set times. The third group used the traditional hands-on approach. Although there was no difference across conditions in performance levels, embodied VR learning seems to better able to support constructivist approach by helping them to see the full picture and build their mental models. One participant commented that the VR affords manipulating "the environment on my own accord. It seems more engaging than the two other methods."

Johnson-Glenberg et al. [116] designed a MR simulation called SMALLab which consists of an interactive physical space to engage multiple sensory systems through floor projection. A study of college students (n = 109) was divided into three groups—MR, interactive whiteboard, and desktop computer. The MR group (high embodiment) performed better on generative knowledge questions. This suggests that embodiment using MR technologies fosters better retention of certain types of knowledge. Virtual science laboratories tapping embodied experiences for inquiry learning have been explored. For example, [119] developed Frame, an innovative sensor augmented virtual lab that uses sensors as physical inputs to control scientific simulations. Results show that the eight-grade students developed an overall understanding of gas behavior with molecular-level explanations. A significant increase from pre-test to pos-test scores with a large effect size of 0.91 was observed.

The virtual experience of embodiment in VR can empower personal growth level [105] through body transfer illusion [120]. Studies have shown that VR is able to produce an out-of-body experience (OBE) [108]. It is the experience in which a person who is awake sees his or her body from a location outside the physical body (Erlsson 2007). Erlsson (2007) and Lenggenhager et al. [121] demonstrated the perceptual illusion whereby healthy participants experienced a virtual body as if it were their own. Maister et al. [122] found that transfer illusion in a body of different race produced a significant reduction of the implicit bias against that race. Yee and Bailenson [123] found that negative stereotyping of the elderly was significantly reduced when participants were placed in avatars of old people, compared with young avatars. Hershfield et al. [124] found that exposing participants to their future-aged selves through VR impacted their attitudes toward monetary saving for the future. These examples show that embodied OBE experiences have the potential to bring about transformational change in individuals through a change in their worldview [125].

4.4 360° Video VR Experience

The scale of experientiality [105] postulates that the greater the immersiveness and presence in a VR experience, the more direct and concrete the experiential learning is in Kolb's cycle. Rupp et al. [126] carried out a study where university students (n = 136) were randomly assigned to watch a six-minute 360° video of the International Space Station (ISS) under one of four conditions—mobile phone, Google Cardboard HMD, Oculus HMD (Development Kit) and Oculus HMD (Consumer). It was found that viewing 360° videos through HMD provided better learning experience, greater immersion, enjoyment than the mobile phone.

The ability of 360^{0} videos to incorporate a first-person perspective induces a greater sense of realism. This is exploited in areas such as medical education. Fukuta et al. [127] filmed an orientation training of operating theatre procedures in 360^{0} video and showed it to medical students (n = 34) as a pre-training before their first operating theatre experience. The 360° camera was mounted on top of the head of a student actor. This gave a first-person perspective which made the viewer's experience more real and personal. Indeed, the cost barrier in producing VR environments has been partially resolved by filming 360° VR videos [128, 129] and viewed through Google Cardboard.

360° VR technologies has also been explored as a viable or more engaging alternative to traditional modes of safety training such as instructional videos or printed materials. In a randomized controlled study by Araiza-Alba [56], children (n = 182) were randomly assigned to three instructional methods—2D video, poster or 360° VR videos. It was noteworthy that the 360° video VR group produced the same level of learning outcomes in terms of improvement of water safety knowledge after going through the training even though they did not have the benefit of interaction with a safety instructor. In addition, participants in the VR group showed greater interest and enjoyment of the learning experience compared to the traditional modes of learning.

In aspects such as environmental conservation, it has always been a challenge to get children and adults to identify with environmental issues. These difficulties can be overcome if we can get these individuals to acquire first-hand experiences [114, 130]. Through personal experiences, people are more likely to see the effects as connected to them and develop positive attitudes to environmental causes. As such, there is scope for the use of 360^{0} videos as a form of experiential learning to bring about improvement in knowledge and attitudes towards environmental conservation. Nelson et al. [131] carried out an interesting 2×2 experimental study whereby they filmed a 360^{0} video of coral reef conservation and sampled (n = 1006) participants by giving them an opportunity to donate to a charity towards reef conservation after watching a 5-min video. They found that 360^0 videos with a negative message garnered larger average donation amounts than 2D videos or 360^0 videos with a positive message. Fokides and Arvaniti [132] also found that 360° videos produced better learning outcomes than webpages or printed material in teaching environmental education.

Calvert and Abadia [133] carried out a comparison study between the use of 360^0 video presented through computer screen (control) and immersive VR with HMD (experimental group) to teach history as a part of the Australian high school curriculum. The topic was about the military campaign between the Japanese and Australian soldiers at the Kokoda, Papua New Guinea in World War Two. The control group participants could only rotate the video using the desktop mouse while the immersive group participants could move their heads and intuitively explore the virtual environment with the HMD. The experimental group students using the HMD VR condition reported higher engagement, presence, empathy, and better knowledge mastery than the desktop 360° video groups. Immersive narrative VR experiences thus have the potential to provide students with deeper experiences due to the sense of realism.

5 Enhanced Visualization

VR/AR technologies transform visualization through **S**patial learning, **I**mmersive simulation, **D**ata visualization and guidance, and **E**nhanced collaboration (**SIDE**) (see Fig. 13). This is adapted from the description of affordances of AR by [134], 12 of illustration of spatial concepts, situated learning in the real environment, intuitive interaction, 3D visualization, and facilitating collaboration.

5.1 Visualization in Surgery and Medical Training

5.1.1 AR Technologies

For more than two decades, AR technologies have been used in data visualization, intra-operative tool guidance and medical training. Enhanced visualization and navigation have led to improved safety and efficacy of surgical procedures [135]. AR can be used to pursue minimal invasiveness and maximum safety by providing real-time updated 3D anatomical details over the real surgical field [136] recent reviews by [137] and [138] have found that AR can enhance the experiences of medical students and improve learning outcomes such as theoretical knowledge and practical skills.

The efficacy of AR in data visualization and navigation is based on its ability to blend imaging datasets into the direct vision of the surgeon [139]. This enhances the accuracy and efficacy of diagnostic and treatment procedures. By augmented images in the operating microscope's eyepiece, an AR system by Cabrilo et al. [140] helped

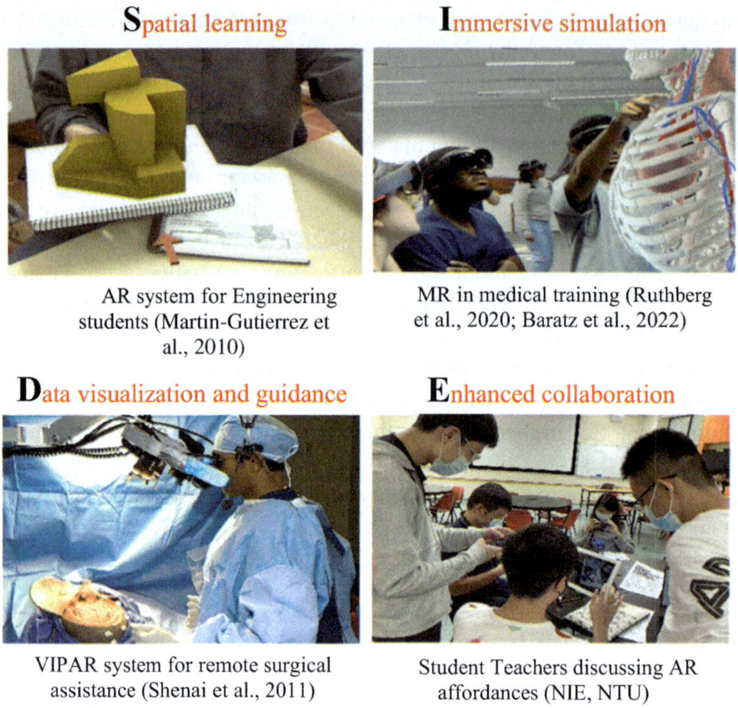

Fig. 13 Enhanced Visualization of VR/AR technologies (**SIDE**)

the surgeons to precisely localize blood vessels during neurosurgery. Abe et al. [141] developed an AR guidance system to visualize the needle insertion point on the patient's skin. The system helped surgeons to prevent inaccurate needle placement that can cause injury to the vertebrae. In tests with 40 spine phantom trials and clinical results with 5 patients, the system was successfully used to assist in needle insertion. Shenai et al. [142] developed a shared AR system called VIPAR that allowed for real-time, virtual interaction between the surgeon and a remote surgeon. The MRI augmentation provided spatial guidance to both surgeons.

The direct projection of virtual images onto the head, skull, and brain surface in real time is a strong advantage to surgeons. Besharati and Mahvash [143] designed and developed an AR system to project a virtual image directly onto the patient's head, skull, and brain surface in real time for neurosurgery. They found that AR visualization of the tumors succeeded in all five cases and there were no significant differences in accuracy in terms of tumor localization and size compared with an existing standard navigation system. AR has also been applied to wide range of orthopedic procedures [144]. For example, [145] developed an AR system to assist surgeons in locking of screws during orthopedic bone surgery. They found that real-time visualization with AR with video guidance improved both the speed and accuracy of the surgeon compared with current navigation tools without AR assistance.

AR technologies can leverage on mobile devices for visualization. Ogawa et al. developed the AR-HIP for orthopedic hip replacement surgery. The AR-HIP system allows the surgeon to view the image of replacement device superimposed in the surgical field through a smartphone. In 56 total hip replacement surgeries, they found the AR system to be more accurate than conventional mechanical guides. In a clinical trial study by Elmi-Terander et al. [146], AR surgical navigation system have been found to offer high accuracy to the placement of thoracic and lumbosacral pedicle screws, and no device-related complications were noticed. Deng et al. [147] developed a Tablet-AR system to directly observe the intracranial anatomical structures with the overlaid projection images during surgery.

In dentistry, immersive technologies have also been found to be beneficial for clinical practice and training [148]. Jiang et al. [149] evaluated the feasibility of an AR system that could overlay 3D images in the real environment for dental implant surgery. They compared the outcomes for the AR-guided navigation method and traditional two-dimensional image-guided navigation method. The results showed that there were smaller errors made and shorter surgery time. By increasing visualisation, AR technologies reduces cognitive load, improves decision making and speeds up task execution [150].

5.1.2 VR/MR Technologies

In surgical training, immersive VR simulations enable improved visualization because it supports stereo vision, viewer-centered perspective and large angles of view [151, 152]. Ghaednia et al. [139] identified the applications of VR as mostly limited to outside of operating theatre in areas such as training and education, remote robotic surgery, and rehabilitation. Silverstein et al. [151] developed an immersive environment for teaching with biomedical models such as virtual models of the liver. The instructor gave a workshop to six surgery residents at two physical locations and found that the workshop produced significant improvements from the pre-test and post-tests scores. Dobson et al. [152] designed a VR-based interactive display to help doctors understand the highly complex anorectal and pelvic floor anatomy. The collaborative VR environment enabled teachers and students to interact and manipulate the components from various locations.

The traditional approach of learning from cadaveric dissections has several limitations. MR technologies may provide an effective tool to visualize complex structures are too small or difficult to view in cadaveric sections [153]. This can lead to a reduction in errors that damage tissues or incomplete procedures [154]. Supplementing cadaveric dissection with MR can improve long-term retention for the anatomical topic of the female breast [154]. Ruthberg et al. [155] studied the use of MR with HoloAnatomy app as an alternative to cadaveric dissection. Medical students using MR took approximately less time to learn the content on limb anatomy than the control group with cadaveric dissection, although no significant difference was found between their exam scores.

In a randomized study by [156], all first-year medical students went through traditional cadaveric dissection teaching. This is followed by VR-based anatomy education with virtual cardiac model or independent self-study. The VR group demonstrated a significant 28% improvement from pre- to post-performance scores. Comments from the VR group were positive and most felt that it offered an immersive experience that helps them understand the size differences and relationships between different cardiac structures. Huang et al. [157] described a VR-enhanced system for intra-cardiac intervention to train medical students and junior surgeons to perform minimally invasive cardiac intervention. The simulations use patient-specific Magnetic Resonance Imaging (MRI) data to provide guidance by means of virtual striped wires to simulate the catheter. Southworth et al. [158] conducted a review of the use of immersive technologies in cardiology.

VR can also be used in pain management. For example, Lahti et al. [159] carried out a randomized clinical trial of the use of VR relaxation before treatment to decrease dental anxiety. The VR group watched 1 to 3.5 min 360° immersion video of a peaceful virtual landscape through HMD while the control group patients could use their own phones if they wished. Dental anxiety decreased more in the VR group. Atzori et al. [160] carried out a within-subjects study on five patients who received VR distraction by putting on Oculus Rift VR HMD during dental treatment compared to "treatment as usual" without VR. The children reported significantly lower "worst pain" and "pain unpleasantness," and had significantly more fun with VR.

Chelsea et al. (2018) showed that an immersive VR environment supplements the learning of neuroanatomy by decreasing the fear of learning neuroanatomy, called neuro phobia, and increases knowledge retention for medical students. The VR group could visualize the 3D relationships in digital structures created from magnetic resonance images from a healthy person's brain, while the control group could only learn from a booklet with color figures. Both groups showed significant performance improvement on the test questions following the intervention.

Immersive VR technologies can provide learners with disabilities with new kinds of access to rehabilitation [1]. For example, the Mind Motion system by Mind Maze (https://www.mindmaze.com) is a VR system that provides intensive physical rehabilitation to adult patients. The device allows patients to control a virtual avatar and visualize the feedback for task-specific training. The system offers interactive VR exercises that engage participants' movements with various levels of difficulty. Perez-Marcos et al. [161] carried out a study using Mind Motion system for ten outpatient stroke survivors who participated in the rehabilitation program. They found that median improvement rate of 5.3% in motor function and of 15.4% at one-month follow-up. For three of them, this improvement was clinically significant.

5.2 Spatial Learning and Immersive Simulation in Education

5.2.1 AR Technologies

In the learning of STEM (Science, Technology, Engineering, Mathematics) subjects, students often fail to make the connection between objects in a real-life and those of the two-dimensional space. For example, students have difficulties in distinguishing geometric solids from flat shapes [162]. AR can assist in the development of spatial abilities and enhance spatial learning [163]. In areas that involve spatial domains such as mechanical machinery, astronomy configurations or human organs, students learn better when with AR compared to printed media or desktop software [65].

AR technologies have been applied in STEM subjects, humanities and arts, astronomy, preschool education, and museum education [164, 165]. VR/AR can be effectively deployed to support skills-based and competency pedagogies, by providing a range of hands-on learning experiences [166]. Research in cognitive psychology show that our minds are inherently embodied, in other words, we create and recall mental constructs based on the way we perceive and move (example—[167–169]). Studies have shown that the users with VR HMDs usually outperformed those in the control group using computer desktop, in areas such as spatial recognition (example [170, 171]) and memory recall (example—[172]). Krokos et al. [173] found that participants using VR HMD have 8.8% improvement in recall accuracy of faces compared to the desktop condition.

In the innovative MagicBook prototype by [174] users can see the graphical content through AR as virtual 3D animated models appearing on pages AR picture books show 3D animated content and provide a means for students to engage in immersive learning experiences with content such as animations, 3D graphics and audio (Dunser et al. 2012). The enhanced visualization of content using AR picture books can reduce stress in children. Alarcón-Yaquetto et al. [175] carried out a randomized experimental design study into the effect of AR book on the salivary cortisol levels of hospitalized pediatric patients. Based on measurement using a visual analogue scale (VAS), ARPB reduced psychological stress better than a standard book. However, ARPB did not diminish cortisol levels to a greater extent than books.

Dünser et al. [176] created three AR books to study the effectiveness of AR for teaching electromagnetism at the secondary school level. They found that the AR group students scored higher than the non-AR group in the immediate post-test and retention test. In addition, students in the AR group did better on questions that require application of spatial abilities, such as drawing field lines on a magnet and questions related to the right-hand rule. Although the sample size was small ($n = 10$), the results show that AR has potential to be effective in helping students to visualize complex 3D scientific concepts such electromagnetism.

Martín-Gutiérrez et al. [177] developed an AR book called AR-Dehaes to help first-year mechanical engineering students perform visualization tasks. They found that the system has a positive impact on students' spatial ability and students rated it easy to use and attractive. Kaufmann and Schmalstieg [178] designed the Construct3D system, a 3D AR geometric construction tool to the improve of spatial abilities and maximize the transfer of learning in geometry education. Besides providing a natural setting for collaboration between teachers and students, Construct3D enables students to see and manipulate objects in 3D space. This enhances their spatial learning abilities.

Estapa and Nadolny (2015) opined that mathematics instruction is a natural fit for AR with benefits in manipulation, visualization, and authentic contexts. Lin et al. [179] developed an AR tool to assist eighth grade students in learning solid geometry. The system utilizes a webcam to photograph the marker and display the 3D object on the computer screen. The AR intervention was effective for students with average and low academic achievement. Students' mathematics scores and spatial perception during the test closely correspond to each other, attesting to the value of AR in improving spatial abilities. This consistent with studies done such as by [180].

In chemistry education, improving spatial reasoning skills has been shown to enhance students' success in learning [181–183]. The ability to visualize a complex structure intuitively is critical because there is a strong relationship between molecular structure and function [184]. Zheng and Waller [184] developed an AR learning experience called Chem Preview for users to interact with molecules using natural hand gestures. The software allows students to build, visualize, modify, and measure molecular structures in a 3D space. Singhai et al. (2012) developed an AR system to help increase the understanding of 3D chemistry modelling and spatial arrangement of molecular structures in space. The AR system allowed students to see actual molecules in the 3D environment, view them from multiple viewpoints and control their interaction. Students also enjoyed the learning experience and improved their spatial intuition.

Smith and Friel [56] developed two mobile AR apps to help students develop 3D cognitive understanding from 2D chemical representations of molecules on slides or textbook images. They found that the use of AR has created an active learning environment and allowing the students to work cooperatively. Most of the students found the AR models easy to use and wanted to see more AR being used in other lessons. Habig [185] carried out a study to determine gender differences in solving stereochemistry problems using AR. While male participants solved items more frequently correct when they were based on AR representations, but females scored higher on items based on 2D ball-and-stick models.

5.2.2 VR Technologies

Fully immersive VR technologies allow users to enter the virtual world and actively participate and interact with simulated 3D objects [186, 187]. An open-source software called VR Chem helps students to visualize and manipulate organic molecules [188]. User studies on using VR Chem to build molecules using a pre-configured set of hand gestures. While they found hand gestures intuitive, its actual use was jarring due to the imperfect accuracy of the tracking device. A VR Enzyme software was developed for some secondary schools in Singapore by NTU. The user can view the enzyme in 3D and use hand gestures to rotate it in such a way to allow the substrate to successfully dock at its active site. The pH of the medium can be changed by intuitively turning the palm upwards to bring up a menu slider for pH value [189–191].

In the science classroom, teachers need to include new ways to teach the complex representations of cells so that students may develop more sophisticated ideas (Celiker 2013). Enhanced visualizations will enable students to make the conceptual leap to a microscopic scale and help them to develop accurate mental models of cellular processes [192]. Thompson et al. [192] developed Cellverse, an inquiry-based VR game to help students learn cellular biology. Two students play together, one using an Oculus Rift with Touch controllers, the other on a tablet or laptop. The game involves a challenge: the players need to use the information given to diagnose the cell which has five possible classes of cystic fibrosis. The students see the 3D cell environment and learn about different organelles such as ribosomes by selecting them to see descriptions. Players can learn about spatial relationships between the micro scale (organelles) and nanoscale (RNA and amino acids). Thompson et al. [192, 193] found that students' post drawings of cells showed greater spatial accuracy and complexity in terms of the types of organelles and their density in the cellular space, indicating a shift towards more accurate mental models. The change in mental models is apparent in their comments, such as "it also helped me put into perspective how many organelles there are in a single cell" and "helped me see where the organelles are in relation to other organelles, for example smooth and rough ER are near (the) nucleus."

Bhattacharjee et al. (2018) found that visualization experiences using immersive VR enhanced students' learning performance and gave them longer term retention of knowledge compared to traditional teaching approaches. In the first experiment, the students could see the functions of the DNA and polymerase from a microscopic level. In the second experiment, students in the VR group learnt chemistry and biology through simulations. The learning performance of attention deficit hyperactivity disorder (ADHD) students increased by 75%. In the third experiment, the students travelled through the simulated human body and its organs. The learning effectiveness of audio processing disorder (APD) students increased by 32%. Tan and Waugh [194] found that immersive VR helped students to overcome frustration and lack of understanding in molecular biology, as traditional media cannot sufficiently represent the structures and dynamic processes taking place in the cell. The study found significant increases in molecular biology achievement in male students. Interviews revealed that visualization enhanced understanding leading to achievement.

6 Learning Motivation

6.1 ARCS Motivational Model

In education, motivation is considered a key determinant of learning [195]. According to Keller [196], before a teacher can begin teaching, he or she must first arouse the learners' attention by making the learning experience "personally relevant". It involves introducing content that taps on the learners' prior experiences and connects to their personal desires and goals. Motivation can influence the level of attention and effort students dedicate to learning activities [195]. Keller's Macro Model of Motivation and Performance [197, 198] provided the foundation and frame of reference for what is known as [199] ARCS motivational model comprising of dimensions of Attention, Relevance, Confidence and Satisfaction. The Macro Model of Motivation and Performance illustrates how motivation combines with their knowledge and skills to influence their overall performance (see Fig. 14).

Motivation can influence initiation, direction, intensity, persistence, and quality of behavior [200] through three feedback loops in dotted lines (Keller 2010). The first loop is that the satisfaction from one's learning experiences will influence the value of the activity and how much attention to give in learning in future. The second and third loops connect performance and outcomes respectively to confidence. The degree to which the learner successfully accomplishes the task and produce outcomes that will affect the expectancies for success in the future.

The following provides explanations of the ARCS domains:

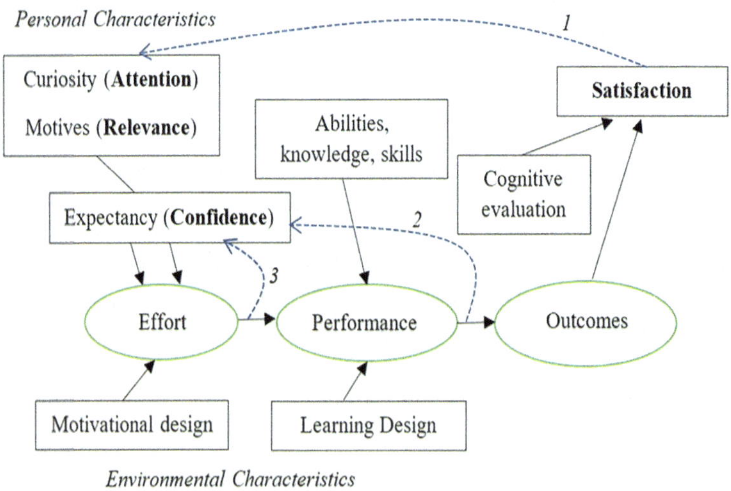

Fig. 14 Macro model of motivation and performance (adapted from Keller [196])

- Attention consists of gaining attention, building curiosity, and sustaining active engagement, and represents a synthesis of several related concepts including arousal theory, curiosity, boredom, and sensation seeking [196].
- Relevance occurs when the knowledge to be learned is perceived to be meaningfully related to a learner's goals, interesting and freely chosen [201] (Keller 2008). Relevance can come by three ways: (a) the link between the task and present and future study and career opportunities, (b) the intrinsic value of learning as something to be enjoyed, (c) the way that something is taught, for example, people who have a high need for affiliation would prefer collaborative work [202].
- Confidence is achieved by helping students to build positive expectancies for success and attribute their accomplishments to their own abilities and efforts rather than external factors [203]. It includes areas such as self-efficacy [204], attribution theory [203], self-determination theory, and goal orientation theory.
- Satisfaction refers to positive feelings such as fun and enjoyment or if that was an interesting activity. It is how they felt about their learning experience that develop a continuing motivation to learn [205]. It is dependent on the user's evaluation and achievement of intrinsic or extrinsic outcomes.

6.2 Motivational Design

The ARCS model is a method for improving the motivational appeal of instructional materials by offering guidance in motivational design. Keller [196] explains that motivation is influenced by the degree to which the instructional materials arouse the attention and sustain the engagement of learners.

The use of a mediatory technology can transform learning activities from passive to active [206]. Pietroni [207] used the term "Experience Design" to describe the integration of virtual and real contents and hybridization of paradigms and immersive technologies to enhance the user experience in museums. AR experiences can maximize user satisfaction and learning outcomes through personalization of experiences [208, 209]. For example, AR can offer museum visitors control what and the pace of learning to find out more about the artefacts that they enjoy most. Paliokas et al. (2020) developed an AR quiz game to increase the time visitors interact with museum artefacts and "offer a playful way of navigation in the 3D space". This is used for the permanent collections of the Silversmithing museum located in Greece.

Teachers can make use of AR to enrich experiences to improve learning motivation through sensorimotor interactions. Bhagat et al. [210] designed an AR-based formative assessment for improving primary students' learning achievement and motivation in a unit of instruction involving butterflies. When the experimental group used AR with iPads, 3D butterflies will pop-up with scientific names next to it and users can rotate and view them from different angles. The system provides immediate feedback for quiz answers and displays the corresponding results at the end of the assessment. The control group used the traditional pen-and-paper method. They found that the

AR-based approach improved not only memory recall, but also the learning motivation in all four domains of attention, relevance, confidence, and satisfaction of [199] model compared to the conventional method.

Huang et. al. [211] designed an outdoor AR learning system for secondary school students to learn about plants and ecology in Taiwan's Science-Botanical Garden based on Kolb's theory. The AR system features real-time prompts with text and images to guide them to explore the surrounding environment. The AR groups showed an enhanced learning experience (Attention) and more positive emotions compared to the control (traditional approach only narration). The second AR group with narration showed the most positive outcome in terms of emotional affect (Satisfaction). They also found that the students' sense of competency (Confidence) was also greater for the AR groups compared to the control.

The scaffolding of instructional materials in the VR/AR learning environment is a type of motivational design to help learners have greater confidence and build their expectancy for success. This in turn can foster intrinsic motivation [197, 198]. This is especially true for subjects such as history which has faced challenges in helping students to relate theory to practice. While it may be possible to organize field trips to historical sites, students are often given an interpretation of history at the site instead of being given the opportunity to explore on their own [212].

6.2.1 Presentation of Instructional Materials

In the learning of STEM subjects, students often fail to make the connection between objects in a real-life and those of the two-dimensional space. Instructional materials can be presented in a way that can capture students' attention and give them greater control over their learning process thus enhancing their sense of satisfaction and confidence. For example, Di Serio, Ibanez and Kloos (2013) carried out a study on the impact of an AR system on the students' motivation for a visual art course. A desktop-based AR system with webcam was used to augment the images of the masterpieces in the visual art with multimedia content. Based on the results of the Instructional Materials Motivation Survey (IMMS), it was found that attention and satisfaction in an AR-based learning environment were better rated than that of a slides-based learning environment.

Khan et al. [213] used IMMS to measure changes in undergraduate student motivation through pre-usage and post-usage questionnaires when using mobile AR to learn human anatomy. They found that the use of the AR has increased the learning motivation in the attention, satisfaction, and confidence domains. Cai et al. (2014) developed an AR software to help junior high school visualize bonding of molecules by providing virtual marker-based ball-and-stick models with electron shells of individual atoms. They found that the AR tool has a significant learning effect for the lower-achieving students compared to the higher-achieving ones and students' sense of satisfaction towards this new way of learning correlated with their evaluation of the AR app. Gunawan et al. [56] reported that chemical engineering undergraduates responded well to an AR application that provided visualization of distillation processes. The application was helpful in increasing students conceptual understanding and even motivated them to self-explore the subject further.

[212] developed a mobile AR system called CI-Spy, to scaffold historical inquiry learning at a local historic site. The Christiansburg Institute was a segregated school for African Americans founded in 1867 in the aftermath of the Civil War. There were three AR-based inquiry activities—Introduction to Inquiry, Classroom Site Orientation and Field Trip. Students were first recruited to become "junior history detectives" and they used handheld tablet devices (iPads) to access 43 historical sources during the exploration experiences. Results show that the students demonstrated a greater understanding of inquiry and gained significant insight into the hidden history of the place. The design of the CI-Spy facilitated a rich AR experience, and the students liked the fact that they could find virtual things in-situ as they explore the different buildings. Students reported that they could mentally connect different categories of sources to each other by remembering at which location they collected them at CI, and they could recall the locations of various sources a week after the visit.

Kaufmann and Dunser [214] compared Construct3D, an AR application with vision-haptic visualization with a standard desktop computer program CAD3D for geometry education. After three rounds of evaluations with more than 100 students, they found that students rated Construct3D as more satisfying than students using the CAD3D program. This is despite the lower ratings on the technical aspects due to cognitive overload in handling the interface and simulator sickness when using the AR HMD. Students found that the AR-based tool increased learning motivation compared to the traditional desktop-based application.

6.2.2 Engagement of Learners

By increasing engagement and motivation, the use of AR/VR can lead to more effective learning and a deeper understanding of concepts [12]. Liu et al. [215] studied sixth-grade students who used VR to learn classification in science and that they were more cognitive, behavioral, and emotional engaged than students in traditional teaching approach. Affordances of the VR system included interaction with the virtual 3D animals, and easy access to touch in-screen links to give information about the animals' habits, features, and life cycles and A robot pedagogical agent that gave instant feedback on the quiz. The VR group obtained higher post-test academic scores than the control group. This study suggests that the VR system enhanced motivation in the domains of attention, confidence, and satisfaction.

Cheng [90] investigated the effects of an AR-book learning activity for 153 university students to learn about an ethnic culture of Hakka group, which comprises of 18% of Taiwan's population. The students can read the paper book while using the AR app and in addition, they can receive augmented information in form of 3D models with audio narration and videos to scaffold their learning. The study reported an increase in the dimensions of attention, confidence, and satisfaction. Out of the three dimensions, the study noted that the attention was the most significantly affected by the AR book. This is attributed to the AR experience offering a new way of interaction that connects the book content with digital information that is more attractive for the students. Multimedia augmentation through AR may have contributed to a reduction of cognitive load, thus increasing motivation through confidence and satisfaction.

Mahadzir [216] designed an AR pop-book to teach primary school children English language grammar based on ARCS model of motivational design [199]. They measured, recorded, and transcribed the motivation level of one student at a time using the observation checklist which was based on the ARCS dimensions. While the sample size was small (n = 5), students perceived the AR pop-up book as being motivational in terms of arousing attention, personal control, success opportunities and intrinsic reinforcement. Three students enjoyed learning so much that they wanted to do more and showed a sense of confidence, pleasure, and satisfaction.

Gopalan also [217] implemented an AR science textbook for 70 secondary school students and found that there is a statistically significant relationship between engagement, enjoyment and fun and students' motivation for science learning. Bacca et al. (2019) reported that students in a vocational school showed increase in attention and confidence when using an AR to learn chemistry compared to the control group with traditional learning approach, as it gave them more control of the learning process. Both AR learning experiences can enhance students' attention, confidence and satisfaction through multimedia augmentations that are activated at their own pace.

Cai et al. [218] found that mobile AR-based learning applications have positive effects on performance and satisfaction in learning mathematics. Junior high school students who used three AR applications to learn probability obtained significantly higher scores in the post-test compared to the control group. One of the AR applications involves providing augmentation of real-time data from the physical tossing of coins with results are displayed in the form of graphs on the screen. This is compared to the traditional approach where they had to manually record data and plot the graphs. after observing the process of flipping coins by teachers. 84% of the students wrote that AR made the class more interesting and fascinating and 50% wrote that AR increased their motivation to learn. The appearance of 3D cartoon figures with animations on top of the coins added to the fun factor and made the learning more enjoyable.

Aydogdu [56] tested an AR software to teach pre-school children about the physical characteristics of animals and their pups. The children could rotate the augmented virtual animals using mobile phones. The post-test scores of the AR group show that indicators such as motivation, social persistence, satisfaction, and general efficacy scales were all significantly higher than the scores of the control group. Hung et al. (2016) found that the use of AR shifted primary school students from passively receiving knowledge to active learning and they expressed greater satisfaction and interest in learning a topic on bacteria.

Contero et al. [219] study of 56 secondary school students randomly assigned to either the high immersion condition using Samsung VR or low immersion using Android tablet to learn a social studies topic regarding events that affect world population distribution found that the high immersion VR group showed better knowledge retention, experienced more positive emotions, and showed greater interest. There was increased satisfaction with good emotions of amazement, satisfaction, curiosity, and inquiry.

Civelek et. al. [220] found a significant effect on the use of VR with haptic force feedback to teach earth science in terms of motivating students, improving attitude towards physics, improving students understanding, raising students' autonomy, and promoting collaborative learning. Parong and Meyer (2018) studied undergraduates who learnt human biology using immersive VR versus learning through didactic teaching with slides. The immersive VR group reported higher motivation, interest, and engagement than those being taught by slides. Allcoat and Muhlenen [221] found that undergraduate students who used immersive VR to learn about plant cells showed better performance gains for recall than those in the textbook or video group. Students using VR reported higher engagement and positive emotions than the control group.

Thompson et al. [192] studied the use of the VR game Cellverse to teach cellular biology. Students reported that the game was more interactive and engaging than traditional ways of learning. This suggested that the motivation design of inquiry-based VR game led to the arousal and sustaining of learner's attention. Students attributed an improved spatial understanding of cells to their game experience, which indicates that satisfaction has also influenced relevance and confidence.

6.2.3 Adding Gamification

Gamification and storytelling are ways to raise intrinsic motivation in learning [222]. Gamification can be defined as "using game-based mechanics, aesthetics and game thinking to engage people, motivate action, promote learning, and solve problems" [223], p. 10). The game design elements that can be integrated into educational contexts include objective and specific rules [224] and reward systems involving trial, error, failure and success through practice, experience, reflection, and learning [225].

Lu and Liu [226] developed an AR game to enhance primary school students' knowledge in marine education. The simulated AR environment enhance students' opportunities to make up for the lack of opportunity for practical observation in coastal and marine environments that are difficult or dangerous for students to visit. They adapted the ARCS Motivation Model as the learning motivation questionnaire. As the study aims to explore the impact the AR on marine education teaching, the questionnaire only adapted the Confidence and Satisfaction dimensions. AR is used to present 3D models of a variety of fish, providing students with vivid immersive teaching materials. Interactive technologies are used to develop the puzzle-based gesture games to increase learning engagement. Results indicate that the students were highly confident by the learning activities and viewed them satisfactorily. This program helped the learners acquire knowledge, especially the lower achievers to improve their learning performance.

Hwang et. al. [227] designed an AR mobile game which comprises a set of gaming missions to help primary school students recognize the predators or natural enemies of the butterflies by combining real-world contexts (the butterfly ecology area) and virtual targets (the predators). Compared with conventional AR-based mobile learning approach, the study found that AR gaming-based learning (ARGBL) improves the students' learning attitudes and their learning performance on the field trip.

AR simulations on mobile devices can augment users' experience of reality by connecting data on the handheld devices to the physical space [228] and enabling ubiquitous, collaborative and situated learning [12]. Klopfer and Squire (2008) developed a mobile location-based AR game called Environmental Detectives to find out the pedagogical and technological affordances of AR in the learning of environmental science.

7 Affordance Analysis for Lesson Design

Bower [229] proposed a five-step affordance analysis design methodology for a technology-enabled lesson to help match task or lesson requirements with the affordances of technologies. It starts with the identification of learning goals, proposing tasks needed to achieve the goals, determining the affordance requirements of the task and the affordances of technologies (**A**) (see Fig. 15).

The next step is to propose the learning tasks (**B**). In technology-mediated lesson for secondary school students in Singapore, the teacher started the lesson by explaining the different kinds of nutrients and introduce the concept of digestion. This is followed by formative assessment tasks comprising of multiple-choice questions (MCQ) hosted in national online platform called Singapore Learning Space (SLS). Based on the results of the quiz, the teacher determines the instructional sequence in the next lesson, which may include scaffolding support for lower- and middle-achievers using AR to help them understand the concept of chemical digestion of nutrients. Based on the **MEEL** model of learning affordances, the requirements are to achieve conceptual understanding by augmentation of the textbook to provide

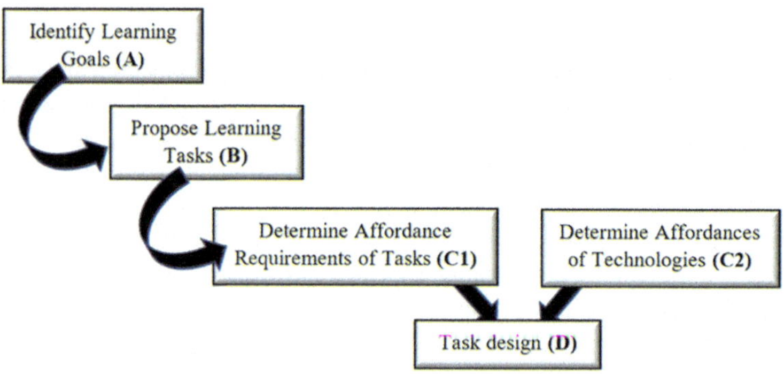

Fig. 15 Affordance analysis design methodology, adapted from [229]

enhanced visualization of the molecules. This should be done in an experiential manner so that the students can learn by doing, and through inquiry, which increases engagement and motivation.

Step **C1** was co-designed by subject teachers as the school had arranged timetabled weekly sessions of professional learning communities (PLC) to allow subject teachers to come together to discuss how to craft technology-enabled lesson.

When it came to step **(C2)** in determining the affordances of the technology, we used the personal edition of the game engine Unity™ with Vuforia plug-in to design the AR app. In view of the short development to deployment cycle, the app was only developed for the Android platform as some students already have Android handphones they could easily download the app for plug-and-play. We also decided to adopt the marker-based AR approach so that the 3D representations could be easily triggered when the student place the camera view of the phone over the specific diagram in the textbook chapter. All students have a copy of the science textbook.

The final step involves integrating the available and required affordances to form the design of the learning task. For task design **(D)**, a design map was produced for the lesson. A design map (Fig. 17) represents the planning process of the teacher to integrate technology and pedagogy to achieve the learning outcomes.

[229] explains that pedagogical and social affordances "serve to illuminate the major foci when designing learning experiences using technology". For example, In the author's lesson design, AR technologies allow collaborative work as the teacher provides scaffolding questions to facilitate the inquiry learning process (see Fig. 16). This is in line with the social constructivist approach of learning.

The process of affordance analysis considers the educational affordances of the technology and the affordance requirements of the task. Chen and Wang [230] designed out a three-stage comprehensive AR-embedded instruction (see Fig. 18) for the secondary school students to learn the earth science phenomena of "day, night, and seasons". They incorporated learning with AR after the teacher presentation stage and then provided opportunities for students to apply and reflect on their learning in the reinforcement stage. They found that the overall learning achievement was significant for the AR-embedded instruction. The students' learning styles and

Fig. 16 AR-based inquiry activity on learning nutrients

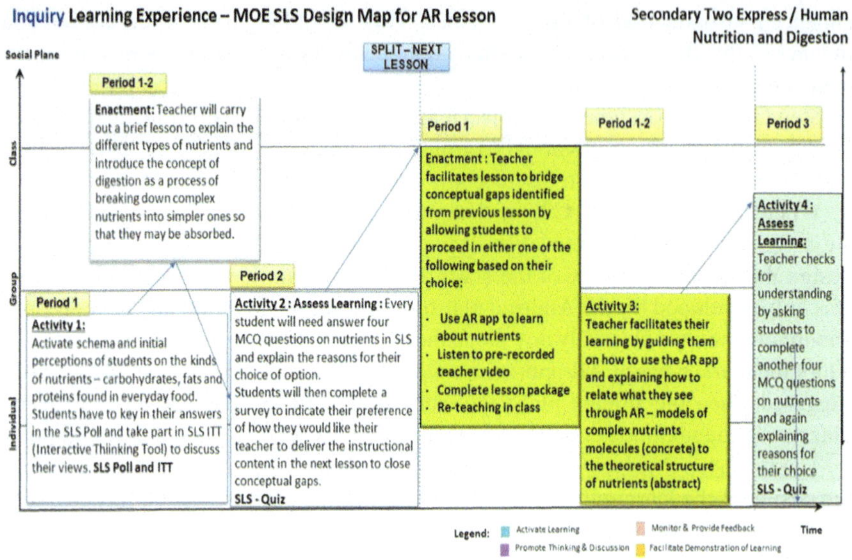

Fig. 17 Design map of proposed learning tasks

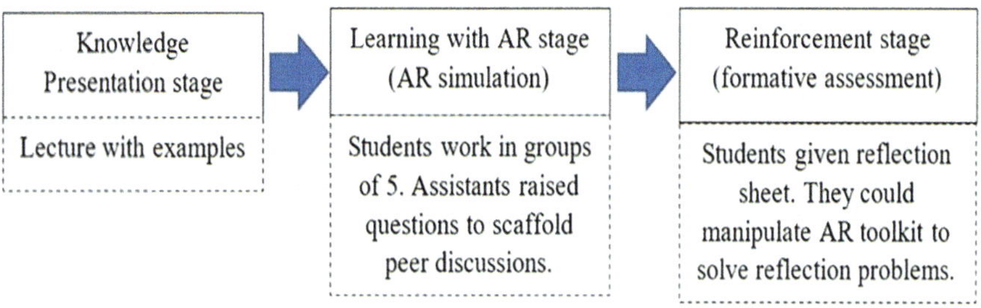

Fig. 18 AR-embedded instruction, adapted from Chen and Wang [230]

ICT competencies did not influence learning achievement, which suggested that the AR-based simulation was well received by the students.

8 Affordance Experiments

A study was carried out in 2021 to determine if 360° VR videos could elicit empathy through personal experience and have a significant effect on students' attitudes towards shark fin consumption. The quasi-experimental design was used with two classes of Secondary Two students (14-year-olds) as the experimental group (n = 80) and control group (n = 80). All four classes have the same ability level being in the Express stream. The experimental group was taught by the author and the control group taught by another science teacher. The teacher would start the lesson by doing a video cum slide presentation on the cruelty of illegal shark finning and decline of shark populations mainly due to demand factors such as shark fin consumption. Students in the experimental group had an additional 30 min for a virtual field trip comprising of two 360° VR videos of sharks swimming in the ocean from a perspective of an underwater diver to elicit empathy (see Fig. 19a). They affixed their own mobile phones to the Google Cardboard. The device-to-student ratio was approximately 2–3. Students completed pre- and post-self-rated surveys comprising of five Likert scale questions to assess changes in their empathy levels towards the conservation of sharks (see Fig. 19b for design).

Analysis of the data using paired samples t-test statistical analysis showed that students in both experimental and control groups experienced statistically significant overall improvement in their attitudes toward conservation ($p < 0.05$). However, the study did not find any significant differences in the amount of empathy change between the experimental and control groups. A handful of students in the VR group even expressed fear when the sharks approached them in the 360° videos!

Some of the possible reasons why there were no significant differences between the experimental and control groups include:

- Design Issues—To reduce the possible effect of external factors, the control group should be allowed to watch the same videos, albeit in 2D format on the class projection screen. There could have been guiding questions specific to the 360^0 content that is given to the students to reflect after watching the video, just like in

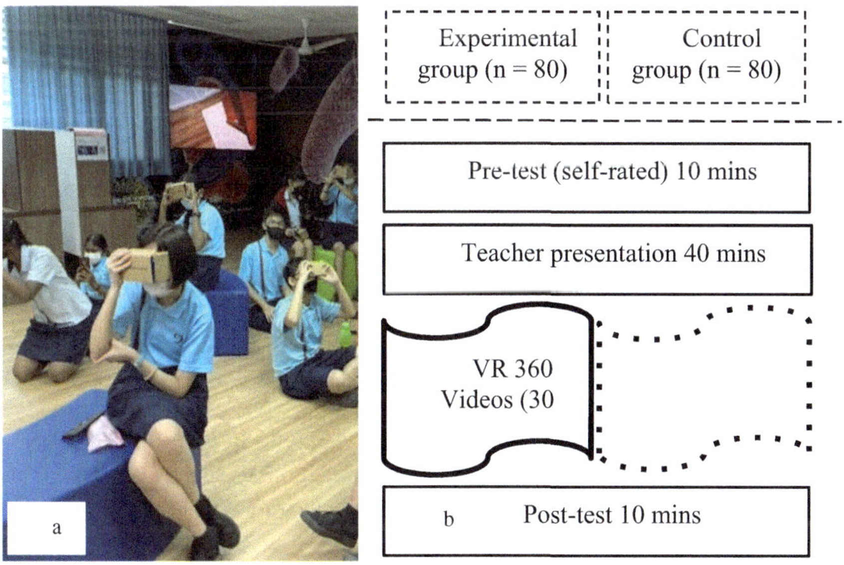

Fig. 19 **a** Students watching 360° videos **b** Design of experiment

the study by [133]. This would ensure that the use of VR was integrated into the lesson instruction.

- Operational Issues—As the ratio of device to students was not 1:1, some students needed to share the Google Cardboard with another person. This reduces the duration of exposure to the intervention. The shorter duration may be insufficient to elicit changes in attitudes. The VR experience was also not uniform as some of them had issues with the viewing the content because lower light settings or poor screen quality in their own handphones.
- Affordance Issues—the 360° videos presented only provided limited immersion and did not allow any form interaction between the user and the content. Based on the scale of experientiality [105], the level of experiential learning may only be limited to simulated or spectator levels. The 360^0 videos may not have elicited a powerful emotional or cognitive response that cause them to see vividly the cruelty of shark finning. In addition, the restricted view of Google Cardboard displays may have hindered collaboration between students [231]

In a separate qualitative study using grounded theory from 2020 to 2021, a total of 25 students from the same school were selected for interviews during the period. The interview questions basically probed for the students' views on the benefits or affordances of VR/AR technologies and how has VR/AR technologies transformed their learning experience in school. The interviews took place either during lessons to capture a timely dip stick of students' perspectives about the VR/AR experience or after lessons, through face-to-face interviews or communication through social media platforms. The duration of the interviews was between 5 and 30 min long and usually 1 to 5 students are interviewed each time. Students used the following VR/AR applications: *1.* Cells and Enzyme VR software (provided by NTU), *2.* InCell VR and The Climb educational VR games, *3.* AR app on nutrients, *4.* Other VR apps used for familiarization purposes such as Beat Saber.

Grounded theory is a qualitative research methodology that was initially developed by [232] to provide a systematic yet flexible approach for theory-constructing inquiry. This methodology generates a theory that is inductively derived from the study of the phenomenon over time at a conceptual level [233, 234]. The constructivist design by [235, 236] emphasizes an interpretivist epistemology that pays more attention to the individual's feelings, opinions, beliefs, principles, and philosophy. Gerunds are used for initial coding in the constructivist approach. Gerunds assist the researcher to move forward analytically and identify actions and processes within the data [237]. After the initial coding stage, the transcripts were placed under each of the four categories of affordances under the MEEL model or PIEE framework (Shin 2017).

During the interviews, students related how immersive VR was able to induce an apparent disconnect between their body and mind due to the strong presence in the virtual world. Students also described positive emotions such as enjoyment and fun during the VR experience, and many recommended for VR/AR to be used more often during lessons to engage learners. In terms of cognitive aspects, students stated that the use of VR/AR has helped them to attain conceptual change as they were previously unable to understand concepts presented in 2D diagrams in textbooks but now can "see" the connections.

9 Discussion

Tamim et al. [238] examined forty years of educational research using a second-order meta-analysis procedure and found a significant positive small to moderate effect size favoring the use of technology to support teaching and learning compared to a no-technology traditional teaching environment. AR technology has been hailed as the transformative technology tool in education [239] and is experiencing broad adoption and mass public acceptance [56]. A meta-analysis conducted by [3] found a moderate effect size of 0.56 affecting learner performance indicating that AR learning environments are effective complementary tools in educational settings.

While AR simulations may afford students with interactive, situated, collaborative problem-solving experiences, teachers may encounter unique technological, managerial, and cognitive challenges when integrating AR during lessons [29]. Interestingly, Kerrawalla et al. (2006) found that the use of AR resulted in 10-year-old children being less engaged than the traditional approach. The AR simulation was to help them understand the interaction of the earth and sun, but teachers used AR only for demonstration purposes and frontal teaching. This is compared to traditional role play sessions in which children were encouraged to move around and they are much more active, both verbally and physically than those in the AR sessions.

The importance of an embedded, constructivist AR instruction is underscored by study by Fokidess and Mastrokoukou [240] of 75 sixth-grade primary school students who used AR to learn about respiratory and circulatory systems app. The students were divided into three groups—Traditional method with printed material, Constructivist learning without AR, Constructivist learning with AR. The third group outperformed students in the other two groups on content recall. The difference in effect sizes between the traditional method and AR with constructivist approach was also large. AR annotation reduces cognitive load, and vision-haptic visualization through embodied interactions helps learners to remember information better [3]. Interactivity in AR supports constructivist learning to bring about conceptual change [241].

As mentioned earlier in the chapter, game-based learning approaches or adding gamification elements is an important strategy in motivational design. Akman and Çakır [242] studied the use of a VR game called "Keşfet Kurtul" to help 4th grade students to learn fractions. The VR group students (see Fig. 20a) produced greater academic achievement and scored higher on the social dimension on engagement compared to the control group The VR game embedded formative assessment quizzes into a 4-part island adventure to be played once a week over four weeks.

While gamification may be used to increase student engagement and participation [223], gamification impacts students with different types of motivation differently [243]. Buckley and Doyle [243] carried out a study on the effect of online gamification

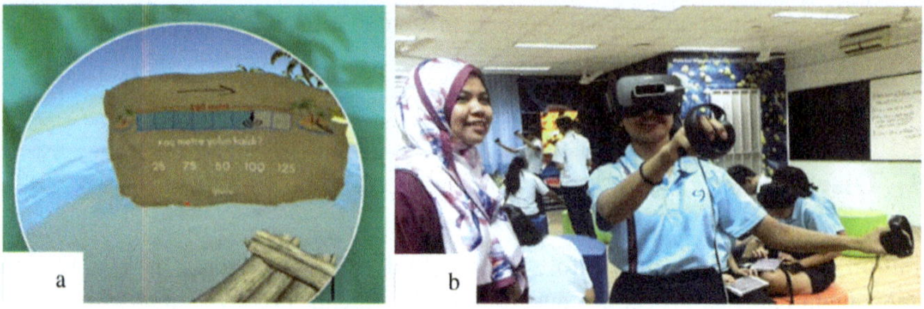

Fig. 20 **a** VR game *Keşfet Kurtul* [242]; **b** *The Climb* VR game

on learning outcomes in an undergraduate taxation module. The results suggest that gamified learning interventions have a larger impact on students who are intrinsically motivated. In the secondary school where the small-scale studies were conducted, a Malay language teacher used. The Climb VR game, whereby students were allowed to play it as a form of cognitive stimulus before their essay writing exercise that involved a topic related to climbing (see Fig. 21b).

Embedding pedagogical problems in a digital narrative that supports social collaboration and immediate feedback is one way to integrate AR/VR technologies into instruction. For example, in the study by [133], the 360° content was integrated closely with the instruction to enhance students' competence and mastery in historical knowledge. The Australian Broadcasting Corporation produced the "Kokoda VR Educational Resource" https://www.abc.net.au/btn/classroom/making-kokoda-vr/10522216 to support teachers in carrying out the VR experience in their lesson. One student said that "it is not like you are in a classroom taking notes... it is like you are doing it with them and you are like experiencing what they (the soldiers) are experiencing." It also contains guiding questions to help students reflect upon the information and apply the knowledge. Lanier et al. [222] created an outdoor game-based platform called Science Spots AR to increase motivation and enjoyment in learning by merging digital technologies, storytelling, games, and outdoor learning.

Media may also interact with learning styles. Safadel and White [244] study of the effect of VR visualization on learning the structure of DNA on undergraduate and graduate students found that immersive VR visualization has a positive impact only for students with lower spatial ability. However, students in the immersive VR group experienced higher satisfaction. This observation is consistent with the results from most studies whereby the students expressed that immersive VR technologies are more hands-on, engaging, and realistic than traditional forms of teaching.

A few main limitations and challenges in using VR/AR current exist, such as the limitation in customizing content, difficulties in using authoring tools, lack of educational content that are aligned to curricular standards, and lack of technical support [245, 246]. Moving forward, feasible solutions include establishing partnerships with universities for students to use the software that had previously been developed for research purposes. For example, a 3D AR wall poster, designed by NTU, is permanently displayed in a secondary school for students to visualize the

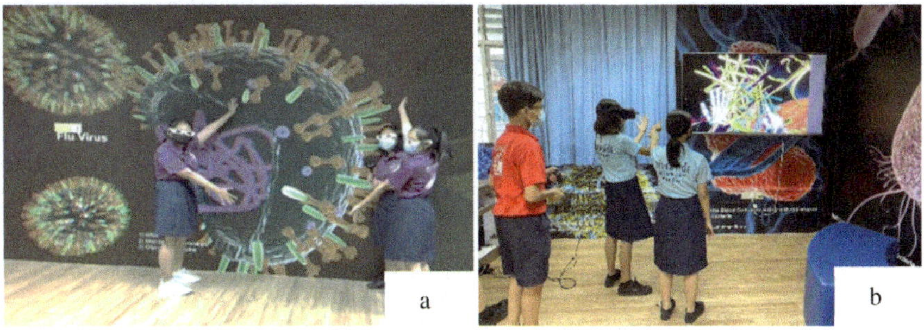

Fig. 21 **a** Coronavirus 3D AR poster **b** Cells VR app © NTU

coronavirus structure and the Cells VR software is used in their science curriculum [189, 190] (see Fig. 21).

An appropriately designed environment can help learners develop skills and knowledge in a more effective way [176]. Appropriate learning structures and instructional scaffolds are crucial to the success of the learning task [12]. Studies such as Estapa and Nadolny (2015), and [218] found that students were distracted by the information presented through the AR technology, suggesting cognitive overload [95, 247]. This necessitates a pre-training approach [248] where students are first taught by the teacher and then given an immersive experience to deepen their understanding [94]. On-screen pedagogical agents can be provided to guide them to learning content [247]. Audio narration to utilize both sensory channels may also help to reduce overload.

VR/AR are novel sensorimotor technologies that have brought about a series of significant changes in multiple fields such as education, medicine, and therapy [15]. The review by Sanfilippo et al. [249] provides a novel perspective on the integration of VR/AR with haptics and the application of learning theories to promote active learning. However, there is still much work to be done to realise the promise that immersive VR is "the learning aid of the twenty-first century" [250]. There is a need for more research on the impact and effects of VR HMDs on learning [14]. [229, 249] proposed a design approach to ensure successful learning based on immersive technologies comprising of the stages of Planning, Design and Evaluation (see Fig. 22). The ability to design immersive learning environments and experiences based on the affordances of the VR/AR can be challenging for educators. Educators cite the lack of competency and lack of understanding of pedagogy to implement VR/AR as key barriers to their implementation in classrooms [245, 251]. One approach is for faculty members in higher education to be trained specifically in creating VR/AR experiences so that they can in turn coach teachers in schools. For instance, the author conducts workshops on using and creating AR and VR applications for faculty members and teachers (see Fig. 23).

More research needs to be carried out using mixed methods designs to give a more holistic overview of VR/AR use in education [252]. While the study by [113] has shown that holistic indicators such as Quality of Experience, Quality of Service

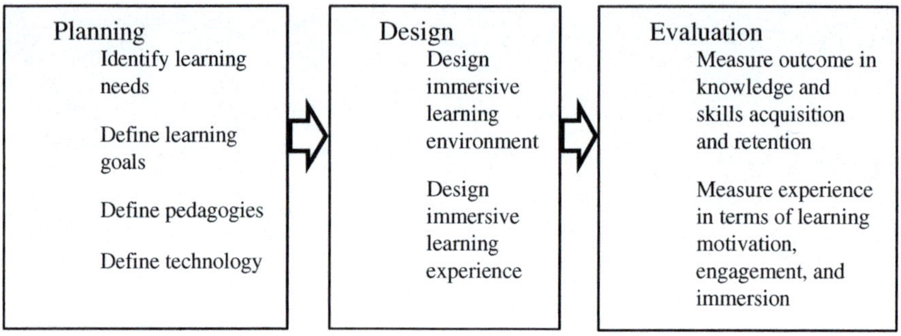

Fig. 22 Design approach for integrating VR/AR into learning process (adapted from [249])

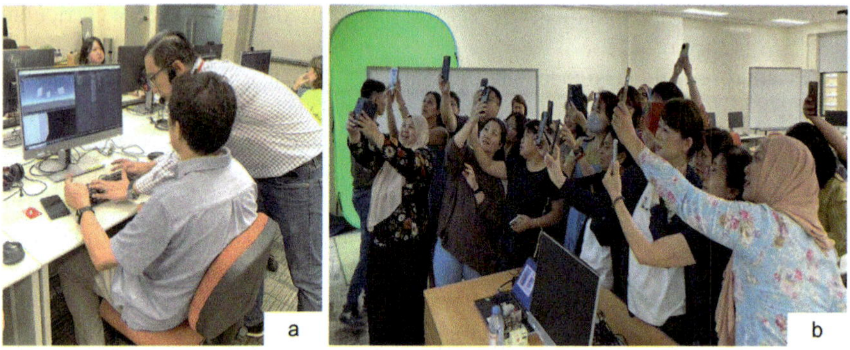

Fig. 23 *a* AR workshop for faculty members **b** Teaching educators how to use AR apps

and Quality of Community to can be used to measure the efficacy of technology-enhanced approaches, it does not seem to be a recognized framework that can be applied to other educational settings. Stylianidou [253] developed an AR game called "Helping Nemo" that draws on the principles of Universal Design for Learning and aims towards higher levels of engagement in learning, including those with learning disabilities or disengaged. The study adopted a combination of quantitative and qualitative methods which included classroom observations, student focus groups, and the teacher's reflective diaries.

10 Conclusion

This chapter has contributed to the body of educational research in VR/AR by identifying the pedagogical, social, and technological affordances of these emerging technologies and exploring how these affordances can be weaved into the cognitive and affective factors in VR/AR learning environments through the application of learning theories such as CTML, ECL and ARCS model of motivational design.

Past review papers tend to focus narrowly on the effect of VR/AR technology to produce learning outcomes without considering the principles of learning theories nor the ecological approach of affordances. Our work makes significant contribution to the VR/AR community by applying the principles of learning theories to the ecological approach of affordances of VR/AR in the learning context. This contribution to a novel application category of VR/AR educational research provides pedagogical and design guidance for future studies. We conclude that schools can transform students learning experiences by using immersive VR/AR technologies as a form of experiential learning based on ECL, and it can be an effective medium for learning if principles such as CTML and Cognitive Load Theory and ARCS model are followed.

Nonetheless, more research needs to be done to further understand how the learning experience occurs in different instructional designs. It is also important to use these new insights for the development of transformative experiences that can support construction of knowledge through higher level critical thinking, problem-solving, collaboration, and self-directed learning. It is still currently unclear as to the exact relationship of the interaction between media and method with immersive VR/AR technologies. For example, the study by [254] to compare e-learning using a desktop computer and m-learning using a mobile AR application to learn geography found that the mobile AR group outperformed the e-learning group in learning performance. However, it is not certain if the learning advantage is due to the use of AR per se or the m-learning approach. We need to strengthen the theoretical contexts and frameworks so that such successes can be replicated in other educational levels and backgrounds. Longitudinal studies with longer-term analysis on changes in user's cognitive and affective levels could provide important insights into impact of VR/AR on learning. More research can be carried out to explore how VR/AR can support the diversity of learners in terms of learning styles, profiles, and needs such as gifted or students with disabilities.

VR/AR technologies are still not mature and therefore pose challenges in the requirements of robustness and reliability [255]. More research needs to be done on how to improve the technical aspects so that the affordances of VR/AR can be maximized. For example, most MR/AR learning environments seem to place an emphasis on virtual elements instead of having active interaction between the virtual and physical objects. Further studies need to explore the pedagogical influence of VR/AR learning environments from a user-centered design approach. In Yilmaz's [256] study of AR trends in education between 2017 and 2018, it was found that just 5.97% out of the 67 studies measured usability or satisfaction, while only 2.99% of studies measure emotions and 1.49% on enjoyment. Further research is also required on improving the user experience in areas such as game-based learning (Pellas et al. 2017), motivation, cognitive load, and learner characteristics [28]. More research needs to be done on how to use VR/AR for online learning and blended environments.

Our book chapter has some limitations. This is not a systematic literature review but a rather more of a meaningful synthesis of relevant peer-reviewed papers that shed light on interactions between the affordances of immersive technologies of VR/AR, learning theories and pedagogical approaches to improve learning outcomes and foster more engaging learning experiences. The complexities of technical implementation coupled with the rapid development of VR/AR hardware and software development makes it hard to achieve a unified approach that considers the technological, psychological, and ecological aspects of VR/AR learning environments [257].

Based on his meta-analysis, [258] asserted that media do not influence learning under any circumstances. This was countered by [259], who concluded that media have distinct capabilities that can synergize with the content in the environment and provide a unique learning experience for the learner [260]. Despite having a large body of experimental and applied research that shows the potential benefits of using VR/AR in learning, VR/AR technologies are still considered largely as an 'alternative' approach to delivering or presenting information [261]. This chapter is a step in the direction to gain a greater understanding of the affordances of immersive VR/AR tools to bring about active, constructivist and transformative learning. This will support pedagogical approaches that develop, implement, and deliver teaching for 21st-century curricula [7]. With the proliferation of learner-centered technologies such as Student Response Systems (SRS) in the form of apps like Kahoot and Mentimeter [262], coupled with the increasing ease of use of emerging technologies such as VR/AR [1], teachers are finding the sweet spot between technological limitations, and pedagogical and social affordances to enhance teaching and learning. Hopefully, the Clark-Kozma debate will be settled once and for all.

References

1. Brown, M., McCormack, M., Reeves, J., Brook, D.C., Grajek, S., Alexander, B., Bali, M., Bulger, S., Dark, S., Engelbert, N., Gannon, K., Gauthier, A., Gibson, D., Gibson, R., Lundin, B., Veletsianos, G., Weber, N.: 2020 Educause Horizon Report Teaching and Learning Edition. Louisville, CO: EDUCAUSE (2020). https://www.learntechlib.org/p/215670/. Last accessed 1 June 2022
2. Dede, C.: The evolution of distance education: Emerging technologies and distributed learning. Am. J. Dist. Educ. **10**(2), 4–36 (1996). https://doi.org/10.1080/08923649609526919
3. Santos, M.E., Chen, A., Taketomi, T., Yamamoto, G., Miyazaki, J., Kat, H.: Augmented reality learning experiences: survey of prototype design and evaluation. IEEE Trans. Learn. Technol. **7**, 38–56 (2014)
4. Dewey, J.: Experience and Education. Macmillan, New York (1938)
5. Dede, C.: Immersive interfaces for engagement and learning. Science **323**(66), 66–69 (2009)
6. Kolb, D.A.: Experiential Learning as the Science of Learning and Development. Prentice Hall, Englewood Cliffs, NJ (1984)
7. Cowin, J.B.: Digital worlds and transformative learning: google expeditions, google arts and culture, and the merge cube. Int. Res. Rev. **10**(1), 42–53 (2020)
8. Yilmaz, R., Kucuk, S., Goktas, Y.: Are augmented reality picture books magic or real for preschool children aged five to six? Br. J. Edu. Technol. **48**(3), 824–841 (2016). https://doi.org/10.1111/bjet.12452
9. Giannakos, M.N.: Enjoy and learn with educational games: examining factors affecting learning performance. Comput. Educ. **68**, 429–439 (2013)

10. Reichhardt, T.: (2021). https://www.smithsonianmag.com/air-space-magazine/new-ar-app-coolest-way-follow-mars-landing-180977007/
11. Challenor, J., Ma, M.: A review of augmented reality applications for history education and heritage visualization. Multimodal Technol. Interact. **3**(2), 39 (2019). https://doi.org/10.3390/mti3020039
12. Wu, H.-K., Lee, S.W.-Y., Chang, H.-Y., Liang, J.-C.: Current status, opportunities and challenges of augmented reality in education. Comput. Educ. **62**, 41–49 (2013). https://doi.org/10.1016/j.compedu.2012.10.024
13. Kwon, C.: Verification of the possibility and effectiveness of experiential learning using HMD-based immersive VR technologies. Virtual Reality **23**, 101–118 (2019). https://doi.org/10.1007/s10055-018-0364-1
14. Radianti, J., Majchrzak, T.A., Fromm, J., Wohlgenannt, I.: A systematic review of immersive virtual reality applications for higher education: design elements, lessons learned, and research agenda. Comput. Educ. **147**, 103778 (2020). https://doi.org/10.1016/j.compedu.2019.103778
15. Freina, L., Ott, M.: A literature review on immersive virtual reality in education: state of the art and perspectives. in the international scientific conference e learning and software for education **1**, 133. "Carol I" National Defence University (2015)
16. Lanier, J., Minsky, M., Fisher, S., Druin, A.: Virtual Environments and Interactivity: Windows to the Future 1989 ACM Siggraph Panel Proceedings (1989)
17. Slater, M., Sanchez-Vives, M.V.: Enhancing our lives with immersive virtual reality. Front. Robot. AI **3**(74) (2016). https://doi.org/10.3389/frobt.2016.00074
18. Scavarelli, A., Arya, A., Teather, R.: Virtual reality and augmented reality in social learning spaces: a literature review. Virtual Reality **25**, 257–277 (2021). https://doi.org/10.1007/s10055-020-00444-8
19. Mezirow, J.: Perspective transformation. Adult Educ. **28**, 100–110 (1978)
20. Mezirow, J.: Transformative learning as discourse. J. Transform. Educ. **1**(1), 58–63 (2003)
21. Bertrand, P., Guegan, J., Robieux, L., McCall, C.A., Zenasni, F.: Learning empathy through virtual reality: multiple strategies for training empathy-related abilities using body ownership illusions in embodied virtual reality. Front. Robot. AI **5**, 26 (2018). https://doi.org/10.3389/frobt.2018.00026
22. Markowitz, D.M., Bailenson, J.N.: Virtual reality and the psychology of climate change. Curr. Opin. Psychol. Dec(42), 60–65 (2021). https://doi.org/10.1016/j.copsyc.2021.03.009
23. Shin, D.-H.: The role of affordance in the experience of virtual reality learning: technological and affective affordances in virtual reality. Telematics Inform. **34**(8), 1826–1836 (2017). https://doi.org/10.1016/j.tele.2017.05.013
24. Klippel, A., Zhao, J., Oprean, D., Wallgrün, J.O., Stubbs, C., Femina, P.L., Jackson, K.L.: The value of being there: toward a science of immersive virtual field trips. Virtual Reality **24**, 753–770 (2020). https://doi.org/10.1007/s10055-019-00418-5
25. Queiroz, A.C.M., Kamarainen, A.M., Preston, N.D., da Silva Leme, M.I.: Immersive virtual environments and climate change engagement. In: Immersive Learning Research Network Conference Montana. 153–164 (2018)
26. Wray, M. (2020). https://globalnews.ca/news/6550977/mom-dead-daughter-virtual-reality/
27. Azuma, R., Baillot, Y., Behringer, R., Feiner, S., Julier, S., MacIntyre, B.: Recent advances in augmented reality. IEEE Comput. Graphics Appl. **21**(6), 34–47 (2001). https://doi.org/10.1109/38.963459
28. Cheng, K.H., Tsai, C.C.: Affordances of augmented reality in science learning: suggestions for future research. J. Sci. Educ. Technol. **22**, 449–462 (2013). https://doi.org/10.1007/s10956-012-9405-9
29. Dunleavy, M., Dede, C., Mitchell, R.: Affordances and limitations of immersive participatory augmented reality simulations for teaching and learning. J. Sci. Educ. Technol. **18**(1), 7–22 (2008). https://doi.org/10.1007/s10956-008-9119-1
30. Caudell, T.P., Mizell, D.W.: Augmented reality: an application of heads-up display technology to manual manufacturing processes. In: Proceedings of Hawaii International Conference on System Sciences, pp. 659–669 (1992)

31. Azuma, R.T.: A survey of augmented reality. Presence: Teleoperators Virtual Environ. **6**(4), 355–385 (1997). https://doi.org/10.1162/pres.1997.6.4.355
32. Sutherland, I. E.: A head-mounted three dimensional display. In: Proceedings of the December 9–11, Fall Joint Computer Conference, Part I on—AFIPS '68 (Fall, Part I) (1968). https://doi.org/10.1145/1476589.1476686
33. Milgram, P., Kishino, F.: A taxonomy of mixed reality visual displays. IEICE Trans. Inf. Syst. **77**, 1321–1329 (1994). http://citeseerx.ist.psu.edu/viewdoc/summary?doi=10.1.1.102.4646. Last accessed 7 June 2022
34. Costanza, E., Kunz, A., Fjeld, M.: Mixed reality: a survey. Lecture notes in computer science. **5440**. 47 (2009). https://doi.org/10.1007/978-3-642-00437-7_3. https://www.researchgate.net/publication/235966290_Mixed_Reality_A_Survey. Last accessed 6 June 2022
35. Leonard, S.N., Fitzgerald, R.N.: Holographic learning: a mixed reality trial of Microsoft HoloLens in an Australian secondary school. Res. Learn. Technol. **26** (2018). https://doi.org/10.25304/rlt.v26.2160
36. Wish-Baratz, S., Crofton, A.R., Gutierrez, J., Henninger, E., Griswold, M.A.: Assessment of mixed-reality technology use in remote online anatomy education. JAMA Netw. Open **3**(9), e2016271 (2020). https://doi.org/10.1001/jamanetworkopen.2020.16271
37. Matsutomo, S., Miyauchi, T., Noguchi, S., Yamashita, H.: Real-time visualization system of magnetic field utilizing augmented reality technology for education. IEEE Trans. Magnetics **48**(2), 531–534 (2012)
38. Cai, Y.Y., Chia, N.K.H., Thalmann, D., Kee, N.K.N., Zheng, J.M., Thalmann, N.M.: Design and development of a virtual dolphinarium for children with autism. IEEE Trans. Neural Syst. Rehabil. Eng. **21**(2), 208–217 (2013). https://doi.org/10.1109/tnsre.2013.2240700
39. Muhanna, M. A.: Virtual reality and the cave: Taxonomy, Interaction Challenges and Research Directions. J. King Saud Univ. Comp Info. Sci. **27**(3), 344–361 (2015). https://doi.org/10.1016/j.jksuci.2014.03.023
40. Huang, L.H., Taib, S.F., Ryan, A., Goh, Z.A., Xu, M.S.: Virtual reality research and development in NTU. Virtual Reality Intell. Hardw. **2**(5), 394–408 (2020). https://doi.org/10.1016/j.vrih.2020.06.002
41. Gibson, J.J.: The theory of affordances. In: Shaw, R., Bransford, J. (eds.) Perceiving, Acting, and Knowing: Toward an Ecological Psychology, pp. 67–82. Erlbaum, Hillsdale, NJ (1977)
42. Norman, D.A.: The Psychology of Everyday Things. Basic Books, New York, NY (1988)
43. Norman, D.A.: Affordance, conventions, and design. Interactions **6**, 38–41 (1999)
44. Turner, P.: Affordance as context. Interact. Comput. **17**, 787–800 (2005)
45. Hochberg, J.: Perceptual theory and visual cognition. In: Ballesteros, S. (ed.) Cognitive Approaches to Human Perception, pp. 223–239. Erlbaum, Hillsdale, NJ (1994)
46. Chong, I., Proctor, R.W.: On the evolution of a radical concept: affordances according to gibson and their subsequent use and development. Perspect. Psychol. Sci. **15**(1), 117–132 (2020). https://doi.org/10.1177/1745691619868207
47. Kirschner, P., Strijbos, J.-W., Kreijns, K., Beers, P.J.: Designing electronic collaborative learning environments. Education Tech. Research Dev. **52**(3), 47–66 (2004). https://doi.org/10.1007/bf02504675
48. Kreijns, K., Kirschner, P.A., Jochems, W., Van Buuren, H.: Determining sociability, social space, and social presence in asynchronous collaborative groups. Cyberpsychol. Behav. **7**, 155–172 (2004)
49. Hutchby, I.: Technologies texts and affordances. Sociology **35**, 441–456 (2001)
50. Zahorik, P., Jenison, R.L.: Presence as being-in-the-world. Presence: Teleoperators Virtual Environ. **7**(1), 78–89 (1998). https://doi.org/10.1162/105474698565541
51. Chittaro, L., Buttussi, F.: Assessing knowledge retention of an immersive serious game versus a traditional education method in aviation safety. IEEE Trans. Visualization Comput. Graph. **21**(4), 529–538 (2015). https://doi.org/10.1109/tvcg.2015.2391853
52. Steuer, J.: Defining virtual reality: dimensions determining telepresence. J. Commun. **42**(4), 73–93 (1992). https://doi.org/10.1111/j.1460-2466.1992.tb00812.x

53. Csikszentmihalyi, M.: Creativity: flow and the psychology of discovery and invention. Harper Collins, pp. 107–126 (1996)
54. Lombard, M., Ditton, T.: At the heart of it all: the concept of presence. J. Comput.-Mediated Commun. **3**(2) (1997)
55. Eckhoff, D., Cassinelli, A., Liu, T., Sandor, C.: Psychophysical effects of experiencing burning hands in augmented reality. In: Bourdot, P., Interrante, V., Kopper, R., Olivier, A.H., Saito, H., Zachmann, G.: (eds.) Virtual Reality and Augmented Reality. EuroVR 2020. Lecture Notes in Computer Science, 12499. Springer (2020). https://doi.org/10.1007/978-3-030-62655-6_5
56. Singh, D., Banerjee, A., Nath, I.: Application of augmented reality and virtual reality in education. In: Russell, D. (ed.) Implementing augmented reality into immersive virtual learning environments, pp. 89–101. IGI Global (2021). https://doi.org/10.4018/978-1-7998-4222-4
57. Dede, C., Jacobson, J., Richards, J.: Introduction: virtual, augmented and mixed realities in education, In: Virtual, Augmented and Mixed Realities in Education. Springer Nature Singapore, 1–16 (2017).https://doi.org/10.1007/978-981-10-5490-7
58. Slater, M.: Implicit learning through embodiment in immersive virtual reality, In: Liu, D., Dede, C., Huang, R., Richards, J. (eds.) Virtual, Augmented and Mixed Realities in Education, Springer Nature Singapore, pp. 1–16 (2017). https://doi.org/10.1007/978-981-10-5490-7
59. Krajčovič, M., Gabajová, G., Matys, M., Grznár, P., Dulina, U., Kohár, R.: 3D interactive learning environment as a tool for knowledge transfer and retention. Sustainability **13**(14), 7916 (2021). https://doi.org/10.3390/su13147916
60. Decety, J.: The neurodevelopment of empathy in humans. Dev. Neurosci. **32**, 257–267 (2010). https://doi.org/10.1159/000317771
61. Decety, J., Meyer, M.: From emotion resonance to empathic understanding: a social developmental neuroscience account. Dev. Psychopathol. **20**, 1053–1080 (2008). https://doi.org/10.1017/s0954579408000503
62. de Vignemont, F., Singer, T.: The empathic brain: how, when and why? Trends Cogn. Sci. **10**(10), 435–441 (2006). https://doi.org/10.1016/j.tics.2006.08.008
63. Singer, T., Lamm, C.: The social neuroscience of empathy. Ann. N. Y. Acad. Sci. **1156**(1), 81–96 (2009). https://doi.org/10.1111/j.1749-6632.2009.04418.x
64. Schutte, N.S., Stilinović, E.J.: Facilitating empathy through virtual reality. Motiv. Emot. **41**, 708–712 (2017). https://doi.org/10.1007/s11031-017-9641-7
65. Radu, I.: Augmented reality in education: a meta-review and cross-media analysis. Pers. Ubiquit. Comput. **18**(6), 1533–1543 (2014). https://doi.org/10.1007/s00779-013-0747-y
66. Mayer, R.E.: Multimedia Learning. Cambridge University Press, New York (2001). https://doi.org/10.1017/CBO9781139164603
67. Mayer, R.E.: Cognition and instruction: on their historic meeting within educational psychology. J. Educ. Psychol. **84**, 405–412 (1992). https://doi.org/10.1037/0022-0663.84.4.405
68. Schunk, D.H.: Learning Theories: An Educational Perspective, 6th edn. Pearson, Boston, MA (2012)
69. Shuell, T.J.: Cognitive conceptions of learning. Rev. Educ. Res. **56**(4), 411–436 (1986)
70. Skinner, B.F.: The origins of cognitive thought. Am. Psychol. **44**(1), 13 (1989)
71. Mayer, R.E.: Thirty years of research on online learning. Appl. Cogn. Psychol. **33**(2), 152–159 (2018). https://doi.org/10.1002/acp.3482
72. Wittrock, M.C.: Generative processes of comprehension. Educ. Psychol. **24**, 345–376 (1990)
73. Wittrock, M.C.: Learning as a generative process. Educ. Psychol. **11**, 87–95 (1974). https://doi.org/10.1080/004615274
74. Sweller, J.: Cognitive load theory, learning difficulty, and instructional design. Learn. Instr. **4**, 295–312 (1994)
75. Mayer, R.E., Moreno, R.: Nine ways to reduce cognitive load in multimedia learning. Educ. Psychol. **38**(1), 43–52 (2003)
76. Mayer, R.E.: Cognitive theory of multimedia learning. In: Mayer, R. (ed.) The Cambridge Handbook of Multimedia Learning, Cambridge Handbooks in Psychology, pp. 31–48 (2005). Cambridge University Press, Cambridge. https://doi.org/10.1017/CBO9780511816819.004

77. Mayer, R.E.: Cognitive theory of multimedia learning. In: Mayer, R.E. (ed.) The Cambridge Handbook of Multimedia Learning, pp. 31–48. Cambridge University Press, New York (2005)
78. Sorden, S.: The Cognitive Theory of Multimedia Learning. Handbook of Educational Theories. Information Age Publishing, Charlotte, NC (2012)
79. Sorden, S.: The cognitive theory of multimedia learning. In: Irby, B., Brown, G., Lara-Alecio, R., Jackson, S. (eds.) Handbook of Educational Theories, pp. 155–168. Information Age, Charlotte, NC (2012)
80. Skuballa, I.T., Dammert, A., Renkl, A.: Two kinds of meaningful multimedia learning: is cognitive activity alone as good as combined behavioral and cognitive activity? Learn. Instr. **54**, 35–46 (2018). https://doi.org/10.1016/j.learninstruc.2018.02.001
81. Mautone, P.D., Mayer, R.E.: Signaling as a cognitive guide in multimedia learning. J. Educ. Psychol. **93**(2), 377–389 (2001)
82. Wang, Y.-H.: Exploring the effectiveness of integrating augmented reality-based materials to support writing activities. Comput. Educ. **113**, 162–176 (2017). https://doi.org/10.1016/j.compedu.2017.04.013
83. Santos, M.E., Lübke, A., Taketomi, T., Yamamoto, G., Rodrigo, M.M., Sandor, C., Kato, H.: Augmented reality as multimedia: the case for situated vocabulary learning. Res. Pract. Technol. Enhanced Learn. **11**(1) (2016). https://doi.org/10.1186/s41039-016-0028-2
84. Sweller, J., Chandler, P.: Evidence for cognitive load theory. Cogn. Instr. **8**, 351–362 (1991)
85. Sweller, J., van Merriënboer, J.J.G., Paas, F.G.W.C.: Cognitive architecture and instructional design. Educ. Psychol. Rev. **10**, 251–296 (1998). https://doi.org/10.1023/A:1022193728205
86. Kirschner, P.A., Sweller, J., Clark, R.E.: Why minimal guidance during instruction does not work: an analysis of the failure of constructivist, discovery, problem-based, experiential, and inquiry-based teaching. Educ. Psychol. **41**(2), 75–86 (2006)
87. Miller, G.A.: The magical number seven, plus or minus two: some limits on our capacity for processing information. Psychol. Rev. **63**, 81–97 (1956)
88. Ginns, P.: Integrating information: a meta-analysis of the spatial contiguity and temporal contiguity effects. Learn. Instr. **16**, 511–525 (2006). https://doi.org/10.1016/j.learninstruc.2006.10.001
89. Artino, A.R.: Cognitive load theory and the role of learner experience: an abbreviated review for educational practitioners. AACE J. **16**(4), 425–439 (2008)
90. Cheng, K.-H.: Reading an augmented reality book: an exploration of learners' cognitive load, motivation, and attitudes. Australasian J. Educ. Technol. **33**(4) (2017). https://doi.org/10.14742/ajet.2820
91. Plass, J.L., Moreno, R., Brünken, R. (eds.): Cognitive Load Theory. Cambridge University Press (2010). https://doi.org/10.1017/CBO9780511844744
92. Klatzky, R., Wu, B., Shelton, D., Stetten, G.: Effectiveness of augmented-reality visualization versus cognitive mediation for learning actions in near space. TAP. **5** (2008). https://doi.org/10.1145/1279640.1279641
93. Tang, A., Owen, C., Biocca, F., Mou, W.: Comparative effectiveness of augmented reality in object assembly. In: Proceedings of the Conference on Human Factors in Computing Systems—CHI '03 (2003). https://doi.org/10.1145/642611.642626
94. Meyer, O.A., Omdahl, M.K., Makransky, G.: Investigating the effect of pre-training when learning through immersive virtual reality and video: a media and methods experiment. Comput. Educ. **140**, 103603 (2019). https://doi.org/10.1016/j.compedu.2019.103603
95. Makransky, G., Terkildsen, T.S., Mayer, R.E.: Adding immersive virtual reality to a science lab simulation causes more presence but less learning. Learn. Instr. (2019). https://doi.org/10.1016/j.learninstruc.2017.12.007
96. Richards, D., Taylor, M.: A comparison of learning gains when using a 2D simulation tool versus a 3D virtual world: an experiment to find the right representation involving the marginal value theorem. Comput. Educ. **86**, 157–171 (2015). https://doi.org/10.1016/j.compedu.2015.03.009
97. Mayer, R.E., Anderson, R.: The instructive animation: helping students build connections between words and pictures in multimedia learning. J. Educ. Psychol. **84**, 444–452 (1992)

98. Fujimoto, Y., Yamamoto, G., Kato, H., Miyazaki, J.: Relation between location of information displayed by augmented reality and user's memorization. In: Proceedings of Augmented Human International Conference. vol. 7, pp. 1–8. ACM, New York (2012)
99. Mayer, R.E.: Multimedia Learning, 2nd edn. Cambridge University Press, New York (2009)
100. Patzer, B., Smith, D., Keebler, J.: Novelty and retention for two augmented reality learning systems. Proc. Human Factors Ergon. Soc. Annu. Meeting **58**, 1164–1168 (2014). https://doi.org/10.1177/1541931214581243
101. Dunlap, J., Dobrovolny, J., Young, D.: Preparing eLearning designers using Kolb's model of experiential learning. Innovate: J. Online Educ. **4**(4), 3 (2008)
102. Merriam, S., Caffarella, R., Baumgartner, L.: Learning in Adulthood: A Comprehensive Guide, 3rd edn. Jossey-Bass, San Francisco (2006)
103. Bonwell, C.C., Eison, J.A.: Active learning: creating excitement in the classroom. In: 1991 ASHE-ERIC Higher Education Reports. Washington, DC, USA, ERIC (1991)
104. Efstathiou, I., Kyza, E.A., Georgiou, Y.: An inquiry-based augmented reality mobile learning approach to fostering primary school students' historical reasoning in non-formal settings. Interact. Learn. Environ. **26**, 22–41 (2018)
105. Gibbons, M., Hopkins, D.: How experiential is your experience-based program? J. Exp. Educ. **3**(1), 32–37 (1980). https://doi.org/10.1177/105382598000300107
106. Conole, G., Dyke, M., Oliver, M., Seale, J.: Mapping pedagogy and tools for effective learning design. Comput. Educ. **43**(1–2), 17–33 (2004). https://doi.org/10.1016/j.compedu.2003.12.018
107. Konak, A., Clark, T., Nasereddin, M.: Using Kolb's experiential learning cycle to improve student learning in virtual computer laboratories. Comput. Educ. **72**, 11–22 (2014). https://doi.org/10.1016/j.compedu.2013.10.013
108. Riva, G., Baños, R.M., Botella, C., Mantovani, F., Gaggioli, A.: Transforming experience: the potential of augmented reality and virtual reality for enhancing personal and clinical change. Front. Psych. **7**, 164 (2016). https://doi.org/10.3389/fpsyt.2016.00164
109. Di Serio, Á., Ibáñez, M.B., Kloos, C.D.: Impact of an augmented reality system on students' motivation for a visual art course. Comput. Educ. **68**, 586–596 (2013). https://doi.org/10.1016/j.compedu.2012.03.002
110. Baus, O., Bouchard, S.: Moving from virtual reality exposure-based therapy to augmented reality exposure-based therapy: a review. Front. Hum. Neurosci. **8**, 112 (2014). https://doi.org/10.3389/fnhum.2014.00112
111. Dale, E.: Audiovisual Methods in Teaching, 3rd edn. The Dryden Press; Holt, Rinehart and Winston, New York (1969)
112. Baukal, C.E., Ausburn, F.B., Ausburn, L.J.: A proposed multimedia cone of abstraction: updating a classic instructional design theory. i-manager's J. Educ. Technol. **9**(4), 15–24 (2013)
113. Blanco-Fernández, Y., López-Nores, M., Pazos-Arias, J.J., Gil-Solla, A., Ramos-Cabrer, M., García-Duque, J.: REENACT: a step forward in immersive learning about human history by augmented reality, role playing and social networking. Expert Syst. Appl. **41**, 4811–4828 (2014). https://doi.org/10.1016/j.eswa.2014.02.018
114. Hsu, W.C., Tseng, C.M., Kang, S.C.: Using exaggerated feedback in a virtual reality environment to enhance behavior intention of water-conservation. J. Educ. Technol. Soc. **21**(4), 187–203 (2018)
115. NUS (2022). https://medicine.nus.edu.sg/nursing/2022/05/keep-calm-and-carry-on-virtual-reality-helps-medical-and-nursing-students-manage-agitated-patients-with-empathy/
116. Johnson-Glenberg, M.C., Megowan-Romanowicz, C., Birchfield, D.A., Savio-Ramos, C.: Effects of embodied learning and digital platform on the retention of physics content: centripetal force. Front. Psychol. **7** (2016). https://doi.org/10.3389/fpsyg.2016.01819
117. Enyedy, N., Danish, J.A., Delacruz, G., Kumar, M.: Learning physics through play in an augmented reality environment. Comput. Supported Learn. **7**, 347–378 (2012). https://doi.org/10.1007/s11412-012-9150-3

118. Madden, J., Won, A., Schuldt, J., Kim, B., Pandita, S., Sun, Y., Stone, T., Holmes, N.: Virtual reality as a teaching tool for moon phases and beyond. In: Paper presented at Physics Education Research Conference 2018, Washington, DC (2018). https://doi.org/10.48550/arXiv.1807.11179
119. Chiu, J., DeJaegher, C., Chao, J.: The effects of augmented virtual science laboratories on middle school students' understanding of gas properties. Comput. Educ. **85**, 59–73 (2015). https://doi.org/10.1016/j.compedu.2015.02.007
120. Slater, M., Perez-Marcos, D., Ehrsson, H.H., Sanchez-Vives, M.V.: Inducing illusory ownership of a virtual body. Front. Neurosci. **3**(2), 214–220 (2009). https://doi.org/10.3389/neuro.01.029.2009
121. Lenggenhager, B., Tadi, T., Metzinger, T., Blanke, O.: Video ergo sum: manipulating bodily self-consciousness. Science **317**(5841), 1096–1099 (2017). https://doi.org/10.1126/science.1143439
122. Maister, L., Slater, M., Sanchez-Vives, M.V., Tsakiris, M.: Changing bodies changes minds: owning another body affects social cognition. Trends Cognit. Sci. **19**(1), 6–12 (2015). https://doi.org/10.1016/j.tics.2014.11.001
123. Yee, N., Bailenson, J.N.: Walk a mile in digital shoes: The impact of embodied perspective-taking on the reduction of negative stereotyping in immersive virtual environments. In: Proceedings of PRESENCE 2006: The 9th Annual International Workshop on Presence. August 24–26, Cleveland, Ohio, USA (2006)
124. Hershfield, H.E., Goldstein, D.G., Sharpe, W.F., Fox, J., Yeykelis, L., Carstensen, L.L., Bailenson, J.N.: Increasing saving behavior through age-progressed renderings of the future self. J. Market. Res. **48**, 23–37 (2011). https://doi.org/10.1509/jmkr.48.spl.s23
125. Gaggioli, A.: Transformative experience design. In: Gaggioli, A., Ferscha, A., Riva, G., Dunne, S., Viaud-Delmon, I. (eds) Human Computer Confluence: Transforming Human Experience Through Symbiotic Technologies, pp. 97–121 (2005). De Gruyter Open, Warsaw
126. Rupp, M.A., Odette, K.L., Kozachuk, J., Michaelis, J.R., Smither, J.A., McConnell, D.S.: Investigating learning outcomes and subjective experiences in 360° videos. Comput. Educ. **128**, 256–268 (2019)
127. Fukuta, J., Gill, N., Rooney, R., Coombs, A., Murphy, D.: Use of 360° video for a virtual operating theatre orientation for medical students. J. Surg. Educ. **78**(2), 391–393 (2021). https://doi.org/10.1016/j.jsurg.2020.08.014
128. Chang, S.-C., Hsu, T.-C., Jong, M.S.-Y.: Integration of the peer assessment approach with a virtual reality design system for learning earth science. Comput. Educ. **146**, 103758 (2020). https://doi.org/10.1016/j.compedu.2019.103758
129. Chang, S.-C., Hsu, T.C., Kuo, W.C., Jong, M.S.-Y.: Effects of applying a VR-based two-tier test strategy to promote elementary students' learning performance in a geology class. Br. J. Edu. Technol. **51**(1), 148–165 (2020). https://doi.org/10.1111/bjet.12790
130. Ahn, S.J.G., Bostick, J., Ogle, E., Nowak, K.L., McGillicuddy, K.T., Bailenson, J.N.: Experiencing nature: embodying animals in immersive virtual environments increases inclusion of nature in self and involvement with nature. J. Comput.-Mediat. Commun. **21**(6), 399–419 (2016). https://doi.org/10.1111/jcc4.12173
131. Nelson, K.M., Anggraini, E., Schlüter, A.: Virtual reality as a tool for environmental conservation and fundraising. PLoS ONE **15**(4), e0223631 (2020). https://doi.org/10.1371/journal.pone.0223631
132. Fokides, E., Arvaniti, P.A.: Evaluating the effectiveness of 360 videos when teaching primary school subjects related to environmental education. J. Pedagogical Res. **4**(3), 203–222 (2020). https://doi.org/10.33902/jpr.2020063461
133. Calvert, J., Abadia, R.: Impact of immersing university and high school students in educational linear narratives using virtual reality technology. Comput. Educ. **159**, 104005 (2020). https://doi.org/10.1016/j.compedu.2020.104005
134. Billinghurst, M., Duenser, A.: Augmented reality in the classroom. Computer **45**(7), 56–63 (2012). https://doi.org/10.1109/MC.2012.111

135. Vávra, P., Roman, J., Zonča, P., Ihnát, P., Němec, M., Kumar, J., Habib, N., El-Gendi, A.: Recent development of augmented reality in surgery: a review. J. Healthc. Eng. 1–9 (2017).https://doi.org/10.1155/2017/4574172
136. Meola, A., Cutolo, F., Carbone, M., Cagnazzo, F., Ferrari, M., Ferrari, V.: Augmented reality in neurosurgery: a systematic review. Neurosurg. Rev. **40**(4), 537–548 (2017). https://doi.org/10.1007/s10143-016-0732-9
137. Dhar, P., Rocks, T., Samarasinghe, R.M., Stephenson, G., Smith, C.: Augmented reality in medical education: students' experiences and learning outcomes. Med. Educ. Online **26**(1), 1953953 (2021). https://doi.org/10.1080/10872981.2021.1953953
138. Eckert, M., Volmerg, J.S., Friedrich, C.M.: Augmented reality in medicine: systematic and bibliographic review. JMIR Mhealth Uhealth **7**(4), e10967 (2019). https://doi.org/10.2196/10967
139. Ghaednia, H., Fourman, M.S., Lans, A., Detels, K., Dijkstra, H., Lloyd, S., Sweeney, A., Oosterhoff, J.H.F., Schwab, J.H.: Augmented and virtual reality in spine surgery, current applications and future potentials. The Spine J. (2021). https://doi.org/10.1016/j.spinee.2021.03.018
140. Cabrilo, I., Bijlenga, P., Schaller, K.: Augmented reality in the surgery of cerebral aneurysms: a technical report. Neurosurgery **10**, 252–261 (2014). https://doi.org/10.1227/neu.0000000000000328
141. Abe, Y., Sato, S., Kato, K., Hyakumachi, T., Yanagibashi, Y., Ito, M., Abumi, K.: A novel 3D guidance system using augmented reality for percutaneous vertebroplasty. J. Neurosurg. Spine **19**(4), 492–501 (2013). https://doi.org/10.3171/2013.7.spine12917
142. Shenai, M.B., Dillavou, M., Shum, C., Ross, D., Tubbs, R.S., Shih, A., Guthrie, B.L.: Virtual interactive presence and augmented reality (VIPAR) for remote surgical assistance. Neurosurgery **68**(1 Suppl Operative), 200–207 (2011). https://doi.org/10.1227/NEU.0b013e3182077efd
143. Besharati, T.L., Mahvash, M.: Augmented reality–guided neurosurgery: accuracy and intraoperative application of an image projection technique. J. Neurosurg. **123**(1), 206–211 (2015)
144. Chytas, D., Malahias, M.-A., Nikolaou, V.S.: Augmented reality in orthopedics: current state and future directions. Front. Surgery **6**, 38 (2019). https://doi.org/10.3389/fsurg.2019.00038
145. Londei, R., Esposito, M., Diotte, B., Weidert, S., Euler, E., Thaller, P., Navab, N., Fallavollita, P.: Intra-operative augmented reality in distal locking. Int. J. Comput. Assist. Radiol. Surg. **10**(9), 1395–1403 (2015). https://doi.org/10.1007/s11548-015-1169-2
146. Elmi-Terander, A., Burström, G., Nachabe, R., Skulason, H., Pedersen, K., Fagerlund, M., et al.: Pedicle screw placement using augmented reality surgical navigation with intraoperative 3D imaging: a first in-human prospective cohort study. Spine **44**, 517–25. https://doi.org/10.1097/BRS.0000000000002876
147. Deng, W., Li, F., Wang, M., Song, Z.: Easy-to-use augmented reality neuronavigation using a wireless tablet PC. Stereotact. Funct. Neurosurg. **92**(1), 17–24 (2014). https://doi.org/10.1159/000354816
148. Fahim, S., Maqsood, A., Das, G., Ahmed, N., Saquib, S., Lal, A., Khan, A.A.G., Alam, M.K.: Augmented reality and virtual reality in dentistry: highlights from the current research. Appl. Sci. **12**(8), 3719 (2022).https://doi.org/10.3390/app12083719
149. Jiang, W., Ma, L., Zhang, B., Fan, Y., Qu, X., Zhang, X., Liao, H.: Evaluation of the 3D augmented reality–guided intraoperative positioning of dental implants in edentulous mandibular models. The Int. J. Oral Maxillofacial Implants **33**(6), 1219–1228 (2018). https://doi.org/10.11607/jomi.6638
150. Porter, M., Heppelmann, J.: Why every organization needs an augmented reality strategy. Harvard Bus. Rev. 1–13 (2017)
151. Silverstein, J.C., Dech, F., Edison, M., Jurek, P., Helton, W., Espat, N.: Virtual reality: immersive hepatic surgery educational environment. Surgery **132**(2), 274–277 (2002). https://doi.org/10.1067/msy.2002.125723

152. Dobson, H. D., Pearl, R. K., Orsay, C. P., Rasmussen, M., Evenhouse, R., Ai, Z., Blew, G., Dech, F., Edison, M. I., Silverstein, J. C., Abcarian, H.: Virtual reality: new method of teaching anorectal and pelvic floor anatomy. Dis. Colon Rectum **46**(3), 349–352 (2003). https://doi.org/10.1007/s10350-004-6554-9
153. Bergman, E.M., van der Vleuten, C.P., Scherpbier, A.J.: Why don't they know enough about anatomy? A Narrative Rev. Medical Teacher **33**(5), 403–409 (2011). https://doi.org/10.3109/0142159X.2010.536276
154. Baratz, G., Sridharan, P.S., Yong, V., Tatsuoka, C., Griswold, M.A., Wish-Baratz, S.: Comparing learning retention in medical students using mixed-reality to supplement dissection: a preliminary study. Int. J. Med. Educ. **13**, 107–114 (2022). https://doi.org/10.5116/ijme.6250.0af8
155. Ruthberg, J.S., Tingle, G., Tan, L., Ulrey, L., Simonson-Shick, S., Enterline, R., Eastman, H., Mlakar, J., Gotschall, R., Henninger, E., Griswold, M.A., Wish-Baratz, S.: Mixed reality as a time-efficient alternative to cadaveric dissection. Med. Teach. **42**(8), 896–901 (2020). https://doi.org/10.1080/0142159X.2020.1762032
156. Maresky, H.S., Oikonomou, A., Ali, I., Ditkofsky, N., Pakkal, M., Ballyk, B.: Virtual reality and cardiac anatomy: exploring immersive three-dimensional cardiac imaging, a pilot study in undergraduate medical anatomy education. Clin. Anat. **32**(2), 238–243 (2018). https://doi.org/10.1002/ca.23292
157. Huang, H., Hwang, G., Chang, C.: Learning to be a writer: a spherical video-based virtual reality approach to supporting descriptive article writing in high school Chinese courses. Br. J. Edu. Technol. **51**(4), 1386–1405 (2020). https://doi.org/10.1111/bjet.12893
158. Southworth, M.K., Silva, J.R., Silva, J.N.A.: Use of extended realities in cardiology. Trends Cardiovasc. Med. **30**(3), 143–148 (2020). https://doi.org/10.1016/j.tcm.2019.04.005
159. Lahti, S., Suominen, A., Freeman, R., Lähteenoja, T., Humphris, G.: Virtual reality relaxation to decrease dental anxiety: immediate effect randomized clinical trial. JDR Clinical Trans. Res. **5**(4):312–318 (2020). https://doi.org/10.1177/2380084420901679
160. Atzori, B., Lauro, G.R., Giugni, A., Calabrò, M., Alhalabi, W., Hoffman, H.G.: Virtual reality analgesia for pediatric dental patients. Front. Psychol. **9**, 2265 (2018). https://doi.org/10.3389/fpsyg.2018.02265
161. Perez-Marcos, D., Chevalley, O., Schmidlin, T., Garipelli, G., Serino, A., Vuadens, P., Tadi, T., Blanke, O., Millán, J.D.R.: Increasing upper limb training intensity in chronic stroke using embodied virtual reality: a pilot study. J. Neuroeng. Rehabil. **14**(1), 119 (2017). https://doi.org/10.1186/s12984-017-0328-9
162. González, N.A.A.: How to include augmented reality in descriptive geometry teaching. Procedia Comput. Sci. **75**, 250–256 (2015). https://doi.org/10.1016/j.procs.2015.12.245
163. Shelton, B.E., Hedley, N.R.: Exploring a cognitive basis for learning spatial relationships with augmented reality. Technol. Instr. Cogn. Learn. **1**(4), 323–357 (2004)
164. Ibáñez, M.-B., Delgado-Kloos, C.: Augmented reality for STEM learning: a systematic review. Comput. Educ. **123**, 109–123 (2018). https://doi.org/10.1016/j.compedu.2018.05.002
165. Sırakaya, M., Sırakaya, D.A.: Augmented reality in STEM education: a systematic review. Interact. Learn. Environ. 1-14 (2020). https://doi.org/10.1080/10494820.2020.1722713
166. Pomerantz, J.: Learning in three dimensions: report of the EDUCAUSE/HP campus of the future project. Louisville, CO: ECAR (2018). https://www.educause.edu/ecar/research-publications/learning-in-three-dimensions-report-on-the-educause-hp-campus-of-the-future-project/executive-summary-key-findings-acknowledgments. Last accessed 15 June 2022
167. Barsalou, L.W.: Grounded cognition. Annu. Rev. Psychol. **59**(1), 617–645 (2008). https://doi.org/10.1146/annurev.psych.59.103006.093639
168. Repetto, C., Serino, S., Macedonia, M., Riva, G.: Virtual reality as an embodied tool to enhance episodic memory in elderly. Front. Psychol. **7**(1839), 1–4 (2016). https://doi.org/10.3389/fpsyg.2016.01839
169. Shapiro, L.: Embodied cognition. Routledge, New York (2010). ISBN 978-0415773423
170. Mania, K., Troscianko, T., Hawkes, R., Chalmers, A.: Fidelity metrics for virtual environment simulations based on spatial memory awareness states. Presence: Teleoperators Virtual Environ. **12**(3), 296–310 (2003). https://doi.org/10.1162/105474603765879549

171. Ruddle, R.A., Payne, S.J., Jones, D.M.: Navigating large-scale virtual environments: what differences occur between helmet-mounted and desktop displays? Presence: Teleoperators Virtual Environ. **8**(2), 157–168 (1999)
172. Harman, J., Brown, R., Johnson, D.: Improved memory elicitation in virtual reality: new experimental results and insights. In: IFIP Conference on Human–Computer Interaction, pp. 128–146. Springer (2017)
173. Krokos, E., Plaisant, C., Varshney, A.: Virtual memory palaces: immersion aids recall. Virtual Reality **23**, 1–15 (2019). https://doi.org/10.1007/s10055-018-0346-3
174. Billinghurst, M., Kato, H., Poupyrev, I.: The MagicBook: a transitional AR interface. Comput. Graph. **25**, 745–753 (2001). https://doi.org/10.1016/s0097-8493(01)00117-0
175. Alarcón-Yaquetto, D.E., Tincopa, J.P., Guillén-Pinto, D., Bailon, N., Cárcamo, C.P.: Effect of augmented reality books in salivary cortisol levels in hospitalized pediatric patients: a randomized cross-over trial. Int. J. Med. Informatics **148**, 104404 (2021). https://doi.org/10.1016/j.ijmedinf.2021.104404
176. Dünser, A., Walker, L., Horner, H., Bentall, D.: Creating interactive physics education books with augmented reality. In: Proceedings of the 24th Australian Computer-Human Interaction Conference on OzCHI '12 (2012). https://doi.org/10.1145/2414536.2414554
177. Martín-Gutiérrez, J., Luís Saorín, J., Contero, M., Alcañiz, M., Pérez-López, D.C., Ortega, M.: Design and validation of an augmented book for spatial abilities development in engineering students. Comput. Graph. **34**(1), 77–91 (2010). https://doi.org/10.1016/j.cag.2009.11.003
178. Kaufmann, H., Schmalstieg, D.: Mathematics and geometry education with collaborative augmented reality. Comput. Graph. **27**, 339–345 (2003)
179. Lin, H.-C.K., Chen, M.-C., Chang, C.-K.: Assessing the effectiveness of learning solid geometry by using an augmented reality-assisted learning system. Interact. Learn. Environ. **23**(6), 799–810 (2013). https://doi.org/10.1080/10494820.2013.817435
180. Saha, R.A., Ayub, A.F.M., Tarmizi, R.A.: The effects of GeoGebra on mathematics achievement: enlightening coordinate geometry learning. Procedia Soc. Behav. Sci. **8**, 686–693 (2010). https://doi.org/10.1016/j.sbspro.2010.12.095
181. Kozma, R., Russell, J.: Students becoming chemists: developing representational competence. In: Visualization in science education, pp. 121–145. Springer, Dordrecht, The Netherlands (2005)
182. Stieff, M., Ryu, M., Dixon, B., Hegarty, M.: The role of spatial ability and strategy preference for spatial problem solving in organic chemistry. J. Chem. Educ. **89**(7), 854–859 (2012). https://doi.org/10.1021/ed200071d
183. Wu, H.-K., Shah, P.: Exploring visuospatial thinking in chemistry learning. Sci. Educ. **88**, 465–492 (2004). https://doi.org/10.1002/sce.10126
184. Zheng, M., Waller, M.P.: ChemPreview: an augmented reality-based molecular interface. J. Mol. Graph. Model. **73**, 18–23 (2017). https://doi.org/10.1016/j.jmgm.2017.01.019
185. Habig, S.: Who can benefit from augmented reality in chemistry? Sex differences in solving stereochemistry problems using augmented reality. Br. J. Edu. Technol. **51**(3), 629–644 (2019). https://doi.org/10.1111/bjet.12891
186. Jimenez, Z.: Teaching and learning chemistry via augmented and immersive virtual reality. In: Technology Integration in Chemistry Education and Research (TICER), vol. 3, pp. 31–52 (2019). https://doi.org/10.1021/bk-2019-1318.ch003
187. Kaufmann, H., Schmalstieg, D.: Mathematics and geometry education with collaborative augmented reality. In: ACM SIGGRAPH 2002 Conference Abstracts and Applications, pp. 37–41 (2002). ACM
188. Pietikäinen, O., Hämäläinen, P., Lehtinen, J., Karttunen, A.J.: VRChem: a virtual reality molecular builder. Appl. Sci. **11**, 10767 (2021)
189. Pang, C.G.: Immersive virtual reality (VR) Classroom to enhance learning and increase interest and enjoyment in the secondary school science curriculum. In: 8th International Conference on Educational Technologies 2021, ICEduTech 2021 and 17th International Conference on Mobile Learning 2021, pp. 99–106 (2021)

190. Pang, C.G., Devi, S., Wong, D., Cai, Y.Y., Ba, R.: The use of immersive virtual reality technology to deepen learning in Singapore schools. In: Cai, Y., van Joolingen, W., Veermans, K. (eds.) Virtual and Augmented Reality, Simulation and Serious Games for Education. Gaming Media and Social Effects. Springer, Singapore (2021). https://doi.org/10.1007/978-981-16-1361-6_5
191. Ryan, B., Yuan, X., Zhang, Y., Faatihah, C.Y.Y., Walker, Z., et al.: Virtual reality enzymes: an interdisciplinary and international project towards an inquiry-based pedagogy. In: Cai, Y.Y., Joolingen, W., Walker, Z. (eds.) VR, Simulations and Serious Games for Education. Springer, Singapore (2018)
192. Thompson, M., Wang, A., Bilgin, C., Anteneh, M., Roy, D., Tan, P., Eberhart, R., Klopfer, E.: Influence of virtual reality on high school students' conceptions of cells. J. Univ. Comput. Sci. **26**(8), 929–946 (2020). https://doi.org/10.3897/jucs.2020.050
193. Thompson, M.M., Wang, A., Roy, D., Klopfer, E.: Authenticity, interactivity, and collaboration in VR learning games. Front. Robot. AI **5**, 133 (2018). https://doi.org/10.3389/frobt.2018.00133
194. Tan, S., Waugh, R.: Use of virtual-reality in teaching and learning molecular biology. In: Cai, Y. (ed.) 3D Immersive and Interactive Learning. Springer, Singapore (2013). https://doi.org/10.1007/978-981-4021-90-6_2
195. Brophy, J.E.: Motivating Students to Learn, 3rd edn. Routledge, New York, NY (2010)
196. Keller, J. M. (2010). The ARCS model of motivational design. Motivational design for learning and performance (pp. 62–68). Boston, MA: Springer.
197. Keller, J.M.: Motivation and instructional design: a theoretical perspective. J. Instr. Dev. **2**(4), 26–34 (1979). https://doi.org/10.1007/bf02904345
198. Keller, J.M.: Motivational design of instruction. In: Reigeluth, C.M. (ed.) Instructional Design Theories and Models: An Overview of Their Current Status. Lawrence Erlbaum Associates, Hillsdale, NJ (1983)
199. Keller, J.M.: The use of the ARCS model of motivation in teacher training. In: Shaw, K.E. (ed.) Aspects of Educational Technology Volume XVII: Staff Development and Career Updating. Kogan Page, London (1984)
200. Maehr, M.L., Meyer, H.A.: Understanding motivation and schooling: where we've been, where we are, and where we need to go. Educ. Psychol. Rev. **9**(4), 371–409 (1997). https://doi.org/10.1023/A:1024750807365
201. Deci, E.L., Ryan, R.M.: Intrinsic Motivation and Self-determination in Human Behavior. Plenum, New York, NY (1985)
202. Keller, J.M.: Development and use of the ARCS model of instructional design. J. Instr. Dev. **10**(3), 2–10 (1987). https://doi.org/10.1007/bf02905780
203. Weiner, B. (ed.): Achievement Motivation and Attribution Theory. General Learning Press, Morristown, NJ (1974)
204. Bandura, A.: Self-efficacy: toward a unifying theory of behavioral change. Psychol. Rev. **84**(2), 191–215 (1977). https://doi.org/10.1037/0033-295X.84.2.191
205. Maehr, M.L.: Continuing motivation: an analysis of a seldom considered educational outcome. Rev. Educ. Res. **46**(3), 443–462 (1976)
206. Wertsch, J.: A sociocultural approach to socially shared cognition. In: Resnick, L., Levine, J., Teasley, S. (eds.) Perspectives on socially shared cognition. American Psychological Association, Washington (1991)
207. Pietroni, E.: Experience design, virtual reality and media hybridization for the digital communication inside museums. Appl. Syst. Innov. **2**(4), 35 (2019). https://doi.org/10.3390/asi2040035
208. Sylaiou, S., Mania, K., Karoulis, A., White, M.: Presence-centred usability evaluation of a virtual museum: exploring the relationship between presence, previous user experience and enjoyment. Int. J. Human-Comput. Stud. (IJHCS) **68**(5), 243–253, Amsterdam: Elsevier, Academic Press (2009)
209. White, M., Liarokapis, F., Darcy, J., Mourkoussis, N., Petridis, P., Lister, P.: Augmented reality for museum artefact visualization. In: Conference: Proceedings of the 4th Irish Workshop on Computer Graphics, Eurographics, pp. 75–80, Ireland (2003)

210. Bhagat, K.K., Liou, W.-K., Michael, S.J., Chang, C.-Y.: To use augmented reality or not in formative assessment: a comparative study. Interact. Learn. Environ. **27**(5–6), 830–840 (2018). https://doi.org/10.1080/10494820.2018.1489857
211. Huang, T.-C., Chen, C.-C., Chou, Y.-W.: Animating eco-education: to see, feel, and discover in an augmented reality-based experiential learning environment. Comput. Educ. **96**, 72–82 (2016). https://doi.org/10.1016/j.compedu.2016.02.008
212. Singh, G., Bowman, D.A., Hicks, D., Cline, D., Ogle, J.T., Johnson, A., et al.: CI-spy: designing a mobile augmented reality system for scaffolding historical inquiry learning. In: 2015 IEEE International Symposium on Mixed and Augmented Reality – Media, Art, Social Science, Humanities and Design. IEEE, Fukuoka, Japan (2015)
213. Khan, T., Johnston, K., & Ophoff, J.: The impact of an augmented reality application on learning motivation of students. Adv. Hum.-Comput. Interact. **2019**, 1–14 (2019). https://doi.org/10.1155/2019/7208494
214. Kaufmann, H., Du¨nser, A.: Summary of usability evaluations of an educational augmented reality application. In: Proceedings of the 2nd International Conference on Virtual Reality, pp. 660–669 (2007). Springer
215. Liu, R., Wang, L., Lei, J., Wang, Q., Ren, Y.: Effects of an immersive virtual reality-based classroom on students' learning performance in science lessons. Br. J. Edu. Technol. **51**, 2034–2049 (2020). https://doi.org/10.1111/bjet.13028
216. Mahadzir, N. N.: The Use of Augmented Reality Pop-up Book to Increase Motivation in English Language Learning for National Primary School. IOSR J. Res. Method Edu. (IOSRJRME), **1**(1), 26–38 (2013). https://doi.org/10.9790/7388-0112638
217. Gopalan, V., Zulkifli, A.N., Bakar, J.A.A.: A study of students' motivation using the augmented reality science textbook. In: AIP Conference Proceedings, vol. 1761, no. 1, p. 020040 (2016). https://doi.org/10.1063/1.4960880
218. Cai, S., Liu, E., Shen, Y., Liu, C.H., Li, S.H., Shen, Y.H.: Probability learning in mathematics using augmented reality: impact on student's learning gains and attitudes. Interact. Learn. Environ. **28**, 560–573 (2020). https://doi.org/10.1080/10494820.2019.1696839
219. Contero, M., Olmos-Ray, E., Ferreira-Cavalcanti, J., Castellanos, M.C., Giglioli, I.A.C., Alcañiz, M.: Mobile virtual reality as an educational platform: a pilot study on the impact of immersion and positive emotion induction in the learning process. EURASIA J. Math., Sci. Technol. Educ. **14**(6) (2018). https://doi.org/10.29333/ejmste/85874
220. Civelek, T., Ucar, E., Ustunel, H., & Aydın, M. K.: Effects of a Haptic Augmented Simulation on K-12 Students' Achievement and their Attitudes Towards Physics. EURASIA J. Math. Sci. Tech. Edu. **10**(6) (2014). https://doi.org/10.12973/eurasia.2014.1122a
221. Allcoat, D., von Mühlenen, A.: Learning in virtual reality: Effects on performance, emotion and engagement. Res. Learn. Technol. 26. https://doi.org/10.25304/rlt.v26.2140
222. Laine, T.H., Nygren, E., Dirin, A., Suk, H.-J.: Science spots AR: a platform for science learning games with augmented reality. Educ. Tech. Res. Dev. **64**(3), 507–531 (2016). https://doi.org/10.1007/s11423-015-9419-0
223. Kapp, K.M.: The Gamification of Learning and Instruction: Game-based Methods and Strategies for Training and Education. Wiley, San Francisco, CA (2012)
224. Smith-Robbins, S.: This game sucks": how to improve the gamification of education. Educause Rev. **46**(1), 58–59 (2011)
225. Lee, J., Hammer, J.: Gamification in education: what, how, why bother? Acad. Exchange Q. **15**(2), 146 (2011)
226. Lu, S.-J., & Liu, Y.-C.: Integrating Augmented Reality Technology to Enhance Children's Learning in Marine Education. Environ. Edu. Res. **21**(4), 525–541 (2014). https://doi.org/10.1080/13504622.2014.911247
227. Hwang, G.-J., Wu, P.-H., Chen, C.-C., Tu, N.-T.: Effects of an augmented reality-based educational game on students' learning achievements and attitudes in real-world observations. Interact. Learn. Environ. **24**(8), 1895–1906 (2015). https://doi.org/10.1080/10494820.2015.1057747

228. Klopfer, E., & Squire, K.: Environmental Detectives—The Development of an Augmented Reality Platform for Environmental Simulations. Educ. Tech. Res. Dev **56**, 203–228 (2008). https://doi.org/10.1007/s11423-007-9037-6
229. Bower, M.: Affordance analysis—matching learning tasks with learning technologies. Educ. Media Int. **45**, 3–15 (2008). https://doi.org/10.1080/09523980701847115
230. Chen, C., Wang, C.-H.: Employing augmented-reality-embedded instruction to disperse the imparities of individual differences in earth science learning. J. Sci. Educ. Technol. **24**(6), 835–847 (2015). https://doi.org/10.1007/s10956-015-9567-3
231. Billinghurst, M., Belcher, D., Gupta, A., Kiyokawa, K.: Communication behaviors in co-located collaborative AR interfaces. Int. J. Human-Comput. Interact. **16**(3), 395–423 (2003). https://doi.org/10.1207/s15327590ijhc1603_2
232. Glaser, B.G., Strauss, A.L.: The discovery of grounded theory: strategies for qualitative research. Transaction Publishers, New Brunswick (Original work published 1967) (2006)
233. Creswell, J.W.: Educational Research: Planning, Conducting, and Evaluating Quantitative and Qualitative Research, 4th edn. Pearson, Boston, MA (2012)
234. Strauss, A., Corbin, J.: Basics of Qualitative Research: Grounded Theory Procedures and Techniques. Sage, Newbury Park, CA (1990)
235. Charmaz, K.: Grounded theory in global perspective: reviews by international researchers. Qual. Inq. **20**, 9 (2014)
236. Charmaz, K.: Constructing Grounded Theory: A Practical Guide Through Qualitative Analysis. Sage, London (2006)
237. Carmichael, T., Cunningham, N.: Theoretical data collection and data analysis with gerunds in a constructivist grounded theory study. Electron. J. Bus. Res. Methods **15**, 59–73 (2017)
238. Tamim, R., Bernard, R., Borokhovski, E., Abrami, P., Schmid, R.: What forty years of research says about the impact of technology on learning a second-order meta-analysis and validation study. Rev. Educ. Res. **81**, 4–28 (2011). https://doi.org/10.3102/0034654310393361
239. Green, M., Challoo, L., Mehrubeoglu, M.: Finding a path forward for integrating augmented reality into K-12 classrooms. In: Russell, D. (ed.) Implementing Augmented Reality into Immersive Virtual Learning Environments, pp. 1–33 (2021). IGI Global. https://doi.org/10.4018/978-1-7998-4222-4
240. Fokides, E., & Mastrokoukou, A.: Results from a Study for Teaching Human Body Systems to Primary School Students using Tablets. Contem. Educ. Tech. **9**(2), 154–170 (2018). https://doi.org/10.30935/cet.414808
241. Tobias, S., Duffy, T.M.: Constructivist Instruction: Success or Failure? Taylor & Francis, New York (2009)
242. Akman, E., Çakır, R.: The effect of educational virtual reality game on primary school students' achievement and engagement in mathematics. Interact. Learn. Environ. 1–18 (2020).https://doi.org/10.1080/10494820.2020.1841800
243. Buckley, P., Doyle, E.: Gamification and student motivation. Interact. Learn. Environ. **24**(6), 1162–1175 (2016). https://doi.org/10.1080/10494820.2014.964263
244. Safadel, P., White, D.: Effectiveness of computer-generated virtual reality (VR) in learning and teaching environments with spatial frameworks. Appl. Sci. **10**(16), 5438 (2020). https://doi.org/10.3390/app10165438
245. Alalwan, N., Cheng, L., Al-Samarraie, H., Yousef, R., Ibrahim, A.A., Sarsam, S.M.: Challenges and Prospects of Virtual Reality and Augmented Reality Utilization among Primary School Teachers: A Developing Country Perspective. Stud. Educ. Eval. **66**, 100876 (2020). https://doi.org/10.1016/j.stueduc.2020.100876
246. Tzima, S., Styliaras, G., Bassounas, A.: Augmented reality applications in education: teachers point of view. Educ. Sci. **9**(2), 99 (2019). https://doi.org/10.3390/educsci9020099
247. Makransky, G., Wismer, P., Mayer, R.E.: A gender matching effect in learning with pedagogical agents in an immersive virtual reality science simulation. J. Comput. Assist. Learn. **35**(3), 349–358 (2019). https://doi.org/10.1111/jcal.1233
248. Mayer, R.E., Pilegard, C.: Principles for managing essential processing in multimedia learning: segmenting, pre-training and modality principles. In: The Cambridge handbook of multimedia learning (2014). Cambridge University Press, New York

249. Sanfilippo, F., Blazauskas, T., Salvietti, G., Ramos, I., Vert, S., Radianti, J., Majchrzak, T.A., Oliveira, D.: A perspective review on integrating VR/AR with haptics into STEM education for multi-sensory learning. Robotics **11**(2), 41 (2022). https://doi.org/10.3390/robotics11020041
250. Roger, S.: (2019). https://www.forbes.com/sites/solrogers/2019/03/15/virtual-reality-the-learning-aid-of-the-21st-century/?sh=3affcbac139b
251. Fransson, G., Holmberg, J., & Westelius, C.: The challenges of using head mounted virtual reality in K-12 schools from a teacher perspective. Educ. Info. Tech 25(4), 3383–3404 (2020). https://doi.org/10.1007/s10639-020-10119-1
252. Hamilton, D., McKechnie, J., Edgerton, E., Wilson, C.: Immersive virtual reality as a pedagogical tool in education: a systematic literature review of quantitative learning outcomes and experimental design. J. Comput. Educ. **8**(1), 1–32 (2020). https://doi.org/10.1007/s40692-020-00169-2
253. Stylianidou, N., Sofianidis, A., Manoli, E., Meletiou-Mavrotheris, M.: "Helping Nemo!"-using augmented reality and alternate reality games in the context of universal design for learning. Educ. Sci. **10**(4), 95 (2020). https://doi.org/10.3390/educsci10040095
254. Joo-Nagata, J., Martinez, A.F., García-Bermejo, G.J., García-Peñalvo, F.J.: Augmented reality and pedestrian navigation through its implementation in m-learning and e-learning: evaluation of an educational program in Chile. Comput. Educ. **111**, 1–17 (2017). https://doi.org/10.1016/j.compedu.2017.04.003
255. Palmarini, R., Erkoyuncu, J.A., Roy, R., Torabmostaedi, H.: A systematic review of augmented reality applications in maintenance. Robot. Comput.-Integr. Manuf. **49**, 215–228 (2018). https://doi.org/10.1016/j.rcim.2017.06.002
256. Yilmaz, R.: Augmented reality trends in education between 2016 and 2017 years. In: State of the Art Virtual Reality and Augmented Reality Knowhow. IntechOpen (2018). https://doi.org/10.5772/intechopen.74943
257. Chebi, H.: Unified approach to augmented reality: taking into account technological, psychological, and ecological Approaches. In: Russell, D. (ed.) Implementing augmented reality into immersive virtual learning environments, pp. 1–33. IGI Global (2021). https://doi.org/10.4018/978-1-7998-4222-4
258. Clark, R.E.: Reconsidering research on learning from media. Rev. Educ. Res. **53**(4), 445–459 (1983). 10.3102%2F00346543053004445
259. Kozma, R.B.: Learning with media. Rev. Educ. Res. **61**(2), 179–211 (1991). 10.3102%2F00346543061002179
260. Becker, K.: The Clark-Kozma debate in the 21st century. In: Canadian Network for Innovation in Education Conference 2010: Heritage Matters: Inspiring Tomorrow. Saint John, New Brunswick (2010)
261. Maas, M.J., Hughes, J.M.: Virtual, augmented and mixed reality in K-12 education: a review of the literature. Technol. Pedagog. Educ. **29**(2), 231–249 (2020). https://doi.org/10.1080/1475939X.2020.1737210
262. Moorhouse, B.L., Kohnke, L.: Using mentimeter to elicit student responses in the EAP/ESP classroom. RELC J. 003368821989035 (2020). https://doi.org/10.1177/0033688219890350
263. Alkhattabi, M.: Augmented reality as e-learning tool in primary schools' education: barriers to teachers' adoption. Int. J. Emerg. Technol. Learn. (iJET) **12**(02), 91–100 (2017). https://doi.org/10.3991/ijet.v12i02.6158
264. Fransson, G., Holmberg, J., Westelius, C.: The challenges of using head mounted virtual reality in K-12 schools from a teacher perspective. Educ. Inf. Technol. **25**, 1–22 (2020). https://doi.org/10.1007/s10639-020-10119-1
265. Aydoğdu, F.: Augmented reality for preschool children: an experience with educational contents. Br. J. Edu. Technol. **53**(2), 326–348 (2021). https://doi.org/10.1111/bjet.13168
266. Bhattacharjee, D., Paul, A., Kim, J.H., Karthigaikumar, P.: An immersive learning model using evolutionary learning. Comput. Electr. Eng. **65**, 236–249 (2018). https://doi.org/10.1016/j.compeleceng.2017.08.023

267. Bolan, C.: Incorporating the experiential learning theory into the instructional design of online courses. Nurse Educ. **28**, 10–14 (2003). https://doi.org/10.1097/00006223-200301000-00006
268. Cai, S., Liu, E., Yang, Y., Liang, J.C.: Tablet-based AR technology: impacts on students' conceptions and approaches to learning mathematics according to their self-efficacy. Br. J. Edu. Technol. **50**, 248–263 (2019). https://doi.org/10.1111/bjet.12718
269. Çeliker, H.D.: Prospective science teacher's levels of understanding and explanation of animal and plant cells: draw-write. J. Balt. Sci. Educ. **14**(4), 501–513 (2013)
270. Chen, M., Chai, C.-S., Jong, M.S.-Y., Chao, G.C.-N.: Modeling learners' self-concept in Chinese descriptive writing based on the affordances of a virtual reality-supported environment. Educ. Inf. Technol. **26**(5), 6013–6032 (2021). https://doi.org/10.1007/s10639-021-10582-4
271. Deci, E.L., Ryan, R.M.: Self-determination theory. In: Van Lange, P.A.M., Kruglanski, A.W., Higgins, E.T. (eds.) Handbook of Theories of Social Psychology, pp. 416–436. Sage Publications Ltd. (2012). https://doi.org/10.4135/9781446249215.n21
272. Ehrsson, H.H.: The experimental induction of out-of-body experiences. Science **5841**(317), 1048 (2007). https://doi.org/10.1126/science.1142175
273. Fujimoto, Y., Yamamoto, G., Taketomi, T., Miyazaki, J., Kato, H.: Relation between displaying features of augmented reality and user's memorization. Trans. Virtual Reality Soc. Jpn. **18**(1), 81–91 (2013)
274. Gandolfi, E.: Virtual reality and augmented reality. In: Kennedy, K., Ferdig, R.E. (eds) Handbook of Research on K-12 Online and Blended Learning, 2nd edn, pp. 545–561. ETC Press
275. Gunawan, P., Kwan, J., Cai, Y.Y., Yang, R.: Augmented reality application for chemical engineering unit operations. In: Cai, Y., van Joolingen, W., Veermans, K. (eds.) Virtual and Augmented Reality, Simulation and Serious Games for Education. Gaming Media and Social Effects. Springer, Singapore (2021). https://doi.org/10.1007/978-981-16-1361-6_4
276. Jeffri, N.F.S., Awang, D.R.: A review of augmented reality systems and their effects on mental workload and task performance. Heliyon **7**(3), e06277 (2021). https://doi.org/10.1016/j.heliyon.2021.e06277
277. Kerawalla, L., Luckin, R., Seljeflot, S., Woolard, A.: "Making it Real": Exploring the Potential of Augmented Reality for Teaching Primary School Science. Virtual Reality **10**(3–4), 163–174 (2006). https://doi.org/10.1007/s10055-006-0036-4
278. Lee, P.: Putting principles into practice: Understanding history. In: Donovan, M.S., Brandsford, J. (eds.) How students learn: history in the classroom, pp. 31–78 (2005)
279. Liarokapis, F., Sylaiou, S., Mountain, D.: Personalizing virtual and augmented reality for cultural heritage indoor and outdoor experiences. In: Proceedings of the 9th International Symposium on Virtual Reality, Archaeology and Cultural Heritage (VAST '08), Eurographics, pp. 55–62 (2008). Braga, Portugal
280. Liu, R., Wang, L., Lei, J., Wang, Q., Ren, Y.: Effects of an immersive virtual reality-based classroom on students' learning performance in science lessons. Br. J. Edu. Technol. **51**(6), 2034–2049 (2020). https://doi.org/10.1111/bjet.13028
281. Maister, L., Sebanz, N., Knoblich, G., Tsakiris, M.: Experiencing ownership over a dark-skinned body reduces implicit racial bias. Cognition **128**(2), 170–178 (2013). https://doi.org/10.1016/j.cognition.2013.04.002
282. Mayer, R.E.: Multimedia aids to problem-solving transfer. Int. J. Educ. Res. **31**(7), 611–623 (1999)
283. Mayer, R.E., Anderson, R.B.: Animations need narrations: an experimental test of a dual-dual coding hypothesis. J. Educ. Psychol. **83**, 484–490 (1991)
284. Mautone, P.D., Mayer, R.E.: Signaling as a cognitive guide in multimedia learning. J. Educ. Psychol. **93**, 377–389 (2001)
285. Mayer, R.E., Moreno, R.: Nine ways to reduce cognitive load in multimedia learning. Educ. Psychol. **38**, 43–52 (2003). https://doi.org/10.1207/S15326985EP3801_6
286. Moreno, R., Mayer, R.: Cognitive principles of multimedia learning: the role of modality and contiguity. J. Educ. Psychol. **91**(2), 358–368 (1999)

287. Nemirovsky, R., Tierney, C., Wright, T.: Body motion and graphing. Cogn. Instr. **16**(2), 119–172 (1998)
288. Noble, T., Nemirovsky, R., Wright, T., Tierney, C.: Experiencing change: the mathematics of change in multiple environments. J. Res. Math. Educ. **32**(1), 85–108 (2001)
289. Roseberry, A.S., Warren, B., Ballenger, C., Ogonowski, M.: The generative potential of students' everyday knowledge in learning science. In: Romberg, T., Carpenter, T., Dremock, F. (eds.) Understanding Mathematics and Everyday Knowledge in Learning Science, pp. 55–80. Erlbaum, Mahwah, NJ (2005)
290. Ryan, R.M., Deci, E.L.: Self-determination theory and the facilitation of intrinsic motivation, social development, and well-being. Am. Psychol. **55**(1), 68–78 (2000). https://doi.org/10.1037/0003-066X.55.1.68
291. Ryan, R.M., Kuhl, J., Deci, E.L.: Nature and autonomy: organizational view of social and neurobiological aspects of selfregulation in behavior and development. Dev. Psychopathol. **9**, 701–728 (1997)
292. Schuemie, M.J., van der Straaten, P., Krijn, M., van der Mast, C.A.: Research on presence in virtual reality: a survey. Cyberpsychol. Behav. **4**(2), 183–201 (2001). https://doi.org/10.1089/109493101300117884
293. Tsatsarelis, C., Ogborn, J., Jewitt and Kress G.: Rhetorical construction of cells in science and in a science classroom. Res. Sci. Educ. **30**(4), 451 (2000). https://doi.org/10.1007/BF02461562
294. Zhang, J., Sung, Y.-T., Hou, H.-T., Chang, K.-E.: The development and evaluation of an augmented reality-based armillary sphere for astronomical observation instruction. Comput. Educ. **73**, 178–188 (2014). https://doi.org/10.1016/j.compedu.2014.01.003
295. Zhu, B., Feng, M., Lowe, H., Kesselman, J., Harrison, L., Dempski, R.E.: Increasing enthusiasm and enhancing learning for biochemistry-laboratory safety with an augmented-reality program. J. Chem. Educ. **95**(10), 1747–1754 (2018). https://doi.org/10.1021/acs.jchemed.8b00116
296. Dalgarno, B., & Lee, M. J.: What are the Learning Affordances of 3-D Virtual Environments? British J. Edu. Tech. **41**(1), 10–32 (2009). https://doi.org/10.1111/j.1467-8535.2009.01038.x

Exploring the Mozilla® HUBS® Platform for Virtual Final Year Project Exhibition

Muhammad Mobeen Movania, Abdul Samad, and Syeda Saleha Raza

Abstract In the current Covid era where all forms of physical engagement with students are severely restricted or stopped, many conventional activities and assessments have been hampered. One particularly important course in any BS(CS) 4 year undergraduate degree program is the final year project. Students are asked to solve a local or global problem under guidance of a faculty member as their final year project supervisor. The final year projects are evaluated multiple times through mid and final presentations and a final year projects exhibition. In these evaluations, the projects are evaluated by both internal and external experts. Based on mid, final presentation and exhibition scores, an aggregate score is generated for the course. Due to strict Covid SOPs, we are unable to host the final year exhibition on campus. So we decided to develop the first virtual final year exhibition at Habib University on the Mozilla® Hubs® platform. This was achieved by creating a virtual replica of the Habib University design thinking and exhibition space (The Habib University Playground). We provided virtual spaces for final year student groups to showcase their posters and presentations. Thanks to the support provided by the Mozilla® Hubs® platform and its built-in Spoke® editor, we were able to do streaming of YouTube® presentations for each group. The students, participants and evaluators had their own virtual avatars in the virtual space. Moreover, distance based sound attenuation enabled talking among participants as if they were meeting physically. This allowed students and evaluators to cross question during the exhibition. Based on our own experience of using Mozilla® Hubs®, we can say with confidence that it is a promising tool for creating engaging virtual events like online exhibitions giving everyone a much more refreshing virtual experience unlike the kind of stereotypic presentations possible through Zoom® or Google Meet®.

M. M. Movania (✉) · A. Samad · S. S. Raza
Habib University, Karachi, Pakistan
e-mail: mobeen.movania@sse.habib.edu.pk

A. Samad
e-mail: abdul.samad@sse.habib.edu.pk

S. S. Raza
e-mail: saleha.raza@sse.habib.edu.pk

Y. Cai et al. (eds.), *Mixed Reality for Education*, Gaming Media and Social Effects,
https://doi.org/10.1007/978-981-99-4958-8_7

Keywords Design thinking · Virtual spaces · Online exhibitions · Virtual experience

1 Introduction

The Covid-19 pandemic has hit hard on everyone irrespective of their geographic position, political or religious beliefs or areas of work. One major blow that Covid-19 did is on the primary, secondary and tertiary level education. Normal modes of engagement are replaced with severely restricted alternatives causing pain and anguish to many students and their families. Everyone had to suffer be it faculty or students or their parents. There is a lot of experimentation at every level. Faculty on one hand are trying to create innovative alternative assessments or completely skipping them. Such adhoc innovations are resulting in more frustration for students and their families who are used to of conventional modes of assessments. This compromise on education will have serious consequences [1].

Being a student centric university, our prime focus has always been on two aspects: student success through engagement and quality content delivery. Like many of our fellow faculty members elsewhere, we try to make our course contents innovative. Sometimes it is through engaging content, which is a mix of in class activities and novel assessments. These are both refreshing and engaging for students. While such innovations work well for most conventional courses, they are difficult to do for a yearlong course like the final year project (also known as the capstone project). Students pick a local or a global problem to solve under supervision of a faculty member as their final year project supervisor. The projects are evaluated multiple times through mid and final presentations and a final year projects exhibition. In these evaluations, the projects are evaluated by both internal and external experts. Based on the mid, final presentations and the project exhibition scores, an aggregate score is generated for the course.

The main reason why such a course is difficult to conduct online is its tight coupling to strict deadlines and the need to continuously monitor student progress via regular assessments. Many of these assessments are carried out by local and external evaluators who judge these projects rigorously. At the end of the year, a final year project exhibition is carried out where the students present their final year projects. Evaluators then visit the capstone project stalls based on their interests and evaluate students by asking questions.

Due to Covid-19, all forms of physical engagement were restricted so we were unable to host the final year project exhibition on campus in the university's design thinking and exhibition space. We initially thought of doing the exhibition online via zoom meetings where students would present their project presentation and videos online. External evaluators would then join in and then evaluate their assigned projects. After watching student project presentations and asking questions and answers (QnA), they would give scores to each project. We felt that this would not be engaging enough for students because,

- Students are already exposed to Zoom® for all other courses.
- Zoom® fatigue is inevitable.
- One-to-one QnA cannot be replicated on Zoom® as we do conventionally in person.
- It is severely restrictive for projects with hardware components, the best students can do is capture videos and share.
- Students lack the freedom to roam around and interact with other projects as they do in in-person final year project exhibitions.

The above restrictions forced us to think out of the box and look out for alternate innovative means of engagement online. We discovered about a promising new tool for online exhibitions and conferences, the Mozilla® Hubs®. We began exploring this new innovative platform and after studying about the platform, we found out that this could very well be a good alternate for hosting final year project exhibitions. The following are some benefits provided by the Mozilla® Hubs® platform,

- Possibility of live interaction with up to 50 participants in one room in the free mode.
- Supports all modern internet browsers.
- Supports all desktop and mobile platforms including smartphones and tablets.
- Supports all AR and VR devices.
- No setup involved, you simply connect through a URL.
- A large library of free assets available (including audio, backgrounds, 3D models etc.).
- Support of world editing through a free online editor (Spoke®).
- Import and export from all major 3D modeling software through the glTF model format.

In addition to the above mentioned benefits, there were some other advantages like streaming of YouTube® videos, ability to roam around in the 3D exhibition space with their own virtual avatars that provided location hints in the virtual space. Distance based sound attenuation enabled participants to talk with each other as if they were meeting physically allowing students and evaluators to cross question during the exhibition. These merits forced us to use the Mozilla® Hubs® platform for the first virtual final year project exhibition. In this chapter, we detail about how we used the Mozilla® Hubs® platform along with a 3D modeling software (3DSMax®) and the Spoke® editor to recreate the virtual replica of the Habib University (HU) Playground for the first virtual final year project exhibition online.

While the Mozilla® Hubs® platform had been used in many virtual exhibitions targeting different domains for e.g. Arts Festival [2], virtual tours for students [3], virtual museums [4] and waste awareness for children [5], to the best of our knowledge it has not been used for final year project exhibition. We elaborate how we developed the exhibition space along with the several design decisions that we made to create a smooth virtual experience for our students, visitors and evaluators. At the end of the virtual final year project, a user study was carried out to investigate the effectiveness of the online tool. We are happy to share that the virtual final year project exhibition was successful in achieving its goals. The details of this user study are also shared in a later section.

The rest of the chapter is organized as follows. Section 2 details the system analysis and design while Sect. 3 details the virtual exhibition, interactivity and user study. Finally, Sect. 4 concludes this chapter.

2 System Analysis and Design

Before working on the first virtual final year project exhibition online, we first tried to analyze the usual plan of activities in the final year project. We then gathered all requirements and identified three main users for this project including their roles:

- Student (To present their project to the project evaluators.)
- Moderator (To ensure the smooth conduct of final year project activities.)
- Evaluator (To evaluate the final year project/s assigned to them and grade students.)

A student may roam around the final year project exhibition hall and interact with props like posters and LED screens. An evaluator may roam around the environment to visit one or more final year projects. He may also interact with props like posters and LED screens. A moderator is a power user who may admit or remove visitors (including students or evaluators). The overall system flow of events is presented in Fig. 1 and a use case diagram highlighting the required uses cases is given in Fig. 2.

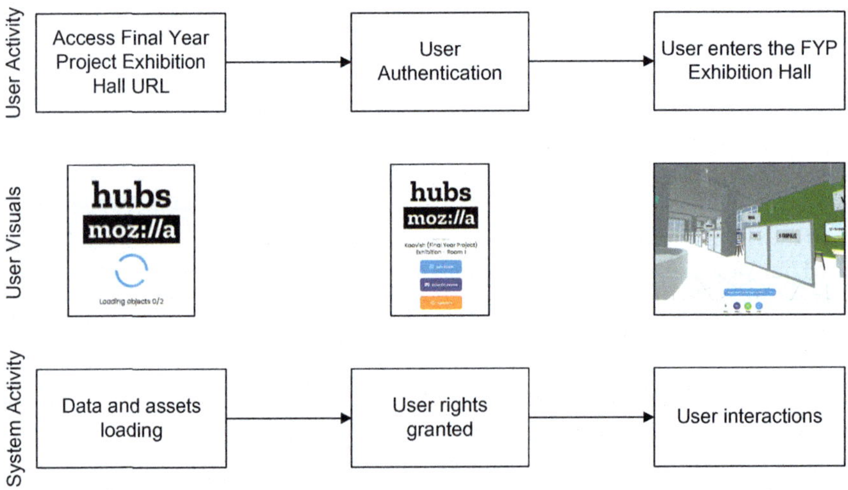

Fig. 1 The flow of events showing the user activities (top row), followed by the visuals user sees (middle row) and the back end processing of the system (bottom row)

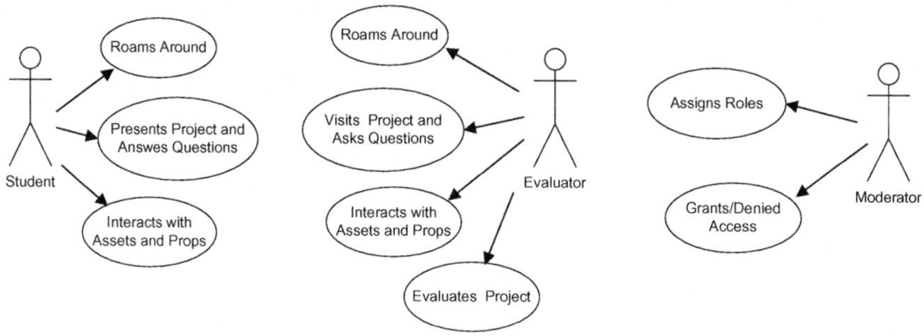

Fig. 2 The three users and their use cases

As can be seen in Fig. 1, a URL to access the virtual final year exhibition room is shared with all users. Power users (moderators) required login so that privileged options may be shared with them. This was to ensure that moderators are able to control options accessible to the students and evaluators. They may also ensure discipline by removing misbehaving students or guests.

2.1 Educational and Design Criteria

Before moving with the design of the virtual final year project exhibition, we had to first identify the educational and design objectives of the application. The educational goals focus on the ability that will help students enhance their learning. The following are the educational goals of the virtual final year project exhibition.

- To allow students to present their final year projects online.
- To allow smoother online evaluation of the final year projects.
- To allow interaction of students and evaluators in a free 3D environment so that they may interact and discuss ideas.

While the educational goals focus on the student learning, the design goals focus on the facilities the application would provide to achieve the educational goals. The following are the design goals of the virtual final year project exhibition.

- To provide smooth presentation of final year projects online.
- To provide hassle free online evaluation of the final year projects.
- To provide space and props for smooth interaction of students and evaluators.

The design ideas presented here have been curated to meet the final year project requirements. But the same could be extended to other courses in the undergraduate curriculum. In the current Covid era, more emphasis is now put on tools for effective student engagement. We believe that the Mozilla® Hubs® platform is a good candidate which has the required tools to do this job effectively. In addition, due to its ease of use, it can be introduced to the students early on in the course. The students may then be asked to use their imagination and generate custom assessment modules by experimentation to fulfill the outcomes of the course.

The effectiveness of these criteria depends on the interactivity and immersiveness of the user experience. This is dependent on what user interactivity options are available and how well the virtual environment is modeled. The latter aspect requires strong 3D modeling and interior design aesthetic sense but fortunately a lot of built-in tools are provided that made this work thoroughly professional and hassle free.

2.2 3D Modeling of the HU Playground

In order to ensure that the students have a similar feeling to the final year project exhibition space, we modeled a 3D replica of the HU playground. The modeling of the whole environment was done in 3DSMax® by using reference photos of the HU playground. All objects were built to scale by using polygonal modeling techniques. Since most objects in the playground were cuboid, the box primitive was first created of the dimension of the object in consideration. The generated cuboid was then edited using the editable mesh modifier to reshape the object to the desired shape and form. The modeling was done in metric units. This was to give a realistic walk through and interactivity experience to the user. The snapshot of the final HU playground 3D model is given in Fig. 3.

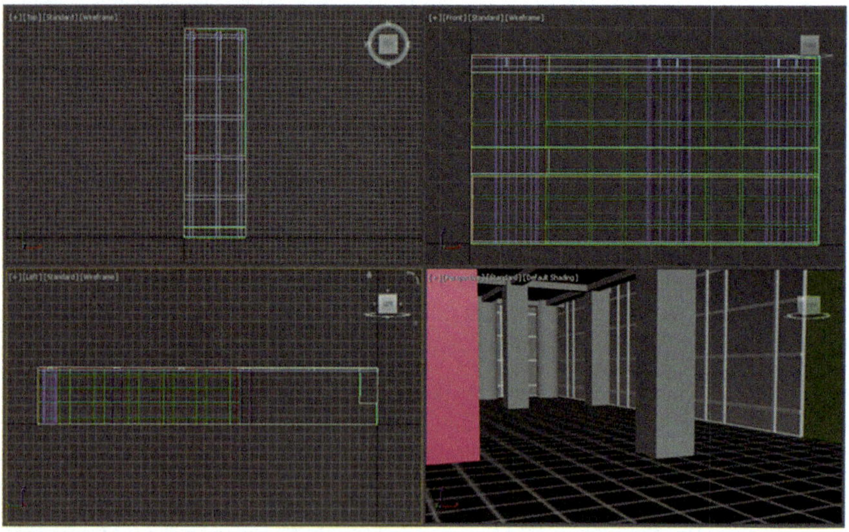

Fig. 3 The top, front, left and perspective view of the 3D model of the HU playground

2.3 *Modeling of Props and Other Assets*

Apart from the HU playground exhibition space model, we required many props which included posters, white boards, title boards, reception desk, chairs, sofa set, rugs, tables, stools, LED TV screens etc. All of these models were built or edited for our needs in-house. Some of the models like whiteboard, chair, sofa set were imported from Sketchfab® but due to the high texture detail and high polygon counts, these models were optimized to suit our need (see Table 1 for model statistics and Table 2 for total savings we obtained by model and texture compression in percentage). We utilized the built-in mesh optimizer modifier in 3DSMax®. This tool provides a quality criterion that preserves sharp edges and corners while optimizing the mesh. Once the desired level of mesh complexity was achieved, the model was saved.

Wherever possible, we replaced textures with colored polygons. Our target was to support all current consumer class devices including smartphones and tablets therefore we had reduced the available texture memory and polygon count budget substantially.

If we went with the same high detailed models with high quality textures, the initial load time and interactivity of the online exhibition were severely hampered. This was evident to us during the trial runs that were carried out initially with the same scene but with highly detailed models with large textures and materials.

Table 1 Statistics of various 3D models

Model	Original			Compressed		
	Vertices	Polygons	Texture memory (MB)	Vertices	Polygons	Texture memory (MB)
Whiteboard	44	68	6	24	22	0
Stool	322	488	8	224	149	0
Table	416	250	8	221	127	4
Bean Bag	460	266	5	460	241	2
White Sofa	2628	1404	6	440	876	0
LED TV	78	26	0	Already compressed		
Red Sofa	82	158	0.505	Already compressed		

Table 2 Savings in percentage

Vertices	Polygons	Texture memory (MB)
66.3%	46.3%	78.2%

2.4 Exporting GlTF Models from 3DSMax®

All 3D models were exported into the glTF binary format (*.glb) [6] using the Babylon.js free glTF exporter for 3DSMax® [8]. This provided additional space savings through compression. Since we had significantly reduced the texture memory required, we could embed textures in the 3D glTF models. This allowed our 3D models to remain portable from one scene to another once the model was uploaded to the Spoke® editor.

2.5 Setup of the Virtual Final Year Exhibition in the Spoke® Editor

After all required assets including the 3D HU playground environment model, the props and audio assets were ready, we imported them into the Spoke® editor (see Fig. 4). Since all our models were built to scale, we were able to place them at exact locations in the environment. In order to provide each final year project group a separate space, we created 5 shelved rooms using white boards. Detailed exhibition floor plan is shown in Fig. 5.

The process of poster placements and LED screen setup took significant time. All final year project groups were asked to prepare a 10 min introductory video and share the YouTube® link with us. These were linked to the LED TV screen through the ability of Spoke® editor to apply YouTube® videos as textures. The users are also provided with the ability to mute or adjust volume of the video by clicking on the LED TV screen (see Fig. 6). The top down view of the floor with all props included is shown in Fig. 7 and a 3D perspective view is given in Fig. 8.

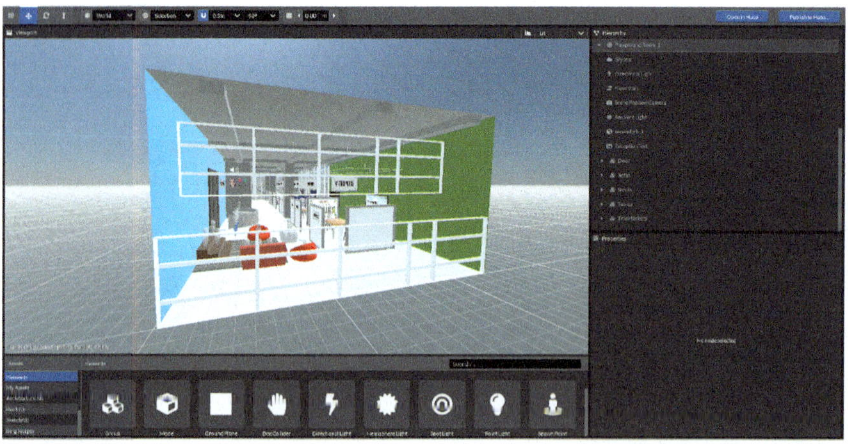

Fig. 4 The Spoke® editor showing the 3D exhibition model and various props

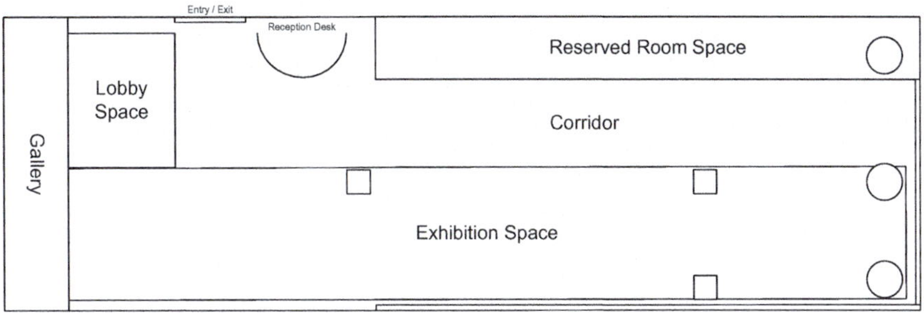

Fig. 5 The 3D virtual final year project exhibition floor plan

Fig. 6 The user controls on the LED screen to control the final year project presentation sound level with decrease volume (- button) and increase volume (+ button) in top row. The video playback buttons are in bottom row showing rewind/pause/forward buttons from left to right. The video current playback and total playback times are visible in the bottom most row

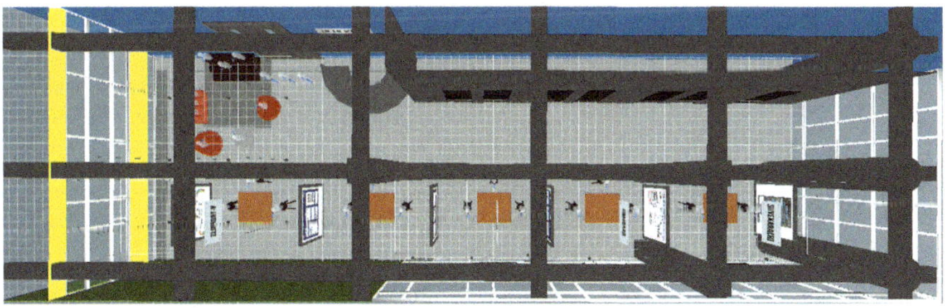

Fig. 7 The 3D virtual final year project top down view showing all props and exhibition spaces

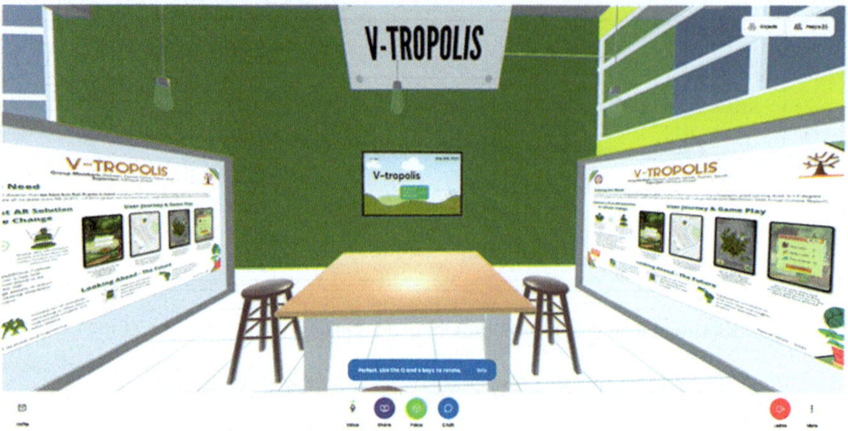

Fig. 8 The 3D perspective view from one of the project exhibition space showing the layout of posters, evaluator table and seating as well as the LED TV screen displaying the project video

2.6 *Publishing the Spoke® Project to Mozilla® Hubs®*

Once the placement of all props was finalized, the rooms were published to host the virtual final year project exhibition. One really nice option that is provided at the time of publishing by the Spoke® editor is the performance check feature. It gives you an overall budget estimate for your project. The Spoke® editor checks all 3D assets, lights, meshes, materials and textures in the scene to ensure that they do not use more resources than what would be available on an average user's platform. This tool separates the required assets into three categories: Low (if the consumption is low), Moderate (if the consumption is on the borderline) and High (if the consumption exceeds the recommended level for an average users platform). This gives a rough idea of the scene complexity which can then be lowered or enhanced as desired. The performance check for one of our rooms is given in Fig. 9.

As can be seen, apart from the 56 unique materials which is clearly higher than the recommended 25 unique materials limit, we were able to reduce the overall 3D mesh and texture memory requirement to keep it at the low level. The large set of unique materials we required were due to the use of multiple unique props. One possible way of reducing the material requirement is to either replace constant color textures with per-vertex colored meshes. Another possibility is to use material atlas to combine multiple materials into a single material. Perhaps we may also introduce material sharing to group similar assets to share a common material. This may help in reducing the overall unique materials budget. Due to time constraints, we were unable to try these experiments but these are some possible avenues of further development.

Fig. 9 The performance check performed by the Spoke® editor on one of our exhibition rooms at the time of publishing of the scene to the Mozilla® Hubs® platform. This option provides a rough estimate of the required memory, lights, materials, textures and polygon budget for the published scene

3 Virtual Exhibition, Interactivity and User Study

In total, we had 15 final year project groups each with 3 student members. The exhibition space was only able to accommodate 5 projects at max. We limited the number of projects to 5 based on our observation that because the exhibition spaces were adjacent, interferences might be introduced when project presentations are run in parallel. So we decided to create three replica rooms following the same floor plan but with a different set of final year projects. This also helped us in catering for the hard limit of 50 visitors posted on the free version of the Mozilla® Hubs® platform. The projects were distributed and so was the audience. This also helped in load balancing as the participants (students and evaluators) were uniformly distributed in the three rooms.

Since most of our users had not used the Mozilla® Hubs® platform before, we prepared and shared a set of video tutorials with all participants to make them comfortable with the new virtual environment. The tutorials helped in getting everyone comfortable with the basic navigation and other controls. The virtual final year project exhibition was attended by around 50 visitors in total including internal/external evaluators and students. The devices on which the virtual exhibition was

Fig. 10 The screen capture from the live virtual final year project exhibition

run concurrently included smartphones, tablets, laptops and desktops. The visitors were distributed in three rooms. The external evaluators were assigned projects based on their area of expertise.

The virtual final year project exhibition room URLs were shared and all attendants were asked to join their respective rooms. Moderators and student assistant were also assigned to ensure smooth execution of the online exhibition. Upon joining, the participants were asked to select a virtual 3D avatar. The whole activity ran for 4 hours. A snapshot from the event is shared in Fig. 10.

The power user's view has a list of participants on the right hand side (see Fig. 10). The moderator names have a red star as a suffix so they can be identified easily by other moderators and power users. Moreover, several buttons are provided to moderators and power users. These are visible in the bottom row (see Fig. 10). These options allow moderators to share their screen or camera, place props dynamically, react (to display a reaction for e.g. applause or happy) and chat (to broadcast messages to all users in the exhibition).

The options available to a guest or student are only limited to chat and react. The reason we restricted the available options to guests or students was to avoid them spoil the generated 3D environment with dynamic content. In special cases when a student wanted to use some extra option (like sharing of screen), a request was submitted to the room moderator. The moderator would then allow the student and then the desired option was available on the student's view instantly.

The final year project students showcased their projects and explained the project inner details through the project posters displayed in their exhibition spaces. Along the way, they also answered questions asked by their evaluators. They presented a short video demo of their project in the LED screens. Some even went on to explain the inner details of their final year project by using the additional features of screen share provided by Mozilla® Hubs®. This helped evaluators in better understanding the students' efforts. At the end of the first virtual final year exhibition, the evaluators were asked to fill the evaluation forms on the basis of which the scores of students were recorded.

3.1 Interactivity and Immersion

The user could navigate in the 3D environment by using the standard first person shooter controls i.e. (keyboard WASD keys to move around and the mouse left click and drag to look around). Right click and drag in any free space (which could accommodate the user) created a teleporting option (see the green curve and ellipse mark in Fig. 11. This allowed a user to quickly hop to the desired location in the 3D space.

One key feature of virtual reality is immersion. To allow users to have an immersive experience, the exhibition space was populated with several props including interactive LED screen (see the 3D perspective view given in Fig. 8). The user is able to skip forward, pause, rewind, or mute playback of videos as desired (see Fig. 6). In addition, a special mode is enabled that creates a clone of the LED screen video near the user's avatar (see Fig. 12). This way the user may watch the video presentation while roaming around. This also helps when an evaluator wants to get a better understanding of some specific aspect of the project presentation.

The sound controls (for example the scene background music, the sound of the LED screen as well as a user's audio) could be muted by moderators by clicking on the available button. This allowed moderators to fully control the sound levels in the virtual exhibition environment. The lobby area had some couches and sofas to allow

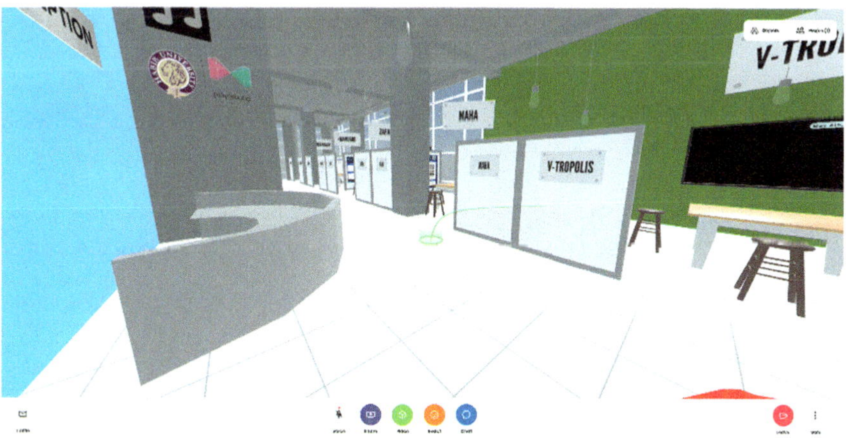

Fig. 11 Right clicking and dragging in any free space allows a user to teleport instantly to the shown location (visible as green ellipse and curve)

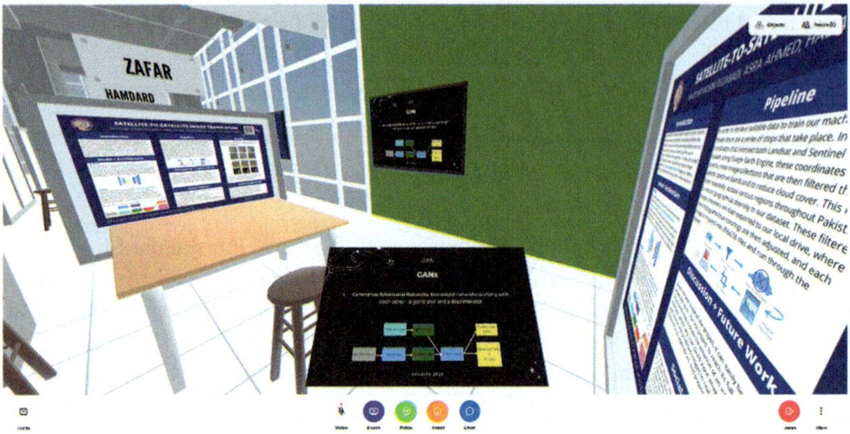

Fig. 12 Clicking on a special props button on the LED screen makes a clone of the LED screen video in front of the user avatar. This comes in handy when some evaluator wants to get some clarity on a specific part of the project presentation

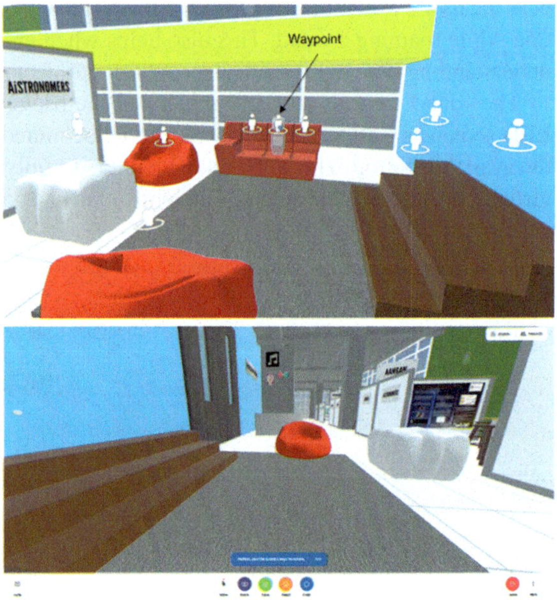

Fig. 13 The lobby area showing waypoint icons on sofa sets (top row) and the view from the shown waypoint when left clicked on it (bottom row)

users to sit and participate in informal discussions. The sitting was enabled through waypoints which were activated by pressing the space bar key and left clicking on the virtual icon. This teleported the user to the waypoint position and oriented him in the waypoint direction as shown in Fig. 13. This feature was really useful in allowing evaluators to sit on the provided stools to watch the project presentations together

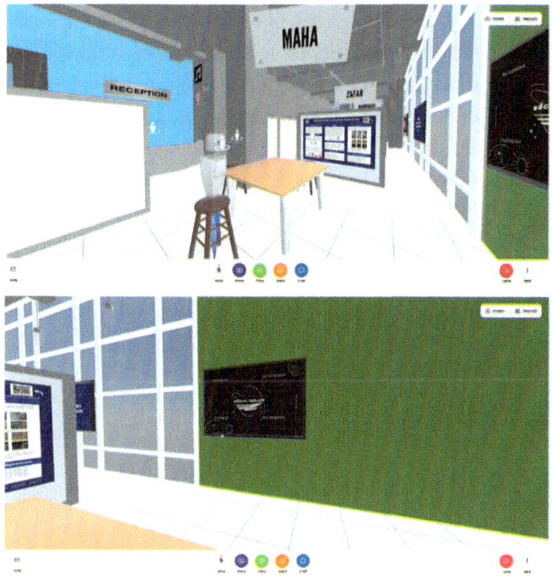

Fig. 14 The project space showing the stools and the discussion table. Pressing the space bar key showed the waypoint positioned on the stools oriented in the direction of the LED screens (top row). Left clicking on the waypoint icon quickly oriented and placed the evaluators on the stool/s to watch the final year project presentation video (bottom row)

and do discussions on the discussion table as shown in Fig. 14. All these features provided a more interactive and engaging experience.

3.2 Scalability

While the pilot study was carried out for final year project evaluations, the ideas learned can very well be applied in other courses in freshman, sophomore and junior levels. For instance, courses that require more interactive discussions between faculty and students or courses that involve design thinking can very well benefit from the expressive freedom provided by the Mozilla® Hubs® platform. Moreover, students may be asked to redesign their course assessments using their expression of ideas through these newer tools. Ideas from design thinking may very well fit here in providing a more engaging student learning experience.

Generating content for this new environment requires good 3D modeling skills and understanding of interior design aesthetics. The Mozilla® Hubs® platform and its provided Spoke® editors are relatively easy for any beginner to work with since they mirror the standard placement options (like translate, scale and rotate) available in all 3D modeling software. In addition, several types of dynamic lights are provided including directional and point lights. Moreover, a big library of readymade assets is also provided which includes sounds and pre-built meshes. Several development kits (like the architecture kit) are also available. These provide a large set of light defaults, sounds and meshes that come in handy for architectural scenes development. Interested readers are encouraged to try the Spoke® editor.

Thanks to these great platform independent tools, it has now become relatively easy for even beginners to generate impressive visuals by just using the built-in tools. In addition, the one click performance analysis and direct publishing of the developed scenes allows the designed scenes to be run on any recent consumer device like smartphone, tablet, laptop or desktop. The only requirement is the availability of a good internet connection and a modern web browser with WebGL support. Such a write once run anywhere capability ensures rapid prototyping and faster experimentation of ideas. Before WebGL, such support required a significant development effort to not only develop the required application but also ensure that it is runnable on all devices through special plugins that required separate installations.

3.3 User Study

One thing that we realized since the beginning was the need to gather all kind of feedbacks from all parties concerned. Therefore, post event, a questionnaire was shared with a subset of attendees asking questions about the utility of the new virtual online final year project exhibition and their overall recommendation. In order to remove any biases, the names of the survey respondents as well as their roles in the final year project exhibition were anonymized. Around 13 respondents observations were recorded. The results of the user study are detailed in Fig. 15.

As can be seen, the new virtual online final year project was liked by 79% of the respondents and around 57% suggested that they would recommend this new way of evaluation. In order to analyze why the recommendations were low, we investigated within the questionnaire about their reason of non-recommendation. We came to know that the user's overall virtual final year project experience is dependent on how fluid the internet connectivity is and how performant the user's platform is. Many of the respondents, their group mates or their evaluators faced connectivity issues which resulted in glitchy responses which caused a low engagement level. This was the main concern which could only be rectified if all parties concerned have a strong internet connectivity. This is something which could be enforced by providing dedicated internet connectivity to all parties concerned.

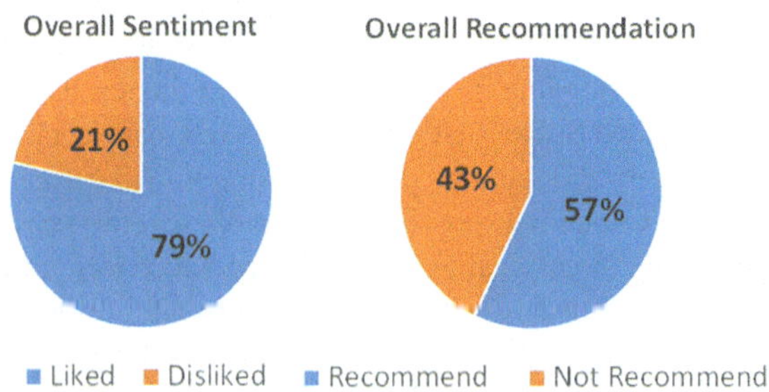

Fig. 15 The results of the user study showing the overall sentiment (left plot) and the overall recommendation (right plot)

3.4 *Attainment of the Design and Educational Goals*

We discussed about three design and educational goals in Sect. 2.1. The design goals were achieved effectively thanks to the versatility of the provided tools. The first educational goal was focused on allowing students to present their final year projects online. We believe this criteria was successfully achieved through our online exhibition. Each student group was given a specified space where their posters and presentation screens were setup.

The second goal was to make the evaluation process smoother. Due to the provided tools in the online exhibition, evaluations were a breeze. Evaluators could clone student presentation near their avatars to seek more clarity about the project or presentation. They could also ask questions if they had any during presentation. The third goal was to allow free interaction of students and evaluators for discussions. We believe the online exhibition design fostered more immersion and engagement between students and evaluators. In addition to dedicated space for each project, a lobby area was also setup to allow informal discussion between students and evaluators to allow more engagement and immersion.

4 Conclusion and Future Work

We have presented the first virtual final year project exhibition that we carried out at our university using the Mozilla® Hubs® platform. Thanks to the flexibility provided by this platform including easy to use 3D editor (Spoke®) and a large library of prebuilt assets, we were able to provide engaging final year project exhibition experience to our participants. There were some technical issues related to internet connectivity that provided glitchy experience to some users but overall around 80% of our participants liked the new way of presenting final year project exhibitions.

Our virtual final year project exhibition used the free version of Mozilla® Hubs® therefore we had limited network bandwidth to access server resources. Moreover, there was a hard limit of 50 participants at max per room in the free version. The Mozilla® Hubs® recommended number of participants in a room is 25. We noticed that the user experience starts to deteriorate at more than 20 participants in one room. Of course there may be several reasons for this which could be internet connectivity or limited bandwidth issues for some participants but a thorough investigation could possibly be carried out in future to find these limits.

The developed rooms required 56 unique materials to support the varied types of props in the exhibition space. We feel that we may reduce the total unique materials requirement by either replacing constant color textures with per-vertex colored meshes or using material atlas to combine multiple materials into a single material. Perhaps we may also introduce material sharing to group similar assets to share a common material. These optimization may help in not only improving the overall experience of the user but also help in making the exhibition space scalable to support a large number of participants. This requires more experimentation and further investigation.

Designing of such virtual exhibits requires a large amount of effort especially to generate 3D models and assets to be consumed in the virtual exhibit. This requires significant 3D modeling and design experience. It always requires knowledge of interior design aesthetics to ensure good placement of props and assets within the virtual exhibition space. In addition, understanding of lighting and shading can help set the mood of the environment. All these enhancements need significant budget (in both time and effort) and this becomes the deciding factor in making a virtual exhibit immersive and engaging for its audience. It will be worthwhile to explore if any automated process could be devised to generate and spawn objects automatically or semi-automatically by limited user intervention.

It will be interesting to extend the ideas learned from this experiment to other courses especially those that have a lab component in them. Such courses could benefit from the large array of built-in assets provided by the Mozilla® Hubs® platform and the Spoke® editor. Students could be guided to develop their own assessments to ensure the course outcomes are met as required. During the process, the students should be allowed to experiment as much as they want. This will certainly help in increasing student interests and enhance immersion and engagement in course contents. This is something that is severely hampered during the Covid' 19 pandemic. Some experimental study could be carried out on other courses which remains to be investigated.

Assuming that the users have a robust internet connectivity, there are several avenues that we would like to explore. The Mozilla® Hubs® platform has built-in support for all modern VR devices. It will be interesting to see how we may make the users experiences more engaging through these VR devices like the Oculus Quest2®/ HTC® Vive®. This could help in providing more immersive and engaging VR experiences. In addition, we are also looking at how we may use the NVIDIA® Omniverse™ [7] and similar cloud based infrastructure for collaborative experiences. This may probably be our future research direction.

References

1. Burgess, S., Sievertsen, H.H.: Schools, skills, and learning: The impact of COVID-19 on education, VOXEU CEPR (2020). https://voxeu.org/article/impact-covid-19-education. Accessed 14 August 2022
2. Brown, C., Scoones, R., Gibson, R., Martelli, B.: WAHAWAEWAO [We Are Here And We Are EveryWhere At Once] VR. [Videography] (2020). Ars Electronica Festival 2020. https://minerva-access.unimelb.edu.au/handle/11343/247760. Accessed 14 August 2022
3. Iglesias, M.I., Jenkins, M., Morison, G.: Enhanced low-cost web-based virtual tour experience for prospective students 2021. IEEE Conference on Virtual Reality and 3D User Interfaces Abstracts and Workshops (VRW), pp. 677–678 (2021). https://doi.org/10.1109/VRW52623.2021.00221
4. Dasgupta, A., Williams, S., Nelson, G., Manuel, M., Dasgupta, S., Gračanin, D.: Redefining the digital paradigm for virtual museums. In: Rauterberg, M. (eds) Culture and Computing.

Interactive Cultural Heritage and Arts. HCII 2021. Lecture Notes in Computer Science, vol 12794. Springer, Cham (2021). https://doi.org/10.1007/978-3-030-77411-0_23
5. Abioso, W.S., Imam, A.I., Maulana, Y.I., Prasetyo, M.J.: Virtual reality utilization as a character–building of children in waste problems. IOP Conference Series: Materials Science Engineering, Vol. 1158(1), 012010, IOP Publishing (2021). https://doi.org/10.1088/1757-899x/1158/1/01201
6. Bhatia, S., Cozzi, P., Knyazev, A., Parisi, T.: The glTF 2.0 Specifications, The Khronos Group Inc (2021). https://github.com/KhronosGroup/glTF/tree/master/specification/2.0. Accessed 14 August 202
7. The NVIDIA Omniverse, NVIDIA Inc. (2022). https://developer.nvidia.com/nvidia-omniverse platform. Accessed 14 August 2022
8. Babylon.js glTF exporter for 3DSMax (2022). https://doc.babylonjs.com/extensions/Exporters/3DSMax_to_glTF . Accessed 14 August 2022

VRLE-Based Teaching for Medical Students: VR-Baby

Grace Ryan, John Murphy, Fionnuala McAuliffe, and Eleni Mangina

Abstract Educational technology plays an important role in supporting mainstream medical education. Virtual reality (VR) is a potentially beneficial learning tool for medical student education, that can enhance the understanding of complex topics in Obstetrics and Gynaecology Education. The objective of this study is to describe the development of three-dimensional (3D) content for a Virtual Reality Learning Environment (VRLE) to teach medical students within specific pedagogical outcomes. The development of the 3D models and the advancement to 3D learning objects within the VR environment is described and the association for further research with the xAPI (Experience API) allowing the learning content and VR learning platform to track the learning experiences. the VR learning solution has been employed and tested for training of medical students based on cost/benefit analysis for the particular application and how it can be used in the future for medical students' education.

1 Introduction

This study explores the use of virtual reality technologies as an alternative to traditional learning methods in understanding the first three trimesters of pregnancy, during teaching of obstetrics and gynaecology. By taking ultrasound images of a fetus provided by the National Maternity Hospital in Dublin, Ireland, the project aimed to create reusable educational accurate models and place the fetus in a 3D virtual environment to assist medical students in their learning. The study evaluated the utility of VR education for medical students in their studies via two randomized controlled trials.

Traditionally medical students would study the development of the fetus using textbooks, lectures, bedside tutorials, and online teaching resources. The use of VR

G. Ryan · J. Murphy · F. McAuliffe
School of Medicine, University College Dublin, Dublin, Ireland
e-mail: grace.ryan@ucd.ie

J. Murphy · F. McAuliffe · E. Mangina (✉)
School of Computer Science, University College Dublin, Dublin, Ireland
e-mail: eleni.mangina@ucd.ie

Y. Cai et al. (eds.), *Mixed Reality for Education*, Gaming Media and Social Effects,
https://doi.org/10.1007/978-981-99-4958-8_8

can create a more active, hands on, learning environment with less distractions. It can be adapted to teach a variety of learning methods such as theoretical, practical, and visual. By blending this new way of educating with traditional means, student can hope to achieve a higher retention of knowledge [1].

According to Sharpley et al. [2], lesson plans that accommodate technology tend towards more innovative and experimental forms of learning and teaching. The recent affordability of VR has now made it possible to view its utility as a legitimate educational tool. The barrier for entry into VR has never been lower. As the technology at a consumer level is relatively new this chapter denotes recent research in medical VR education and focuses on its advancements, pros and cons, and implementations.

The ultrasound images utilised for the case study presented in this chapter have been provided in DICOM file format from the experts and converted into the STL format so they can be used in a 3D VR environment. VR medical software is at the cutting edge of medical education and practice. Some of the applications available today will also be assessed. Ultrasound images themselves are quite noisy, and the implementation of ultrasound modelling for VR environments has been challenging. Most research of 3D anatomy modelling takes place using CT and MRI scans. For the purposes of the the teaching and learning activities within the lesson plans described in this case study, the 3D models developed, take advantage of Learning Object standards [3] with the aim of contributing to the research on 3D Learning Objects.

Section 2 describes the background work with a variety of methods researchers utilised to construct 3D models of anatomy. The problem domain in terms of lesson plans within the purposes of the pedagogical VR scenes are also discussed in Sect. 3 along with the data considerations for the case study. Section 4 details how the '3D Baby' virtual environment was implemented. Section 5 provides information of the study design and the learning experience evaluation. Section 6 discusses the findings of this study, although the focus of this chapter is to describe the virtual learning environment. Section 7 provides the conclusions of this work.

2 Background Work

Current research has focused on the effectiveness of VR for medical education. For example, Nicholson et al. [4], have constructed a fully interactive 3D model of the middle and inner ear and tested its educational usefulness in a randomized controlled study and tested 57 students using a web-based tutorial. 28 students completed the tutorial, which included the interactive model, while 29 completed it without it. By the end of each tutorial, each group was given a quiz to determine their knowledge of the 3D structure of the middle and inner ear. They concluded that the group who took the quiz had a mean score of 83% as opposed to those that didn't with a score of 65%. Although the model was not used with a VR headset, it did allow for the student to manipulate the model within a web browser. The significant statistical improvement ($P < 0.001$) showed that students can benefit from 3D-based interactive learning as a result, this engagement can lead to positive results.

Allcoat and von Mühlenen [5] explored the use of VR education by testing it against two other mediums, video and traditional textbook. The study assessed 99 first-year psychology students. The model that they were to examine was a plant cell, and be given a questionnaire on how the learning type affected them. 34 of the participants were tested using VR, another 34 were tested using video (in a VR environment), and the remaining 31 were tested on textbook. Of those who did the VR, they encountered a floating plant cell in the middle of the environment, which they were able to navigate, and interact with using controllers. To try and mitigate the effects of wearing a headset in relation to tainting the study, those who did the video test did it in a VR environment. Although they were in a VR environment, the same information as the VR test group was presented as a 2D video. While they were in VR, this group lacked the interactivity and immersion that a traditional VR environment would have. The remaining 31 participants used a traditional textbook method where they saw screenshots of the 3D model along with a PDF of the same information the other two groups received. Using pre and post questionnaire the VR group saw a significant increase in attitudes towards remembering and understanding. Both groups were significantly better than the one with the traditional textbook. The study noted this may have been due to some students having to learn how to use the VR headset and controller hampering their learning but there are many factors to consider. Although VR is becoming more ubiquitous, it still isn't fully mainstream yet. A more suitable study would allow time for users to become acquainted with VR before testing. Regardless, the VR-based intervention had a positive reaction and the authors noted it was likely due to its interactivity and immersion, rather than graphics or visuals.

Maresky et al. [6] assessed the use of VR for teaching cardiac anatomy with 42 students in total. 14 students were place in a control group. Both groups performed the same quiz with the same set of questions (5 cardiac, and 5 visual-spatial related questions). Both groups undertook standard cadaveric dissection training in relation to heart anatomy. However, the intervention group also underwent an immersive cardiac VR experience. The experience allowed the student to closely examine all parts of the heart in a non-destructive manner. The students who were in the intervention group performed better overall with a performance increase of 24.6% over the control group. The study also highlighted the re-usability of 3D models, which is not feasible with cadavers. A student may make a mistake in VR without worrying about ruining a human cadaver organ.

Augmented Reality (AR) enables users to interact with virtual objects in the real environment. Chien-Huan Chien et al. [8] noted that's its use would help students learn complex anatomy. Its advantageous to tradtional forms of teaching anatomy as it can represent complex 3D structures that can't be as accurately portrayed by a medical atlas. They concluded that it would help students to learn anatomy faster and better compared to traditional methods.

The role of VR to teach can be shown to be very beneficial to students and teachers. However, they're are some adverse effects. Its immersion can in some cases be its downside. It can trick the brain causing nausea and dizziness. Moro et al. [9] studied whether learning through virtual reality was as effective as tablet based learning. They took 59 participants and allocated them into 3 groups in terms of VR, AR, and tablet based learning. They completed quizzes on skull anatomy and their results were measured. While they found no significant difference in students learning outcomes, they did find that VR users were more likely to feel unwell with headaches, dizziness, and blurred vision. Hsin-Kai Wu et al. [10] researched the challenges currently facing the adoption of virtual reality in education. Aside from nausea they found that students may feel overwhelmed when using it. Those who were less adept at technology may suffer adverse learning outcomes as they would encounter multiple technological devices. This may be mitigated by prolonged use to VR while students get used to the technology [7]. However both reports did note that despite these adversities, they found largely positive responses from users even if they suffered adverse effects while using VR technology.

Another barrier for adaptation of VR in education could be the cost of the VR headset, which is redused over the years. Ray and Deb [7] studied Google cardboard's use in the classroom as a low cost means of VR in education. Students were split into two groups A and B. Those in A received teaching content provided by a projector and slides. Group B received the same teaching content, but presented in a 3D panoramic view with VR. After two months of testing students in both groups, it was shown that group B was performing significantly better. This showed that low cost of entry options into VR are viable routes for teaching. therefore, tools created for medical VR training could be utilised regardless of impact on costs.

2.1 3D Modelling of Anatomy

Researchers have adapted different methods to construct 3D models of anatomy. While some did not place the models into a VR headset environment, their methodology is of vital importance for good quality medical content development.

Mohankumar and Yie [11] presented a way of displaying 3D models of vertebrae suffering from scoliosis. CT and MRI scans were utilised to obtain DICOM data. Taking the DICOM data, they converted to an STL model using InVelasius software, which allowed them to edit mask, surface, and threshold properties accordingly. From there it was transferred to Magics software to remove noise and smooth the model. This method of 3D modelling creation may be useful to researchers and eliminate dependance on cadavers.

Mangina et al. [12] presented 3D models of cardiac defects placed in VR for medical teams to rehearse surgical procedures. To create a more realistic model of a heart model, 3 software tools, Blender, Netfabb, and AutoDesk Meshmixer were evaluated. The user would select sections of the heart so the mesh had to be separated into sections. Blender and Netfabb were cumbersome as they only allowed for planar cuts, cutting the model in straight lines. MeshMixer allowed to perform more intricate cuts and included more features. InVelasius and MeshLab were also tested for cleaning the mesh and noted their pros and cons. At the time of implementation Oculus SDK didn't provide headset tracking so that limited the cababilities of interacting with the 3D models.

Cross and Bauer [13] looked at different open-source software for the purpose of 3D modelling organs. They attempted to construct a 3D model of the heart and a left ventricular assist device. Taking data from CT scans in the form of DICOM and JPEG they aimed to "virtually fit" their model into a patients left ventricle. They used 3D slicer and its "Grow from Seeds" algorithm to perform auto and manual segmentation. Segmentation is the process of partitioning the meshes into something easier to manipulate and analyze. The 3D model was then exported into MeshMixer to be divided along its long axis. Finally it was imported to Blender to take advantage of its Boolean subtraction. They noted the steps they took could be transferred to other forms of human anatomy 3D models.

Khalil et al. described [14] to conversion of DICOM to a 3D object in Unity3D for a medical training system. 3D Slicer was utilised to extract the DICOM data and export it as an STL file. As Unity cannot interact with STL files, Blender was used to convert it into an object file that Unity could import. Once imported Unity's physics rules "Rigidbody" were applied to make it behave realistically. A "Sphere Collider" was also added to define the perimeter of the object so the users hand could detect where it is. Users were able to interact successfully with the 3D objects they had created.

The quality of the 3D content and the adaption of VR learning environments in education are highly depended on the technology available for the creation of accurate and interactive 3D educational content. The related software/technology for 3D content development includes the following:

- DICOM (Digital Imaging and Communication Software in Medicine) is the international standard that different manufacturers use to convey medical information. Its main aim is to facilitate communication of diagnostics, images, and associated data. It allows for DICOM interfaces to communicate with other DICOM equipment. Its use is now ubiquitous in medicine for transferring imagery. It can handle most forms of 3D medical data such as Ultrasound Scan (USS) data, Computerised Tomography scans (CTS), and Magnetic resonance imaging (MRI). For USS it supports image acquisition and processing, as well as providing any other relevant data such as patient name and condition. As there is a vast variety of medical tools and types of data acquisition in the medical field, the DICOM standard has facilitated open source software to present medical data [15].
- STL (Stereolithography) is a file format that describes the surface geometry of 3D objects as triangular meshes ordered by the right hand rule in a 3D Cartesian coordinate system. While it doesn't support colours or textures it is widely used for rapid-prototyping in 3D printing. The files are created by using a CAD (Computer-aided design) process as the final step in modelling 3D objects. They can be used for generating 3d models from DICOM data with the aid of medical software. Although STL files are generally used for 3D printing they provide the perfect basis on which to be constructed and refined for use in a VR environment [16].

- 3D Slicer is an open source software that allows users to process and view medical images in 3 dimensions. Developed in MIT in 1999 as a result of a Masters thesis, today its currently supported by engineers collaborating with algorithm designers and scientists. It supports DICOM data in MRI, CT, and USS, as well as other medical file formats. It supports plugins which can add algorithms and applications to its functionality. It provides auto image segmentation and some noise reducing abilities that can be used before trying more specialized software. Its capable of taking 3D dimensional ultrasound images and converting them to the STL format using the SlicerHeart extension. The easy to use GUI, and with great documentation from the online community made it a great choice for converting DICOM data to STL for the needs of thes study described in this chapter [17].
- MeshMixer is a free 3D sculpting program by AutoDesk. It allows the user to import STL files and offers an elegant UI, supports re-meshing, auto-fixing holes, creating face groups, allows for plane cuts, provides model smoothing. Mostly catering to 3D printing, its tools are ideal and catered for working with STL files. As its role is specialized its tools allow for rapid fixing and segmentation of meshes before they are imported into Unity or Blender. However as it is more catered to 3D printing it, its missing some features that may be necessary for the construction of the 3D models in this case study. If the models need further processing, Blender will be used after processing with MeshMixer [18].
- Blender is an open-source 3- = D modelling software released in 2002 that works on all major operating systems. Its capable of modelling, sculpting, and rendering meshes as well as a range of features but most importantly it allows for STL models to be imported. It allows for STL files to be converted to obj files necessary for Unity. Its large community means there's lots of guides for working with STL files and meshes. The large amount of tutorials for Blender make it a valid option for this project [19].
- Unity is a cross platform game engine created in 2005. It allows the user to create realistic virtual 3D environments. It is lightweight, fast and has an efficient UI experience making it easy to use for beginners. It exists as a free version with lots of professional tools for rapidly generating scenes. It takes advantage of prefabs, pre-configured game objects that reduce errors during development. It contains an asset store for various add-ons and facilitates the creation of virtual reality environments with the intention of using virtual reality headsets. It supports HTC Vive as well as Oculus Rift by using a base API that supports multiple devices [20].
- VRTK is an open-source framework that allows developers to quickly create interactivity within the virtual environment using a variety of scripts. Its primary use is for providing functionality with touching, grabbing, and using objects within a VR environment. Its hardware agnostic so it runs of a variety of different headsets and controllers. It provides a VR simulator allowing for development without a headset [21].
- BabysliceO is specialized standalone software that allows users to transfer USS images in DICOM, vol. and .mvl file formats, and then convert them to STL files for the purpose of 3D printing. Its powerful editor allows users to clean up areas

around the baby before exporting to STL. The paid-for software also allows users with VR headsets to view the STL file in a VR environment. It is offered as a free service in demo mode but does not allow the saving of any files. As this software only works with USS, it would not work with any MRI data that might be obtained during the course of this study [22].

3 Problem Domain: Lesson Plans

The lesson plans for the "VR Baby" case study, focused on student's relevant curriculum and the immersive qualities of virtual reality. User's walk around virtual environments discovering new models and new information. Each part of the lesson contained a wall and a model with appropriate information in the form of text or video. To make it more interactive, the lesson plans incorporated gamification. Via scripting in Unity, users would use the Oculus Quest controls to select models and annotated information would appear. In other scenarios, users would be asked written questions presented in 3D space. When users find new information or answer correctly they will hear a sound as well as receive visual feedback. Two virtual lessons were created and developed for Obstetrics and Gynaecology and Midwifery university level education.

With regard to the first lesson, it was developed from content in the Obstetrics and Gynaecology curriculum for medical students. The area of Obstetrics and Gynaecology has a unique need for immersive technologies to enhance learning and understanding. Medical students are required to complete a six-week module in Obstetrics and Gynaecology as part of their clinical rotations, during their final two-years of the undergraduate or graduate medical programme at University College Dublin. Our experience has lent us a valuable insight into the requirements of the students when they undertake their Obstetrics and Gynaecology rotation via formal and informal feedback. Students often struggle at the beginning of the placement with regard to trying to remember basic anatomy and embryology essential for understanding the development of the fetus during pregnancy, as they would have undertaken anatomy and embryology modules in their first two-years of the programme. With that in mind and as this was the first attempt at developing a resource tool for Obstetrics and Gynaecology education, a simple lesson on the topic of "Stages of fetal development during pregnancy" was developed. The intervention involved the development of a lesson in the form of an immersive virtual reality learning environment (VRLE), which was created to explore the stages of fetal development during pregnancy. The VRLE was designed based on documentation in the literature regarding VRLE development [23–26].

The interactive VRLE involved a "treasure hunt" for information, linking images to key learning points relating to fetal development. Using the controllers, participants could select internal organs on the fetal images being visualised in the HMD that were relevant to the stage of development (an example shown at Fig. 1). Doing so revealed

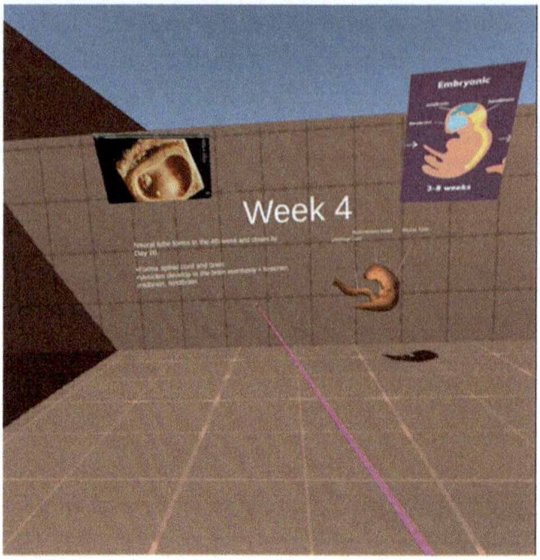

Fig. 1 Stages of fetal development—a VRLE scene at week 4 of gestation

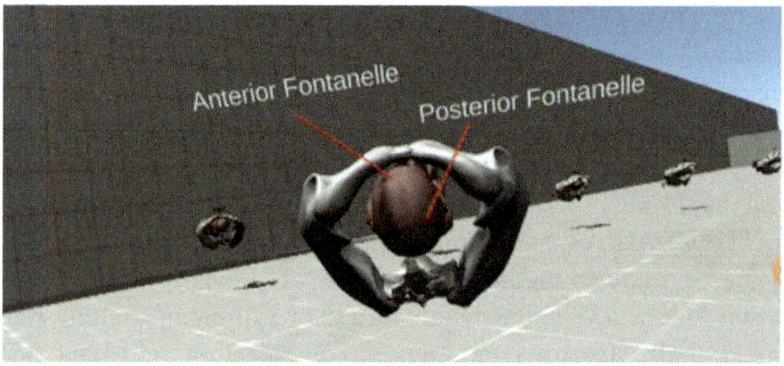

Fig. 2 VRLE scene for fetal head presentation, with the fontanelle highlighted

learning points necessary for knowledge assessment after the VRLE experience. The bespoke designed VRLE was given the name of "VR Baby".

The VRLE created for the Midwifery students was a review lesson on the topic of "Fetal lie, position, and presentation in the third trimester of pregnancy". The development process was similar to the first lesson as described above (an example shown at Fig. 2) the underlying value of VR technology for healthcare education is the ability for VR to create an environment for practicing basic clinical skills and to visualise 3D objects for enhanced understanding of complex topics. Examples of complex concepts related to pregnancy and midwifery include fetal anatomy and circulation, and the orientation of the fetus within the uterus and pelvis. These concepts present unique difficulties for medical and midwifery students' learning. Midwifery and nursing education has a unique need for VRLE's as an educational tool to enhance learning and understanding of the fetus during pregnancy and associated pelvic anatomy.

3.1 Data Considerations

The data compiled for use in this study included antenatal ultrasound images and magnetic resonance (MRI) images (Fig. 3). Informed patient consent was obtained prior to collection of the data. Patients were given a detailed patient information leaflet to read through prior to their scan appointment, consent was then obtained from the research study participants. They were given the patient information letter prior to commencement of the scan, allowing them time to read and understand the research project. On entry to the scan room, informed consent was either obtained or declined. The number of participants approached was recorded for data purposes. This was recorded in paper format and stored securely. A copy of the consent was electronically scanned into the patients online chart. The patient was given a hard copy of the information leaflet. The research participants were pseudo-anonymised for the study. The data was stored in a protected research area in the hospital which required electronic hospital card swipe access.

The DICOM data were collected and processed with software capable of turning it in an STL file (Fig. 4). Data were likely to be noisy and needed to be cleaned using software before exporting to a VR environment. The process of cleaning and refining the models continued until they're seen as being accurate enough to be used. Within the VR environment the user was able to walk around and view different parts of the model's anatomy. The models were presented in different conditions. The first 3 trimesters were represented as well as another model suffering from spina bifida in the 2nd trimester. Its inclusion is to highlight the difference between an abnormal and normal pregnancy.

4 3D Baby VRLE Implementation

VR Baby, comprises two learning experiences which incorporate key learning points within the obstetrics and gynaecology curriculum. Obstetrics and gynaecology speciality module is undertaken in the 5/6th year of undergraduate entry medicine and 3/4th year of graduate entry medicine. It is a 6 week module comprised of theory and practical learning.

The early stages of embryonic development is a core concept in the field of obstetrics. It builds the foundation of knowledge to help students understand the physiological process, the effects of disease, and their treatments. Students learn

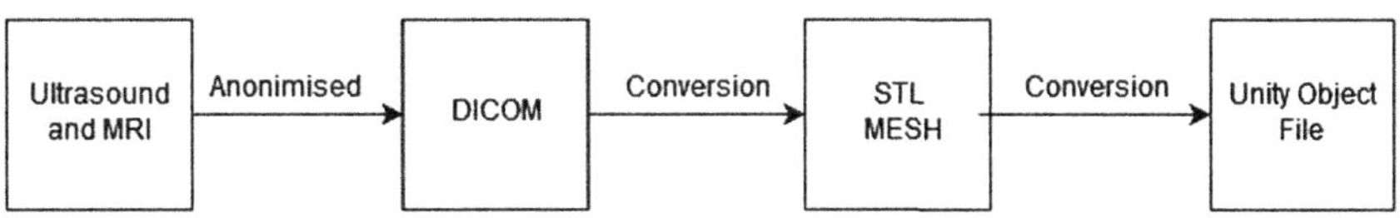

Fig. 3 VR baby 3D content format

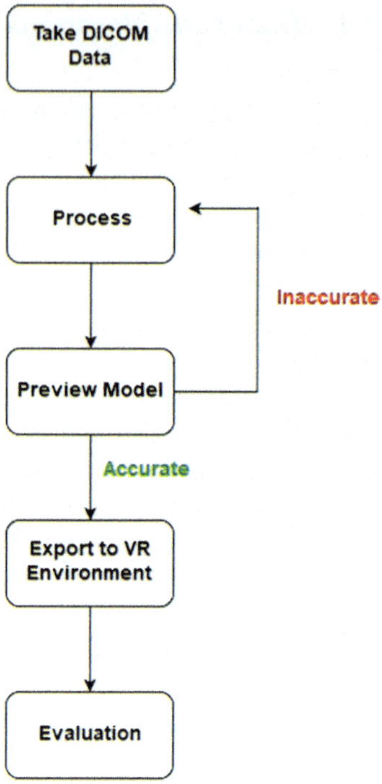

Fig. 4 VR baby: outline of approach

about tissues and early development as part of their anatomy education. When a student has completed a module in obstetrics and/or gynaecology the should have the following knowledge:

- Ability to demonstrate understanding of the female reproductive physiology from early embryonic life to birth.
- Understand the early development of the mammalian embryo according to key divisions, events and time course.
- Know where the basic tissue types originate from during embryonic development.
- Describe developmental disturbances that can arise during the early development of the mammalian embryo and present as malformations in the newborn child.

For this lesson plan 3D models were acquired that showed fetal development from week 4 to week 40 (Fig. 5). The models indicated the key aspects of fetal development that medical students should learn. In the various weeks of development, different life signs appear. These are often shown to students via ultrasound or MRI images, so to give students an immersive 3 dimensional experience of these life signs would greatly assist their learning. Images and video would be placed inside the virtual environment alongside the appropriate model, to give the learner something to relate to in the real world. Taking into account what student should know when taking an obstetrics module, the lesson plan with intended learning outcomes was created.

Intended learning outcomes for this lesson:

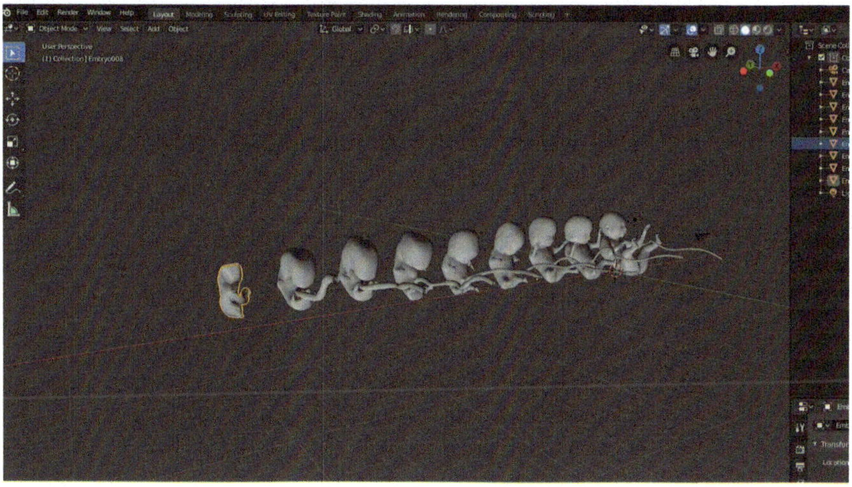

Fig. 5 VR baby: stages of development models in blender

- Know the division of pregnancy into 3 trimesters.
- Students should be able to understand the development of the fetus in each trimester, specifically the first trimester and organogenesis.
- Understand the key development points in the first trimester and what happens if neural tube doesn't close e.g. spina bifida.
- When the foetus starts to move, viability, and delivery.

To accomplish these learning outcomes, a linear learning experience was implemented. Students would start at a 4 week fetus and move through the virtual environment until they reach a fully grown fetus at 40 weeks. As the fetus grows, different life signs emerge, and complications can be detected. These would be shown to the user as they make their way through the environment. Some details would be hidden, and it would require interaction for the user to find them.

Fetal Lie/Position and Presentation in Pregnancy lesson plan also focused on the immersive learning experiences. During pregnancy, babies will move around, twisting, stretching, and tumbling in the womb. Before labour begins, they usually settle into a position that allows them to be delivered head first through the birth canal. This occurs in the third trimester of pregnancy, and sometimes complications can arise. This topic introduces medical students to the pathological aspects of reproduction and and their management. Students that study this topic should be able to:

- Know the physiological changes of pregnancy and puerperium.
- Know normal labour and parturition.
- Demonstrate satisfactory ability to perform clinical assessment in the abdominal and pelvic examination of non-pregnant women, and abdominal examination of pregnant women.
- Conduct of a normal vaginal delivery.

For this lesson plan we acquired a model of a pregnant women and fetus in the third trimester of pregnancy (Fig. 6). The model of the fetus was perfect as it contained a rigging armature. Not unlike an artist's manikin, this allowed the model to put into different positions and lies of pregnancy. Again, images of MRIs and ultrasounds would be placed inside the virtual environment that are relevant to the course work students would learn. The experts formulated the lesson plan with intended learning outcomes.

Intended learning outcomes for this lesson:

- Definitions and Differences of Fetal Lie/Presentation and Position
- Types of Lie
- Types of presentation
- Breech—types and management options
- Appreciate the risks associated with malpresentation
- Understand the Cephalo-pelvic relationship.

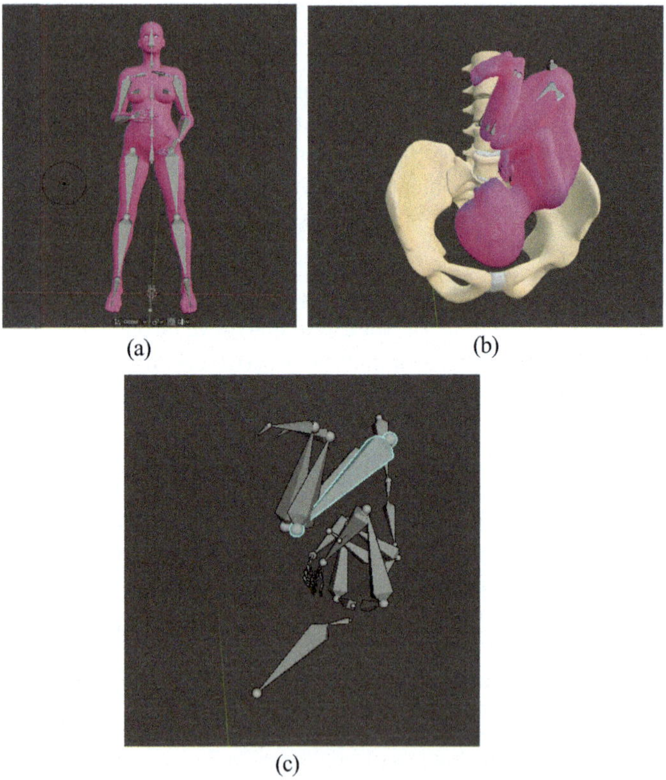

Fig. 6 Model used for Fetal Lies, showing the armature of the fetus that could be used to manipulate its position **a** Model of pregrant woman **b** Fetus position **c** Fetus Pose within Unity

The accomplish the learning outcomes, different types of presentations and lies were arranged in the virtual environment. Similar to the other lesson plan, the user would navigate sections of the environment, with each relating to a part of the learning outcomes.

4.1 3D Models

Implementing the lesson plans consisted of three stages. First pre-processing of the models ready in Blender, then importing into Unity. Unity set up followed up so that it would work on the Oculus Quest Headset. Finally, creating the virtual environment inside Unity and compiling towards the Oculus Quest. It was mostly the same procedure for both lesson plans, except for the Fetal Lies and Positions lesson plan that required more editing in Blender than the content for the Stages of Development.

4.1.1 3D Models

Prior to embedding the models into Unity, editing had to take place in Blender. The Oculus Quest being standalone, and not tethered to a computer, meant the models would need to be optimized for the platform. Initially, the vertex count on each model meant the Quest would come into difficulty when the user would view them. The frame rate would drop and the user would lose their immersive experience of virtual reality. To combat this, the decimate tool in Blender was utilised, which reduced the vertex count. By editing the ratio parameter, it would reduce the face count and the models were re-usable. For both sets of models needed to be exported into a file format Unity would accept. Blender allows for models to be exported in the FBX file format which Unity accepts. The textures for those were added within Unity environment. For the Lies and Presentations lesson plan, it required editing the models further within Blender.

4.1.2 Changing Poses Within Blender

The lesson plan for the fetus position required the user to view different types of breeches and the Blender's pose mode was utilised. The model of the pregnant women was hidden through Blender's hiding feature. To get the fetus into a correct pose, first the model's armature was selected, and then the pose mode from the drop down menu. Furthermore, different actions based on trial and error moved each part of the skeleton into the pose that needed to be create (Fig. 7). Experts provided images of different breech positions, and they were used as a guide for manipulating the model. In total 15 versions of the model, relating to head presentations, breeches, lies, regular presentations, and a placenta previa were posed using this method.

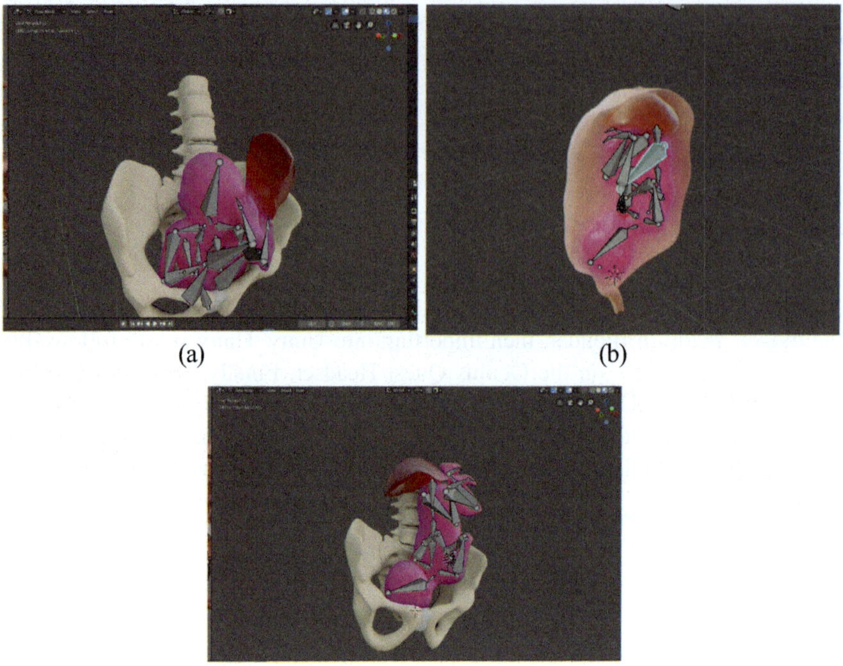

(a) (b) (c)

Fig. 7 Model arranged into different poses in Blender **a** Breech **b** Lobgitudal **c** Presentation

4.1.3 Sculpting

For the head presentation section of the "Fetal Lies and Presentation" lesson sculpting was necessary to show the gap between the bones of presenting baby. As the skull of a newborn isn't fully formed, a gap can be seen on the fetus' skull. There is two main gaps called posterior and anterior fontanelles, with a thin gap connecting both. These can be used by a doctor to determine which way the fetus is facing when presenting. Blender has a sculpt mode with numerous brushes that fitted the task (Fig. 8). The tool allowed gaps and fontanelles to be created. By following along with images the experts provided, the necessary features were created to show to medical students. There would be five types of head presentation, but this relates to the orientation of the baby, so this allowed to use the same sculpted model, and orientate it using Blender's pose mode to retrieve all five models.

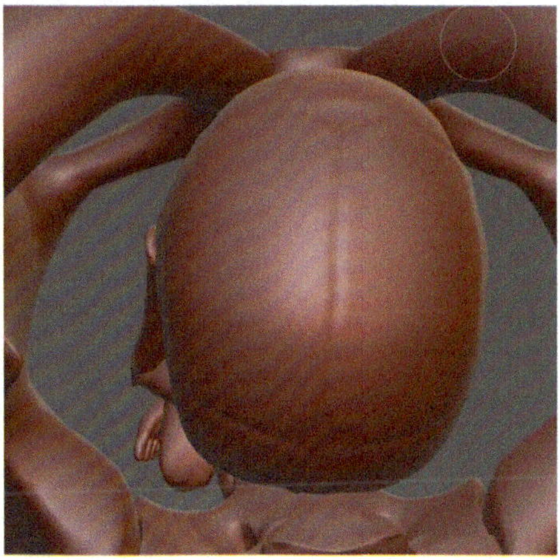

Fig. 8 Sculpted skull with the posterior (top) and anterior (bottom) fontanelle

4.2 *VR Environment*

Once the models were edited in Blender, confirmation received from the experts on their accuracy in order to be used in the lesson plans. Next step involved the the Oculus Quest set up and the virtual environment creation. Working in unity consisted of 3 stages.

- Enabling development for Oculus Quest.
- Creating a virtual environment with the models
- Scripting.

4.2.1 Creating the Virtual Environment

While Unity has standard tools to create an environment they're quite rudimentary and the ProBuilder package was utilised instead. It allowed to rapidly create floors and walls, and had a more streamlined workflow. For the Stages of Development lesson plan each model was given a section. Further sections were hidden behind walls for the user to walk around and discover. Each section with the model contained relevant textual or visual information. Using Unity's canvas scheme, to display text and video in the relevant locations. The canvas acts as placeholder, allowing sub user interface elements to be contained inside it (Fig. 9). Where the canvas needed to be interactive (Fig. 10), an OVR physics ray caster script was placed on it. This allowed the user to point and select buttons, and the canvas to receive its input.

The models were imported from Blender and scaled using Unity's scale tools. Textures were applied to give the models a more lifelike appearance (Fig. 11). In the Stages of Development lesson plan, anatomical points of interest would be shown of each internal organ and were coloured so the user could distinguish between them. Some models had transparent textures to see into the anatomy of the developing fetus.

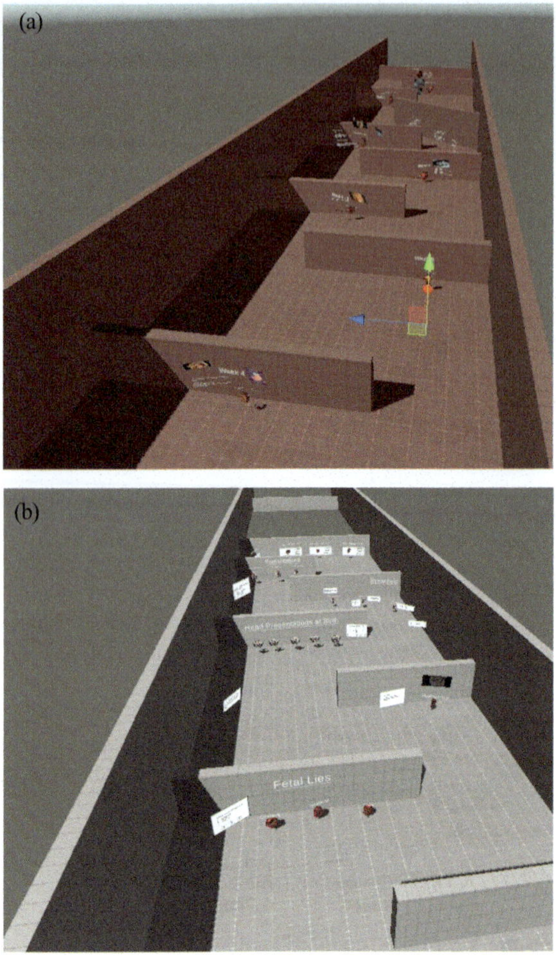

Fig. 9 Overview of model arrangement in each lesson plan **a** Stages of embryonic development **b** Fetal Lie/Position and Presentation in Pregnancy

In Week 6 of fetal development, the fetal heartbeat is detectable. To highlight this, an animation was created of a pulsing heart. Close lit lighting was used to present the model's features better.

For the Stages of development lesson, audio was incorporated. The audio mimicked the sound a baby would hear in the womb of the muffled heartbeat of the mother and the amniotic fluid swirling around the womb. When the user interacts with questions or reveals information, a sound would also play to tell them they're right or discovered something of interest.

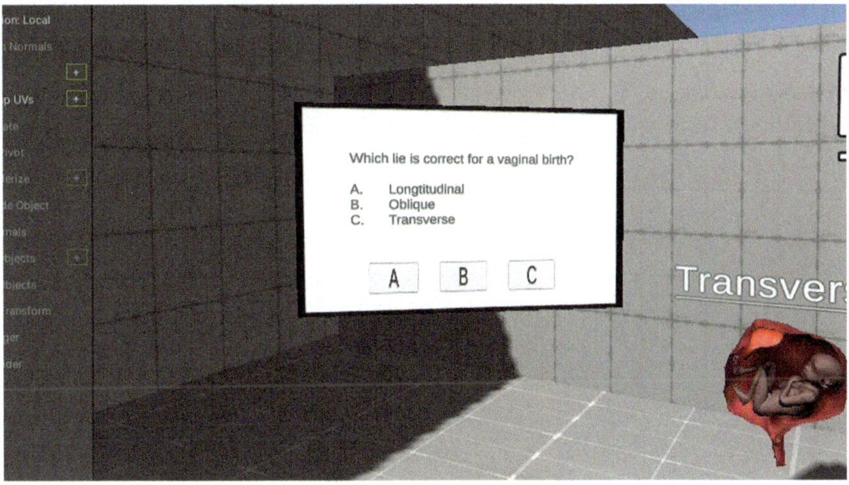

Fig. 10 Question using Unity's canvas

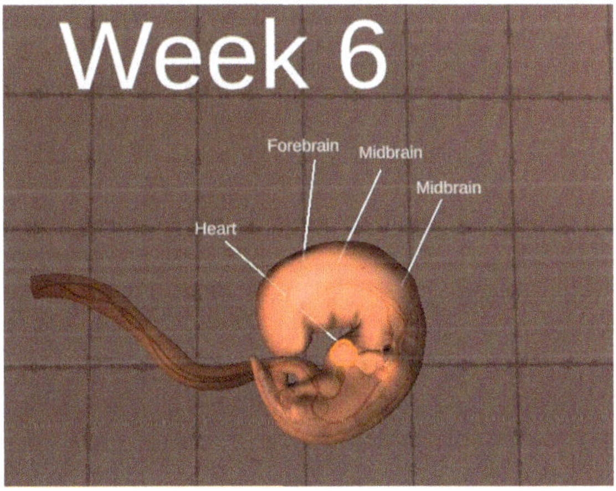

Fig. 11 Week 6 model showing internal fetal anatomy

4.2.2 Scripting

In order to enhance interaction to the experience, aspects of the environment were gamified. For some models users would need to select a model or a piece of text which would reveal more information. For others, questions were asked and selecting the right one would reveal a sound, along with visual feedback. As this required logic, scripting was implemented using Visual Studio Code. The scripts were placed on the relevant canvas. Using the OVR physics ray caster along with a script, meant it was able to detect when the user was pointing at the canvas and when they select a button. By using the Serialization on private variables, it allows objects to remain encapsulated but assignable in Unity's Editor. The scripts allowed for different prototyping.

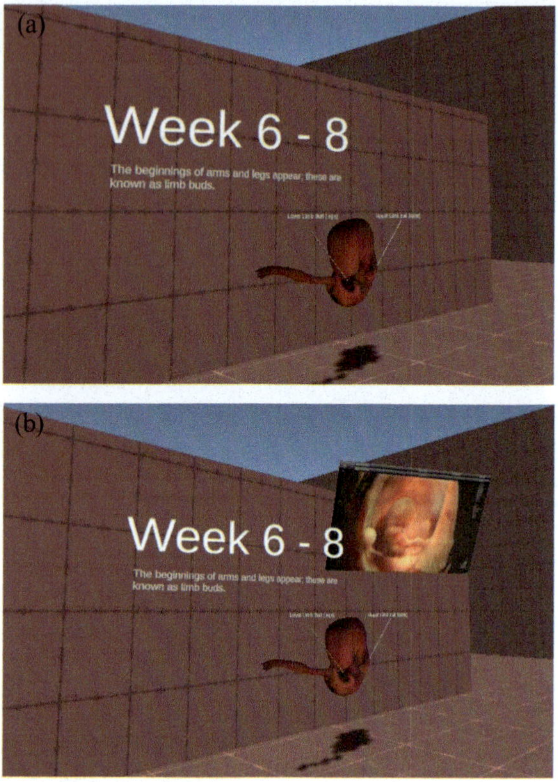

Fig. 12 **a** VR scene without activated image **b** VR scene with activated image at the Week 6–8 stage content (**a**)

Below is an example of one script used to show an image when a canvas is clicked. Most of the scripts followed this pattern in each lesson plan (Figs. 9 and 12). It was just a matter of tweaking it slightly depending on the use case.

Further work within the 3D development for medical education can include the production of standards for the interoperability of learning environments. Learning resources that have used these standards can be reused, and adapted for the learner. The 3D learning objects would require standardised metadata that describe then with a vision to allow users to search and access objects that are relevant to their subject much easier. An open source digital repository for 3D learning objects for medicine would advance the teaching and learning practicies embedded in VRLEs. Through the use of metadata harvesting and repositories, learning objects can facilitate more streamlined education.

5 Study Design

Ethical approval for the study was granted by the University College Dublin Research Ethics Committee (Reference: LS-20–09-McAuliffe) on the 26th May 2020 and registered with the ISRCTN trial registry (ISRCTN No. 39107). We investigated whether a virtual reality learning environment (VRLE) enhanced student understanding and retention of knowledge compared with a traditional tutorial. Students studying in the undergraduate (6-year programme) or postgraduate (4-year programme) medical degree programme at University College Dublin were invited to participate. Students were invited to take part in the study via announcements on Brightspace (university e-learning platform), class announcements and medical student societies. Interested students contacted the study team via the email address provided with the announcement and were sent an information leaflet and consent form to review. All medical students except those under the age of 18, non-English speaking, or with a medical condition including cardiac (e.g. pacemakers) and binocular vision abnormalities, psychiatric disorders or epilepsy, were eligible to take part in the study. In addition articipating students were required to have completed the embryology module on the stages of development of the fetus (Fig. 13).

Figure 14 provides the CONSORT Diagram for the pilot study representing the randomisation process of the participants. Assuming a normal distribution with a standard deviation of 1.75 taken from literature (Nicholson et al., 2006), the sample required to demonstrate a difference of 2 between the intervention and the control groups at a type I error rate of 0.05 (adjusted for multiple comparisons) at 90% power with an allocation ratio of 1:1. After an adjustment to allow for 20% drop-out, the total sample size required is 40 for 90% power (Fig. 15).

The interactive VRLE involved a "treasure hunt" for information, linking images to key learning points relating to fetal development. Using the controllers, participants could select internal organs on the fetal images being visualised in the HMD that were relevant to the stage of development. Doing so revealed learning points necessary for knowledge assessment after the VRLE experience. The bespoke designed VRLE was given the name of VR Baby. The control group underwent a traditional learning experience consisting of a face-to-face teaching tutorial on the above mentioned topic. Due to COVID-19 restrictions, the tutorial was conducted via an online video conferencing platform called Zoom (Zoom Video Communications, version 5.0.1) using a Microsoft PowerPoint (Microsoft Office, version 16.43) presentation. Two small group tutorials were conducted to encourage increased engagement within groups, and consisted of 10 students each. The tutorial lasted approximately 30 min in total, 15 min allocated to explaining the study and requirements for participation, and then a 15-min tutorial on the study topic.

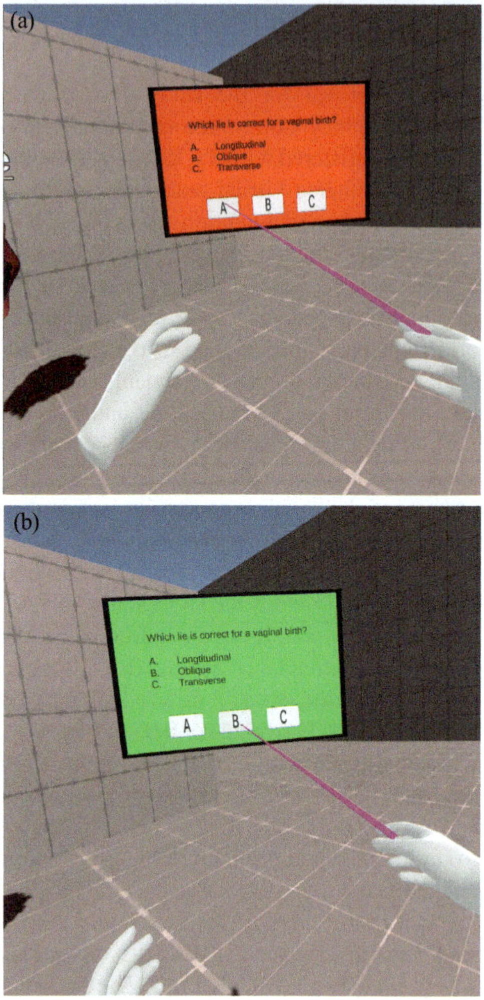

Fig. 13 Example of a question script **a** Wrong selected answer **b** Correct selected answer

5.1 Learning Experience Evaluation

Student knowledge was evaluated using a multiple choice questionnaires (MCQ) consisting of 10 questions related to the study topic. The MCQ was developed and peer reviewed by experts in the field of anatomy and embryology at the UCD School of Medicine and Obstetrics and Gynaecology professionals from the National Maternity hospital. MCQs were taken prior to intervention (baseline) and at two time points post intervention; immediately and after one week. The MCQ at one week was intended to evaluate knowledge retention.

In addition to the study MCQ, participants were also asked to complete two scales on the learning experience and a VR design questionnaire (intervention group only)

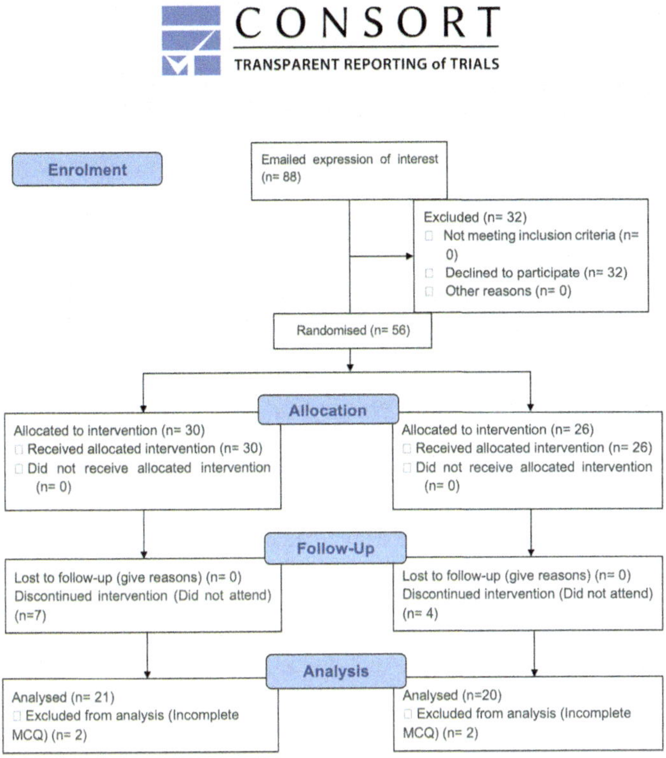

Fig. 14 CONSORT Diagram representing the randomisation process

	Sample Size per group		**Total sample size**	
	Power = 80%	*Power = 90%*	*Power = 80%*	*Power = 90%*
Difference = 2 (score)	13	17	26	34

Fig. 15 Power analysis for sample size

immediately post intervention. The two scales were designed to evaluate teaching methods, and included the Student Satisfaction and Self-Confidence in Learning Scale (SLSC) and a Simulation Design Scale (SDS; renamed and adapted for the purpose of this study as Virtual Reality Design Scale) [27] (Fig. 16). Both scales are validated and were originally designed for evaluating simulation studies. The VR design questionnaire investigated the acceptability and side effects of the VRLE using the SDS scale adapted for this study.

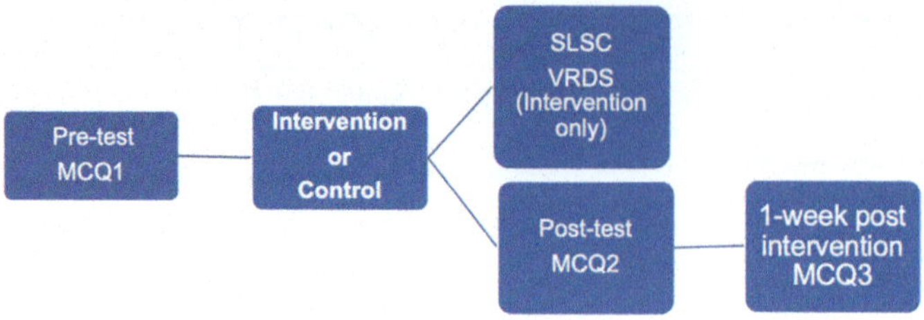

Fig. 16 Flowchart of evaluation timeline

The 13-item SLSC measured student satisfaction with the teaching activity (5 items) and self-confidence in learning (8 items). The SLSC was developed and validated by the National League for Nursing in order to implement a standardised approach to assessing learning outcomes, and is reported to have a Cronbach alpha of 0.94 and 0.87 for satisfaction and self-confidence, respectively [32]. SLSC is used for simulation-based studies, which we have adapted for our assessment of the VRLE. Total scores for the scale were obtained by adding responses to all items, with a maximum total score of 65. Higher scores denote higher levels of satisfaction and self-confidence [30]. The 20-item SDS, also developed by the National League of Nursing, was used to measure constructs from the framework for simulations [31]. The SDS is based on a Likert scale and divided into 5 sections, which are rated by the student. These included objectives and information (5-items), student support (4 items), problem solving (5 items), guided reflection or feedback (4 items), and fidelity (2 items). The reported Cronbach's alpha for the SDS is 0.92 [32], and adaptation for use within our VRLE was reasonable as parameters of evaluation were applicable [30]. We further assessed the reliability of both scales in the context of our study using reliability analysis for further validation. Cronbach alpha scores were 0.87 and 0.89 for the SLSC and SDS respectively (Fig. 17).

Scales	Original version, Cronbach alpha	Correlation inter-item total	VR Version study sample Cronbach alpha	Correlation inter-item total (Mean)	Correlation inter-item total (Range)
Student Satisfaction in learning and self-confidence scale	0.87-0.94	0.34-0.75	0.87	0.35	0.19-0.75
Simulation Design scale	0.92	0.35-0.69	0.89	0.36	0.34-0.75

Fig. 17 Cronbach's Alpha for satisfaction and self-confidence in learning scale and the simulation design scale and the adapted version for our study

The primary outcome was knowledge (measured by MCQ score) post intervention. Pre-specified secondary outcomes included attitudes on the learning experience assessed using the SLSC and the VDS. Side effect profiles of the VRLE were also evaluated to determine the usability of VRLE as an educational tool.

MCQ scores were compared between groups immediately post intervention (MCQ 2) and after one week (MCQ 3) using one-way analysis of covariance (ANCOVA). Both models were adjusted for baseline score (MCQ 1), and also MCQ 2 in the latter test. There were no significant differences between groups, so individual one-way analysis of variance (ANOVA) and post hoc Tukey's HSD tests were undertaken to explore the impact of time on knowledge scores for each study group independently. ANCOVA was used to assess between group differences for total SLSC scores, correcting for previous virtual reality use. The intervention group reported higher mean scores (54.19 (7.54)) than the control group (50.45 (7.20)) (Fig. 18).

Mean VDS scores are reported for each design feature (Fig. 19). Objectives and information (4.67 (0.44)) and support (4.80 (0.30)) were rated the highest among the design features. In reference to fidelity and resemblance to real life, students also reported high levels of realism (4.19 (0.86)). This strengthens the support for using the VRLE by providing content and face validity.

Scale	Intervention (n = 21)	Control (n = 20)	Corrected P-value	95% CI
Total SCLS Score *	54.19 (7.54)	50.45 (7.20)	0.21	47.14, 54.24
Satisfaction in Learning	21.57 (3.67)	19.55 (3.27)	-	-
Self-Confidence in Learning	32.62 (4.42)	30.90 (4.64)	-	-

Fig. 18 Student satisfaction and self confidence in learning Scale (SCLS) scores for intervention and control groups

Fig. 19 Virtual reality design scale scores (VRDS) in intervention group on

Questionnaire Item	Score*
Objectives & Information	4.67 (0.44)
Support	4.80 (0.30)
Problem Solving	4.35 (0.73)
Feedback	4.44 (0.60)
Fidelity (Realism)	4.19 (0.86)

6 Discussion

Following the implementation of the Virtual environment, the first virtual lesson was used to conduct a Randomised Control Trial (RCT) with medical students in their penultimate years at university level. Participants were randomised to an intervention (VRLE involving a 15-min learning experience on the stages of fetal development) or control (PowerPoint tutorial on the same topic) group. Multiple choice questionnaires (MCQ) assessed knowledge at three time points; pre-intervention, immediately post intervention and one week post intervention. Primary outcomes were differences in MCQ knowledge scores post intervention between groups. With regard to the RCT, 56 medical students volunteered to participate in this study and 41 completed the study with 21 and 20 participants in the intervention and control groups respectively. With regard to the primary outcome testing the effect of intervention on MCQ knowledge scores, there were no differences in post intervention knowledge scores. There were significant knowledge scores among time points for both the intervention ($P < 0.01$; CI 5.33, 6.19) and control ($P = 0.02$; CI 5.74, 6.49) groups. Within the intervention group, knowledge scores were significantly higher following the intervention, and persisted at one week. Mean levels of satisfaction and self-confidence in learning were higher in the intervention (54.19 (7.54)) compared to the control group (50.45 (7.20), $P = 0.21$).

The second virtual lesson was used to conduct a cohort study with university level midwifery students. 41 midwifery students participated in the study, with a 100% response and participation rate. Repeated measures one-way analysis of variance (ANOVA) revealed no statistically significant differences in knowledge scores pre and post-intervention. Participants rated high satisfaction and self-confidence scores with a mean score of (69.73 (6.81)). Side effects most commonly experienced by participants included dizziness (49%), disorientation (30%) and symptoms similar to motion sickness (32%). The VRLE had no impact on knowledge gain, however high levels of satisfaction and self-confidence indicate a positive response to the VRLE amongst midwifery students.

A novel and innovative educational tool consisting of a VRLE on the topic of "fetal lie, position, and presentation in pregnancy" has been developed and assessed. Although the VRLE did not change knowledge levels of participants in this study, the learning experience was highly rated by students. Additionally, higher levels of engagement and enhanced understanding of fetal and pelvic anatomy were self-reported.

7 Conclusions

A bespoke design VRLE on topics pertaining to the Obstetrics and Gynaecology and Midwifery curriculum has been developed in the work described in the chapter. We conducted an RCT with medical students that found knowledge retention improved with a VRLE compared to a more traditional learning modality i.e. a PowerPoint lecture. Knowledge increased after both learning modalities however it was retained in the intervention group at one week follow up. Results of the questionnaire reflected the degree of satisfaction with this learning tool by medical students. Participants also rated the design features of the VRLE highly in the questionnaire despite having experienced some side effects. This strengthens the support for using the VRLE by providing content and face validity. This study demonstrates that VRLEs are a useful educational tool for invisible concepts, such as those found within embryology by enhancing student's retention of knowledge, satisfaction and self-confidence in learning.

However it is important to note that immersive technologies may not replace traditional lecture-based teaching by experts but may serve as a valuable educational tool, enhancing and reinforcing knowledge, clinical reasoning and skills. A meta-analysis of 36 studies, including 28 RCTs and 2226 participants, showed that 3D visualization technology was associated with higher factual knowledge, better results in spatial knowledge acquisition, increased user satisfaction and growth of the learners' perception of tool effectiveness [28]. Therefore VR can be considered as a valuable supplementary educational tool, which is a very exciting prospect for both students and curricula developers alike [29].

Acknowledgements This project was based on the knowledge provided from the experts at National Maternity Hospital in Ireland. The lesson plans, study design and protocols were supervised from Prof. Fionnuala McAliffe, Professor of Obstetrics and Gynaecology at National Maternity Hospital Dublin and Director of the UCD Perinatal Research Centre. The technical requirements and supervision of the VR software development was supervised from Pro. Eleni Mangina, School of Computer Science, University College Dublin.

References

1. Prince, M.: Does Active Learning Work? A Review of the Research (2004).https://doi.org/10.1002/j.2168-9830.2004.tb00809.x
2. Shapley, K., Sheehan, D., Maloney, C., Caranikas-Walker, F.: Effects of Technology Immersion on Middle School Students' Learning Opportunities and Achievement (2011). https://eric.ed.gov/?id=EJ934897. Accessed 20 November 2021
3. Mangina, E.: 3D learning objects for augmented/virtual reality educational ecosystems. 2017 23rd International Conference on Virtual System Multimedia (VSMM), pp. 1–6 (2017). https://doi.org/10.1109/VSMM.2017.8346266
4. Nicholson, D.T., Chalk, C., Funnell, W.R., Daniel, S.J.: Can virtual reality improve anatomy education? A randomised controlled study of a computer-generated three-dimensional anatomical ear model. Med Educ. **40**(11), 1081–7 (2006). https://doi.org/10.1111/j.1365-2929.2006.02611.x. PMID: 17054617
5. Allcoat, D., Mühlenen, A.: Learning in virtual reality: effects on performance, emotion and engagement. Res. Learn. Technol., 26 (2018).https://doi.org/10.25304/RLT.V26.2140
6. Maresky, H.S., Oikonomou, A., Ali, I., Ditkofsky, N., Pakkal, M., Ballyk, B.: Virtual reality and cardiac anatomy: exploring immersive three-dimensional cardiac imaging, a pilot study in undergraduate medical anatomy education. Clinical Anatomy (New York, N.Y.) **32**(2), 238–243 (2019). https://doi.org/10.1002/ca.23292

7. Ray, A.B., Deb, S.: Smartphone based virtual reality systems in classroom teaching—a study on the effects of learning outcome. 2016 IEEE Eighth International Conference on Technology for Education (T4E), pp. 68–71 (2016). https://ieeexplore.ieee.org/abstract/document/7814797. Accessed 3 October 2021
8. Chien, C.-H., Chen, C.-H., Jeng, T.-S.: An interactive augmented reality system for learning anatomy structure. Proceedings of the International Multi-Conference of Engineers and Computer Scientists 2010, Vol. 1, IMECS (2010)
9. Moro, C., Štromberga, Z., Raikos, A., Stirling, A.: The effectiveness of virtual and augmented reality in health sciences and medical anatomy. Anat. Sci. Educ. **10**(6), 549–559 (2017). https://doi.org/10.1002/ase.1696
10. Wu, H.-K., Lee, S.W.-Y., Chang, H.-Y., Liang, J.-C.: Current status, opportunities and challenges of augmented reality in education. Comput. Educ. **62**, 41–49 (2013), ISSN 0360-1315.https://doi.org/10.1016/j.compedu.2012.10.024
11. Mohankumar, P., Yie, L.W.: 3D Modelling With Ct and Mriimages of ascoliotic vertebrae 2015. Journal of Engineering Science and Technology, 4th EURECA 2015 Special Issue February (2016) 188–198, © School of Engineering, Taylor's University
12. Mangina, E., De Oliveira Ranito, O., Campbell, A., McMahon, C.J.: 3D Stereolithographic models placed in virtual reality to assist in pre-operative planningin. Proceedings of the 10th E AI International Conference on Simulation Tools and Techniques (ACM, Hong Kong,China), pp. 87–92 (2017). ISBN: 978–1–4503–6388–4. https://doi.org/10.1145/3173519.3173522.13
13. Cross, D. E., Bauer, T. M., Tchantchaleishvili, V.: 3D organ modeling with open-source software. Artif. Organs **43**, 596–598 (2018).https://doi.org/10.1111/aor.13395.
14. Seif, M., Umeda, R., Uehara, Y., Higa, H.: A Data Conversion for Medical Training Systemin (2016)
15. Mustra, M., Delac, K., Grgic, M.: Overview of the DICOM standard. 50th International Symposium ELMAR, pp. 39–44 (2008)
16. All3DP 2020. STL File Format (3D Printing)—Simply Explained https://all3dp.com/what-is-stl-file-format-extension-3d-printing/.17.3D. Accessed 20 November 2021
17. SLICER. (2020). https://www.slicer.org/ Accessed 20 November 2021
18. . AutoDesk (2020) MeshMixerAutoDesk. http://www.meshmixer.com/
19. Blender Online Community.Blender—a 3D modelling and rendering package, Blender Foundation. http://www.blender.org.20. Accessed 22 April 2021
20. Unity (2021) Unity Technologies.https://unity.com/.21. Accessed 15 February 2020
21. VRTK.VRTK Library https://www.vrtk.io/ Accessed 15 February 2020
22. BabySlice0 (2020). https://3dprintedultrasounds.com/baby-slice-o-software/.23 Accessed 2 January 2020
23. Allen, L.K., Ren, H.Z., Eagleson, R., De Ribaupierre, S.: Development of a web-based 3D module for enhanced neuroanatomy education. Studies Health Technol Inf **220**, 5–8 (2016)
24. Höhne, K.H., Bomans, M., Pommert, A., et al.: 3D visualization of tomographic volume data using the generalized voxel model. Vis. Comput. **6**, 28–36 (1990). https://doi.org/10.1007/BF01902627
25. Advincula, W.D.C., Choco, J.A.G., Magpantay, K.A.G., Sabellina, L.A.N., Tolentino, J.G.M.F., Baldovino, R.G., Bugtai, N.T., See, A.R. and Du, Y.C.: Development and future trends in the application of visualizationtoolkit (VTK): The case for medical image 3D reconstruction. AIP Conference Proceedings, 2092, 020022 (2019)
26. Werner, H., Castro, P.T., Matos, A.P., Lopes, J., Ribeiro, G., Daltro, P.: Ultrasound in Obstetrics Gynecology, **54**, 23–24 (2017). https://doi.org/10.1002/uog.20483
27. Jeffries, P.R., Rizzolo, M.A.: Designing and Implementing Models for the Innovative Use of Simulation to Teach Nursing Care of Ill Adults and Children: A National, Multi-Site, Multi-Method Study. National League for Nursing and Laerdal Medical, New York (2006)
28. Yammine, K., Violato, C.: A meta-analysis of the educational effectiveness of three-dimensional visualization technologies in teaching anatomy. Anat. Sci. Educ. **8**(6), 525–38 (2015). https://doi.org/10.1002/ase.1510. Epub 2014 Dec 31. PMID: 25557582

29. Moro, C., Štromberga, Z., Raikos, A., Stirling, A.: The effectiveness of virtual and augmented reality in health sciences and medical anatomy. Anat. Sci. Educ. **10**(6), 549–559 (2017). https://doi.org/10.1002/ase.1696. Epub 2017 Apr 17. PMID: 28419750
30. Unver, V., Basak, T., Watts, P., Gaioso, V., Moss, J., Tastan, S., Iyigun, E., Tosun, N.: The reliability and validity of three questionnaires. the student satisfaction and self-confidence in learning scale, simulation design scale, and educational practices questionnaire. Contemp. Nurse. **53**(1), 60–74 (2017). https://doi.org/10.1080/10376178.2017.1282319. Epub 2017 Feb 10. PMID: 28084900
31. Jeffries, P.R.: A framework for designing, implementing, and evaluating simulations used as teaching strategies in nursing. Nurs. Educ. Perspect. **26**(2), 96–103 (2005). PMID: 15921126
32. Jeffries, P.: Simulation in nursing education: from conceptualization to evaluation, Lippincott Williams Wilkins (2020)

Immersive and Interactive VR for Lifting Training of Tower Cranes

Zonghong Yu, Lihui Huang, Ching Hui Ooi, Yi Gong, and Yiyu Cai

Abstract Virtual Reality (VR) is a rapidly evolving technology that provides an experience involving most of the senses and creating an immersive synthetic world. It is an enabling technology that is entering our everyday life, from our homes to workplaces. Various sectors look towards using technology to achieve better safety, and better management in terms of energy, time, and cost. Being a complex and competitive sector, the construction industry can take advantage of VR's ability to provide an immersive and interactive simulation. This is especially favourable for the industry since simulations can be used for training workers in the construction sector, decreasing the cost and providing a much safer training environment. The simulation also serves an educational purpose for teaching novices about the construction industry without physically visiting the site, benefiting the industry by allowing the typical layman to have exposure to the industry with fewer safety hazards in place. This chapter reviews the recent developments of VR installations, specifically the case study of tower crane lifting through innovative VR interface designs. VR will be discussed in detail for fidelity modelling, immersive visualization, and real-time interaction of tower crane lifting in a simulated construction environment.

Keywords Virtual reality · Tower crane · Lifting

1 Introduction

1.1 VR History

The essay, The Ultimate Display [20] by Dr. Ivan Sutherland in 1965 started the emergent of VR technology. During his visit to Nanyang Technological University, Singapore, in February 2016, Dr. Sutherland gave a talk on Sixty Years of Computing: A Personal Recollection as shown in Fig. 1.

Z. Yu · L. Huang · C. H. Ooi · Y. Gong · Y. Cai (✉)
Nanyang Technological University, Singapore, Singapore
e-mail: myycai@ntu.edu.sg

Y. Cai et al. (eds.), *Mixed Reality for Education*, Gaming Media and Social Effects,
https://doi.org/10.1007/978-981-99-4958-8_9

Fig. 1 Head mounted display (HMD) invented by Dr. I. Sutherland

Today, VR, Augmented Reality (AR), and Mixed Reality (MR) are much more accessible due to the rapid growth in the industry. Facebook's acquisition of Oculus VR, Inc for a total of approximately $2 billion in 2014 is a notable milestone in VR technology and commercial development [9]. In March 2016, Microsoft released the company's venture plan into MR with HoloLens [15]. The VR market size is estimated to grow to more than 12 billion U.S. dollars by 2024 [1], with the increased usage from both enterprises and consumers. VR, AR, and MR are disrupting numerous sectors, being able to provide an effective, immersive experience in a safe environment. They are widely adopted in various applications in medicine [2, 7], education [3, 4, 6, 12], and training [5].

This chapter discusses VR application in crane lifting training. Section 2 introduces crane lifting as a case study with a focus on productivity and safety. Section 3 describes the VR enabled simulation of tower cranes, encompassing fidelity modelling, immersive visualization, and real-time interaction for crane lifting. The summary and further developments of the work is then provided in Sect. 4.

2 Tower Cranes and Crane Lifting

A crane is a machine shown in Fig. 2, commonly used to lift and transport heavy cargo in various sectors, such as the construction and maintenance industries. There are various types of cranes meant for the different functions and environments, where the two main categories are static cranes and mobile cranes. The crane that we will be focusing on, is the tower crane which is a type of static crane.

Fig. 2 Tower crane in construction [25]

According to GlobalData's report [10], the construction industry is predicted to increase by 5.2% globally in 20021, 2.5% higher than the level reported in 2019. This is a conservative prediction with the COVID-19 crisis that has been ongoing since the end of 2019 and the observed recovery at the end of 2020. Deloitte has also reported their positive outlook towards the construction industry, with the shift of the field towards digital transformation which will increase efficiency [14]. The Independent Commodity Intelligence Services (ICIS) also expects the industry sector to grow to about 35% from 2020 to 2030 despite the gloomy short-term outlook due to the COVID-19 pandemic [8].

Tower cranes are an essential part of the construction industry. Despite the relatively low accident rates of around 15 incidents a year globally [19], the aftermath of a tower crane-related accident is severe. Hence with the predicted growth in the construction sector, there is an increasing need for crane operators to go through rigorous training to acquire the necessary experience to decrease accident rates.

2.1 Tower Crane Basic Structure

A tower crane generally consists of several key components: the base, mast, jib, slewing unit, operating cabin, and counterweight [11] as shown in Fig. 3.

1. **Base**: The base of a tower crane controls the stability of the entire crane by forming a connection between the ground or building that the crane is built on, and its towering structure.

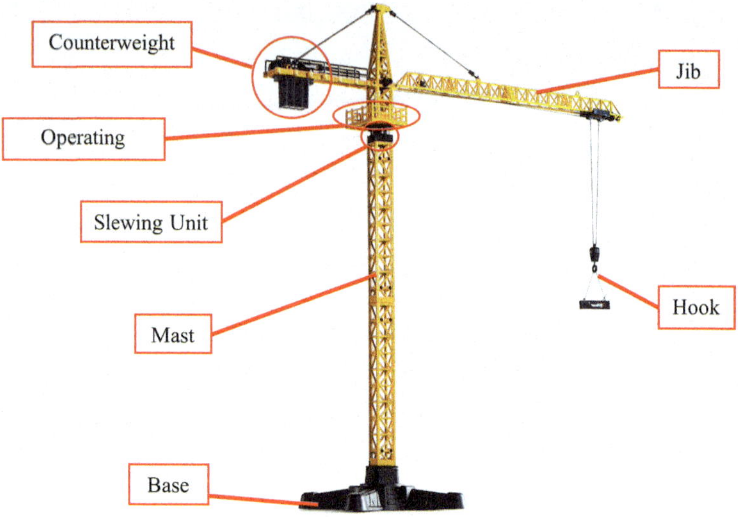

Fig. 3 Basic structure of a tower crane [21]

2. **Mast**: The mast is the structure between the base and the jib of the tower crane. It also determines the height of the tower crane by the number of masts stacked on top of each other. Apart from its structural function, it also has guard and guide rails for the elevator meant for the operator and technician's transportation up the crane.
3. **Jib**: The jib acts as the working arm of a tower crane that extends horizontally. A trolley moves along the jib and is used to carry loads via a hook attached to the trolley. The jib is made up of separate sections which allow for the length to be adjustable, allowing it to be adapted to suit the environment of the working area.
4. **Counterweight**: The counterweight is found on the other end of the jib, to balance the weight of the load it will be lifting. It is generally adjustable and dependent on the weight of the load it is meant to carry.
5. **Slewing Unit**: The slewing unit is the axis of a tower crane's rotation, containing a set of huge gears and motors, enabling the tower crane's jib to rotate almost 360°.
6. **Operating Cabin**: The operating cabin where the crane operator works from, located above the slewing unit to obtain the best possible field of vision.

In general, the working area of a tower crane is almost a full circle, with the slewing unit as the centre, and the length of the jib as the radius. This is due to the slewing unit providing the ability to rotate almost 360° and the trolley for being able to move along the jib.

The location of where the tower crane is set up requires careful consideration about various factors such as the approximate weight of the load, the distance of

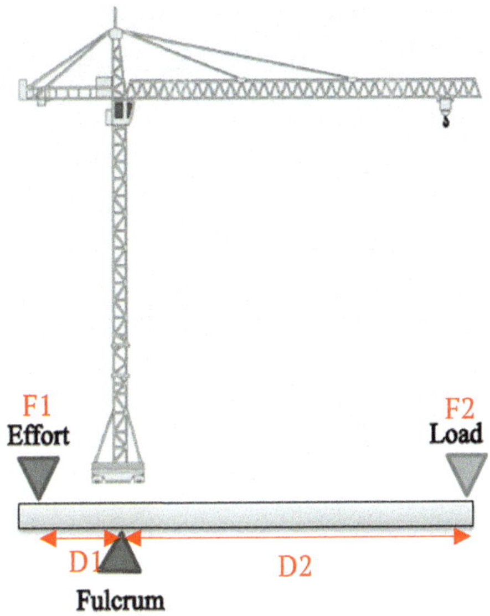

Fig. 4 Force distribution in a tower crane

carrying the load, and environment factors such as the distance and height of nearby buildings, and distance from other tower cranes.

2.2 *Tower Crane Force Distribution*

A tower crane can be simplified into a lever as shown in Fig. 4. The jib is the beam where the point of intersection of the mast on the jib is the fulcrum of the lever. The load attached to the tower crane can be taken to be the load on the lever and the counterweight to be the effort. Assuming that the distance between the effort and the fulcrum is D1, and the distance between the load and the fulcrum to be D2, we can further take the effort to have a force of F1 and the load to be F2. Using conservation of momentum, we obtain the following equation:

$$D1 \times F1 = D2 \times F2 \tag{1}$$

Different tower crane models have their specific load capacity, limited by the restriction of the counterweight of each crane. The load capacity also varies with the trolley's distance along the jib. An example of various tower cranes' load capacity is shown in Fig. 5, where the general trend can be deduced that the load capacity decreases exponentially with the increase in the jib's length. The load capacity can also be seen to have both an upper and lower limit due to the restriction of the tower crane's counterweight.

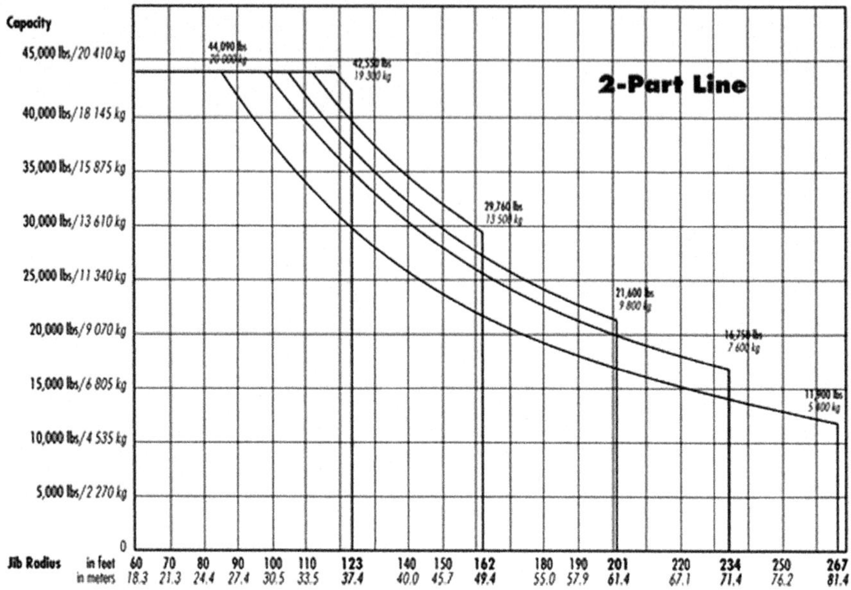

Fig. 5 Load capacity of hammerhead tower crane [16]

2.3 *Tower Crane Operation*

The operating of a tower crane may seem simple; however, it requires the combined efforts of various roles such as the crane operator, lifting supervisor, rigger, and signalmen [26]. Crane operators are required to check through functional and integrity checks, where an example of the checklist is shown in Fig. 6, before operating the crane. In the situation of identifying deficiencies of the crane, adverse weather conditions, and when the dynamic environment around the crane is not in accordance with safe practices, crane operators are given the responsibility to cease crane operations.

Lifting supervisors is to coordinate, brief and supervise works of the crane operator, rigger, and signalmen, ensuring that the lifting operations are performed safely. Riggers must check if the load meant for lifting is within the load range of the crane and ensure that the tower crane is properly equipped and secured before rigging up the load. Signalmen act as the eyes and ears of the tower crane operator from the ground by providing clear and accurate signals. The signals provide cues to inform the crane operator if it is safe to lift the load and guide the load to its destination. An example of the signals is shown in Fig. 7.

	1st	2nd		1st	2nd
Seat adjustment			Hoist limiter		
Control identification (levers and switches) and operation			Slew lock pin		
Warning devices (including horn)			Anti-collision device operation (where fitted)		
Slew lock			Maximum wind speed		
			Electricity isolating switch		
Load charts					
Load chart (capacity at radius)			**Safety inspection:**		
Load chart (variations for different boom lengths)			• ropes • rams		
Load chart (variations for different rope falls)			• wire rope wear of pads/sheaves • boom		
			• grease points		
Rated capacity limiter			• fluid leaks		
Rated capacity limiter – features			• fluid levels		
Rated capacity limiter – set up and operation			• brakes • cabin visibility		
Rated capacity limiter – relationship to load chart			• lights • tower bolts		
Rated capacity limiter – override procedures			• slew ring bolts • fall-arrest harness and lanyard		
Rated capacity limiter – digital display			• boom pins and retainers		
Slew function					
Luff function (where applicable)					
Trolley function (where applicable)			Platform and tower edge protection		
Hoist function			Cable drum and cable guide		
Slew brake			Hoist, travel and trolley brakes		
Slewing ring ball races			Fluid couplings		
Slipring unit					
Wheel bogies			Fire control procedures and equipment		

Fig. 6 Example of a checklist for tower crane operation [22]

Fig. 7 Example of signal guides [17]

2.4 *Safety of Tower Crane Lifting*

The most common crane accidents are the collapsing or overturning of the tower crane, workers and loads falling from the crane, and workers being trapped between the crane's mast and jib according to Occupational Safety and Health Administration (OSHA) [18]. These accidents often occur due to the overloading of the crane and falling of loads [13]. The overloading compromises the structural integrity of the crane. It is usually caused by hoisting loads beyond the crane's capacity, usage of defective components, swinging or sudden drop of the load. The falling of loads is hazardous, where it can result in fatal injuries and lead to an increase in both cost and time required. The cause of the falling of loads is often due to loads not being secured properly, mechanical failure of components, and two-blocking.

Hence a safety warning system that can detect and anticipate critical risks is paramount to prevent accidents caused by human error. Such a system includes collision path protection, zone protection, boundary protection, early earning control, and fault detection. However, these safety warning systems can only help with the detection and relying on them can simply reduce accident rates. Hence, to achieve a safe tower crane working environment, proper operation training for operators and signalmen are of the essence.

3 Immersive and Interactive VR for Tower Crane

This work creates a simulation to provide various tower-crane related lifting experiences. The assembling of a tower crane, job scopes of a crane operator and a signalman is mimicked. This provides trainees with a better understanding of the construction industry while staying in a safe environment. It also acts as a possible training ground for crane operators and signalmen, imparting commensurate knowledge to trainees through the simulation. The replication of the physical experience is only feasible through the usage of various interfaces.

The simulation is developed with Unity [24], a cross-platform game engine that can be used to create three-dimensional interactive simulations that also allows for leap motion applications. It can be done easily by setting up leap motion drives and assets for Unity which can be directly downloaded and imported into Unity 3D. Several assets provide different models, prefabs, and scripts that fulfil basic functions such as detecting the extension of fingers or palm orientation. This provides convenience to the developer, by simply dragging the preferred prefab into the program to invoke basic settings or functions.

3.1 Fidelity Modeling and Assembly of Tower Cranes

A tower crane consists of multiple components. In this simulation, major components are modelled by combining various small components into several parent objects. The combination of parts allows for trainees to apply functions on parent objects, effectively affecting multiple components simultaneously. The tower crane is divided into OT, masts, front jibs, back jibs, and basic mast structures whose components are classified into different parents according to their position and function. The base section of the tower crane model is named OT in this simulation for simplicity. This is shown in Fig. 8.

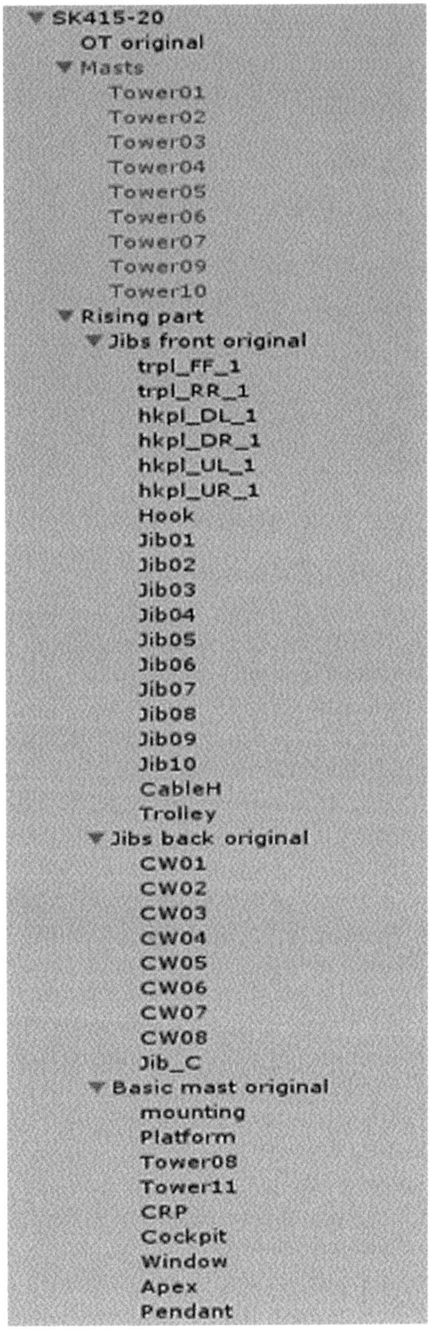

Fig. 8 Component grouping in simulation

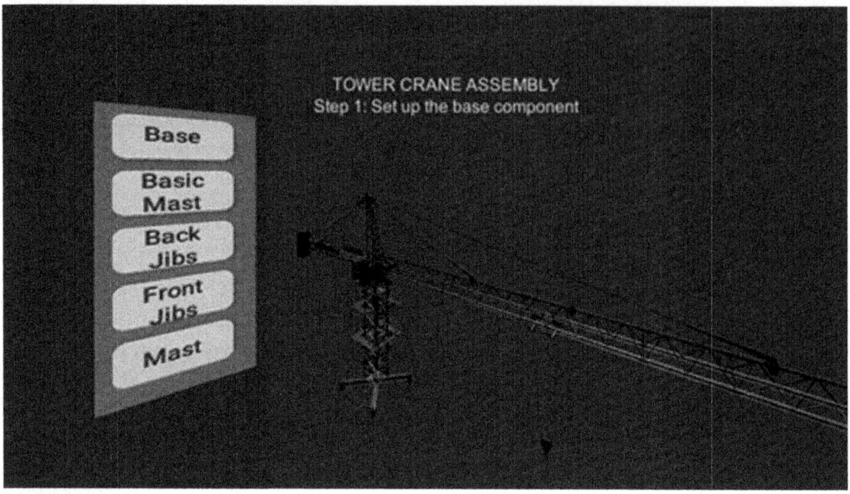

Fig. 9 Tower crane assembly interface

The overall assembly process of the tower crane has three milestones. The first is the assembling of the base structure, followed by the subassembly process of the mainmast structure, and the third is the raising of tower cranes through assembling of the mast. With considerations about the interface design, the author chose GUI buttons for the selection and activation of components, as shown in Fig. 9, before proceeding with the assembling of the tower crane. Instructions are provided on the screen, for trainees to follow the directions and complete the tower crane assembly simulation.

The instructions provided are based on an actual tower crane's assembly process. The OT base is first set on the ground, followed by the setting up of the basic masts, made up of the cabin, slewing unit, and mounting structures. The front and back jibs are then hoisted up the crane for installation before finally installing the masts to raise the tower crane to its appropriate height.

To achieve the assembly function of the simulation, the entire tower crane model is duplicated, where the parent groups of the copied model are set to a fixed position. The colour of the copied model is then selected to be transparent, to differentiate between the fixed and operational models. For the collision detector to work, all models have the box collider and rigid body components added to them. Event scripts are then attached to their respective components, which are activated through the pressing of their designated buttons. Events include the changing of the colour of the target position for assembly, locating the position of the moving component to the fixed position, stopping the controller's work and model's movement once collision between proper components have been detected, and activating the next instruction. The script is shown below:

Example of an event script for components

```
using Unity Engine;
using System.Collections;
using UnityEngine.Events;
using Leap;
using Leap.Unity;
public class collision : MonoBehaviour {
    public GameObject original;
    public GameObject text1;
    public GameObject text2;
    public GameObject explaintext;
    [HideInInspector]
    public Color[][] previousChildColors;
    [HideInInspector]
    public Color previousColour;
    private MeshRenderer[] renderers;
    void Start () { }
    void Update()
    {
        original.GetComponent<MeshRenderer>().material.color = Color.red;
        for (int i = 0; i < original.GetComponentsInChildren<MeshRenderer>().
            Length; ++i)
        {
            MeshRenderer childRenderer = original.GetComponentsInChildren<
                MeshRenderer>()[i];
            for (int j = 0; j < childRenderer.materials.Length; ++j)
            {
                childRenderer.materials[j].color = Color.red;
            }
        }
    }
    void OnTriggerEnter(Collider other)
    {
        if (other.gameObject==original)
        {
            transform.position = original.transform.position;
            transform.eulerAngles = original.transform.eulerAngles;
            GetComponent<LeapRTS>().enabled = false;
            original.SetActive(false);
            text1.SetActive(false);
            text2.SetActive(true);
            explaintext.SetActive(false);
        }
    }
}
```

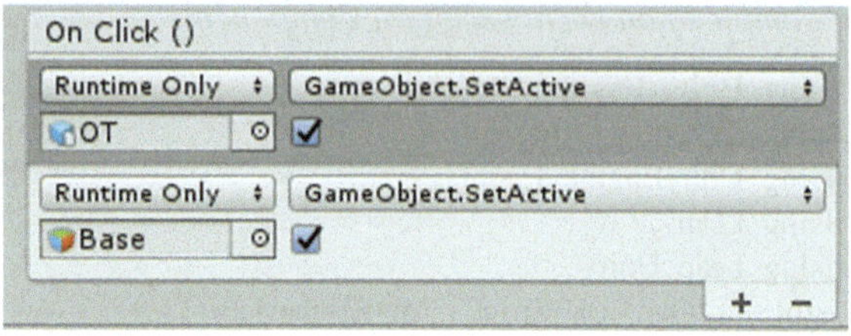

Fig. 10 Button settings (on click)

3.2 *Leap Motion Enabled Interaction*

This section introduces the functions of the fidelity modelling of the simulation. The assembly section of the simulation uses a Leap Motion Controller, where trainees engage with the interactive environment via hand motions, tracked by the controller. LeapRTS [23] is a basic script available from leap motion's asset base which can be switched on easily as shown in Fig. 10. It allows trainees to move objects via the pinching motion. By attaching the LeapRTS script to components in the simulation, the component's movements are performed by replicating the movements of trainees' pinched fingers.

When trainees click on the appropriate button by following the instructions provided, the selected movable component will become active and appear on the screen with a brief introduction about the component. The position that trainees should move the selected component to, will be highlighted by changing the colour of the corresponding fixed component to red. The assembling of components can be seen in Fig. 11, where trainees use a pinching motion to move the activated component to its respective position highlighted in red. When the collision between the movable and fixed component in the simulation is detected, the movable component will take on the position, rotation, and scale of the fixed component. The LeapRTS script and introduction text will be disenabled simultaneously. The instruction for the next part of the assembly sequence of the simulation will replace the previous instructions, guiding trainees to complete the simulation.

To enhance trainees' impressions about the specific details of the essential basic components of the tower crane, the simulation allows trainees to assemble them. The button 'Basic Mast' leads trainees to the subassembly process of assembling the components. When the button is selected, the camera providing trainees' field of view will move to a different scenario. trainees can then proceed with the subassembly of the basic mast structure which includes the apex mast, operating cabin, mounting system and platform around the centre unit mast that is fixed at a central position as shown in Fig. 12.

The component selection for the basic mast subassembly uses ray casting instead of controls via buttons. Ray casting is an interaction technique that imitates pointing with a laser pointer, where a ray is cast from the centre of the camera. The selection of component is done by having the projected ray touches the component. The name of the component will be displayed on the screen and the LeapRTS script attached

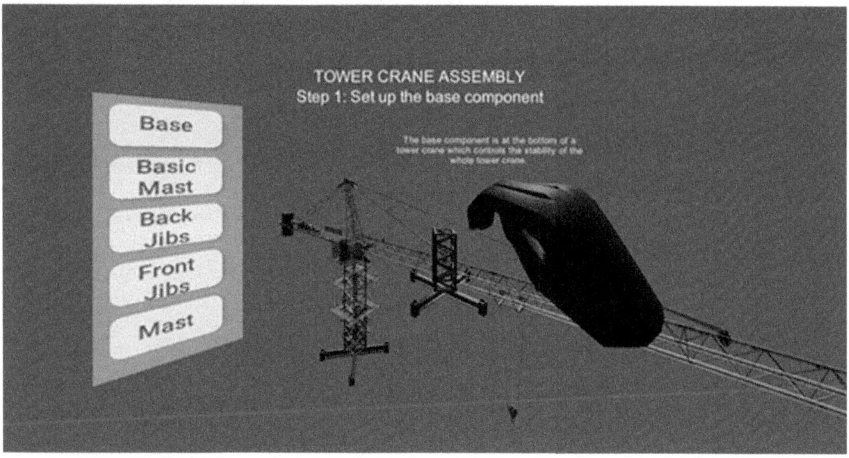

Fig. 11 Assembling base components in simulation

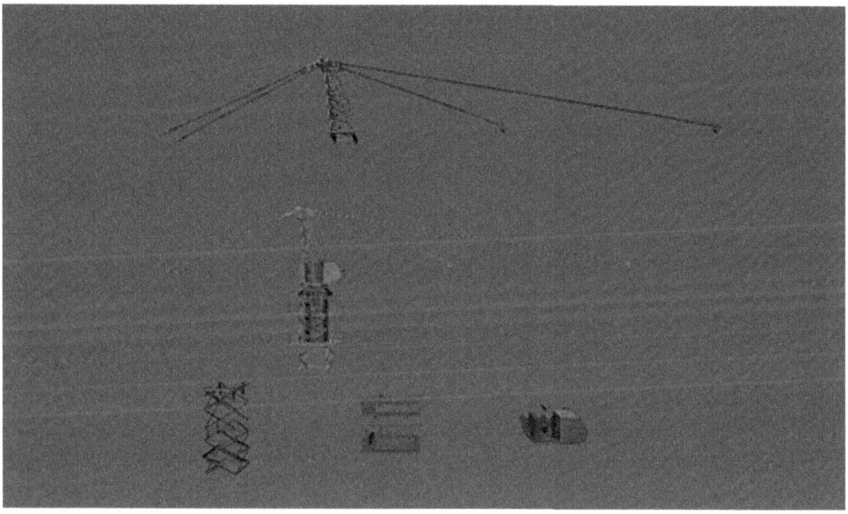

Fig. 12 Basic mast subassembly in simulation

will be activated. trainees can then assemble the components via the pinching action. The following script creates the ray cast from the camera:

Example of ray cast script

```
Ray ray = mycamera.ViewportPointToRay(new Vector3(0.5F, 0.5F, 0));
```

After performing the subassembly process, the thumbs-up gesture, shown in Fig. 13, is used to return to the main assembly scenario to continue the procedure. Gestures

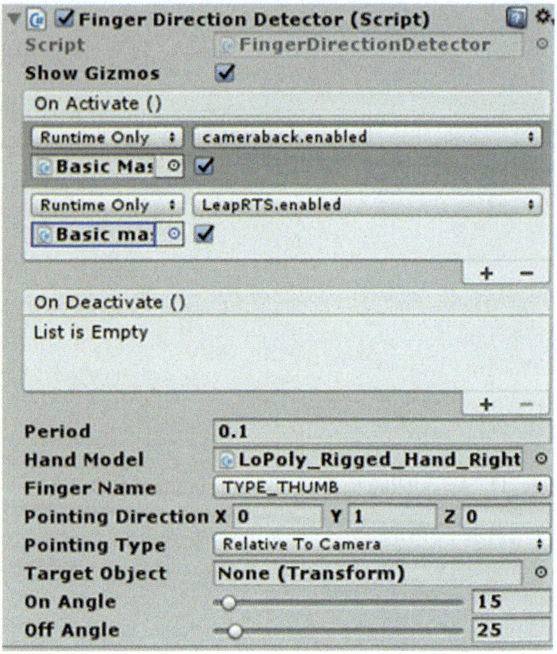

Fig. 13 Thumbs up gesture detector setting

are controlled by the Finger Detection Detector script which will be further explained later in the chapter. Upon completing the basic tower crane structure's model, trainees are left with the final assembly simulation of raising the tower by installing the tower crane's mast.

The method of activating the mast is alike the other assembly processes, via the selection of the indicated button from the main assembly's panel. The assembly method simulates the raising of real cranes, using a mounting platform to hold the mast, followed by raising the crane with hydraulics before inserting the mast into the gap. In this simulation, the collision detector is placed at the mounting model. When trainees collides the mast with the mounting platform, the mast will move into the centre of the tower crane, increasing the height of the crane. trainees will then receive the next piece of the mast.

For the VR application's basic settings, the camera view uses the prefab, *LMHead-MountedRig* to replace the main camera in a scene to suit the perspective for VR. The hand model used is *Capsule Hand*, shown in Fig. 14, which has the prefab *LeapHandController* attached to it to simulate hand motions in VR.

On the assumption that the process proceeds smoothly, and that all collisions are detected accurately, the overall assembly process will proceed as stated below:

1. Hint 1: "Step 1: Set up the base component." and press the "Base" button to activate the base model.
2. Pinch to assemble the base model. Hint 1 will be replaced by the next hint.

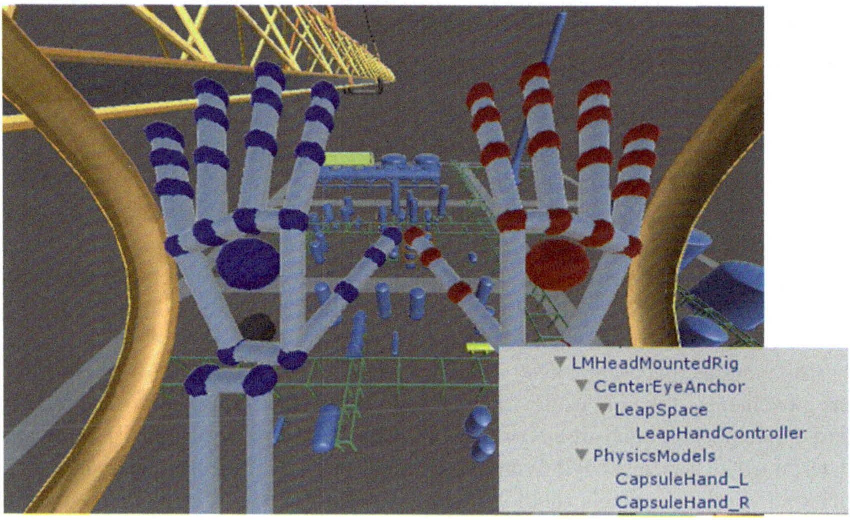

Fig. 14 Thumbs up gesture detector setting

3. Hint 2: "Step 2: Assemble mounting structure with the basic unit of a mast." and press the "Basic Mast" button to switch the scenario to the subassembly.
4. Complete the subassembly process of the mainmast structure and use a thumbs-up gesture to return to the main assembly scenario.
5. Pinch to assemble the basic mast component. Hint 2 will be replaced by the next hint.
6. Hint 3: "Step 3: Fix the backward jibs for balancing." Press the "Back Jibs" button to activate the back jibs model.
7. Pinch to assemble the back jibs. Hint 3 will be replaced by the next hint.
8. Hint 4: "Step 4: Install the front Jibs to finish the basic structure assembly." Press the "Front Jibs" button to activate model front jibs.
9. Pinch to assemble part front jibs. Hint 4 will be replaced by the next hint.
10. Hint 5: "Step 5: Install the masts to raise the tower crane." and press the "Mast" button to activate masts.
11. Pinch to assemble all the masts to raise the tower crane and complete the assembling of tower crane.

3.3 *Simulated Tower Crane Lifting*

The simulation also provides trainees with the perspective of being a tower crane operator from the operator's cabin. This allows trainees to experience controlling a tower crane from a safe environment and can even be used for training tower crane operators. Compared to traditional training methods, training in a virtual environment has several advantages.

In a virtual environment, the cost of training a tower crane operator is much lower. without the need to assemble and disassemble a tower crane, as well as the reduced need for electricity and power supply. Using virtual reality also increases the convenience of training since it only requires a head-mounted display. Without using an actual tower crane, trainees eliminate the chance of exposing themselves to the safety risks of operating a heavy industrial machine. This simulation can also be modified to be suited to other crane models, allowing trainees to experience controlling the various models. This flexibility is only available to the virtual reality simulation as opposed to when training using actual cranes.

The motions that are involved in the simulation of controlling a tower crane, are the movements of the jib, trolley, and hook. These three parts are controlled by their parent group where each group corresponds to the controls, and degrees of motion are determined by trainees. For the camera view to move with respect to the first person's perspective of the crane's rotation, the camera controller *LMHeadMountedRig* is placed under the parent of the tower crane model. The mesh and material of the window model are then adjusted to transparent for a better perspective. The motion of moving the jib is named "top". It includes moving all components above the slewing unit that can rotate around the axis. This allows trainees to rotate the jib via a keyboard. The script for rotation control is:

Example of rotation control script

```
using UnityEngine;
using System.Collections;

public class rotate : MonoBehaviour {
    public GameObject crane;
    public float velocity;
    void Start () { }
    void Update () {
        if (Input.GetKey(KeyCode.LeftArrow))
        {
            transform.Rotate(0, -1, 0*velocity);
        }
        if(Input.GetKey(KeyCode .RightArrow))
        {
            transform.Rotate(0, 1, 0 * velocity);
        }
    }
}
```

When trainees press the left arrow on the keyboard, the tower crane will rotate anticlockwise. When the right arrow is pressed, the tower crane rotates clockwise instead. The script also has a rotating speed that can be varied depending on trainees' preference as seen in Fig. 15.

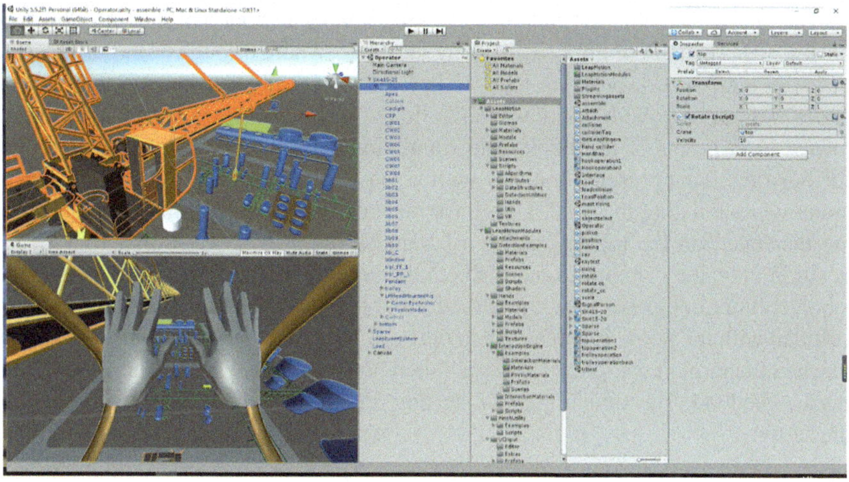

Fig. 15 Tower crane operation in simulation

The trolley on the other hand has a motion of sliding along the front jib via pulleys. The movement of the trolley is similarly, controlled with a keyboard via its up and down arrow keys. The up-arrow key moves the trolley forward and the down-arrow key moves the trolley backwards. Since the jib has a definite length, the script includes values that act as the farthest and nearest limit points as endpoints to restrict the movement of the trolley. The script below controls the trolley's movement with every update frame:

Example of trolley movement script

```
void Update () {
    Z = gameObject.transform.localPosition.z;

    if (Z <= max && Z >= min − 100)
    {
        if (Input.GetKey(KeyCode.UpArrow))
        {
            transform.Translate(Vector3.forward * velocity);
        }
    }
    if (Z <= max+100 && Z >= min)
    {
        if (Input.GetKey(KeyCode.DownArrow))
        {
            transform.Translate(−Vector3.forward * velocity);
        }
    }
}
```

The final movement provided in the simulation is the motion of moving a hook to hoist loads. The path of the hook is along the cable, where through a similar function of using a keyboard, pressing the 'w' key moves the hook upwards and pressing the 's' key moves the hook downwards. It also has two values as the highest and lowest limit points of the hook. The simulation also includes a load as the target for hoisting. This presents trainees with a sense of realism of controlling a tower crane while hoisting the load. For convenience, the positions of the hook and load are displayed on the top left corner of trainees' field of view, similar to a GPS's position indicator. The hook and load components in the simulation are attached to a box collider component for collision detection. When the load is detected to have collided with the hook, it will attach to the hook and move with it. After the load has reached the target position, trainees can then press the space bar on the keyboard to release the load. The attaching and releasing functions of the load are controlled by the script:

Example of attaching and releasing function of load script

```
using UnityEngine;
using System.Collections;
public class Attach : MonoBehaviour {
    public GameObject hook;
    public GameObject load;
    private float X;
    private float Y;
    private float Z;
    private bool attach = false;
    void Start () { }
void OnTriggerEnter(Collider other)
    {
        if (other.gameObject.tag == "hook")
        {
            attach = true;
        }
    }
    void Update () {
        X = hook.transform.position.x;
        Y = hook.transform.position.y;
        Z = hook.transform.position.z;
        if (attach == true)
        {
            load.transform.position = new Vector3(X,Y-400,Z);
        }
        if(Input.GetKey(KeyCode.Space))
        {
            attach = false;
        }
    }
```

In addition, the simulation has a scoring system that indicates any undesirable collisions made by trainees. All environmental objects have a box collider component attached to them, such that when the hook is detected to have collided with an environmental object, 20 marks will be deducted from the total of 100 marks. Hence a total of four collisions are tolerated. Upon the fifth collision, the simulation will end with the text "Game Over" appearing on the screen as seen in Fig. 16. The scoring system provides an accuracy test in hoisting and releasing the load from the tower crane hook, which will be suitable for training a tower crane operator. The scoring and ending script for the simulation are shown below:

Example of scoring and ending script

```
using UnityEngine;
using System.Collections;
using UnityEngine.UI;

public class loadcollision : MonoBehaviour {
    public Text Score;
    private int count;
    public Text gameover;
    void Start () {
        count = 1;
    }
    void OnTriggerEnter(Collider other)
    {
        if (other.gameObject.tag == "building")
        {
            Destroy(other.gameObject);
            count = count + 1;
            if(count>=6)
            {
                gameover.text = "Game Over";
            }
        }
    }
    void Update () {
        Score.text = "Score:" + (120 - count * 20);
    }
}
```

When operating a real tower crane, the operator controls the crane using a panel and rocking bars to achieve every degree of motion. In comparison, the simulation uses a keyboard as the form of input, where the left and right arrow keys are used to rotate the jib, up and down arrow keys for moving the trolley along the jib, and 'w' and 's' keys for moving the hook. This provides a sufficiently close experience that will be useful for novice tower crane operators to get a feel of how it is like to operate a tower crane while staying in a safe environment.

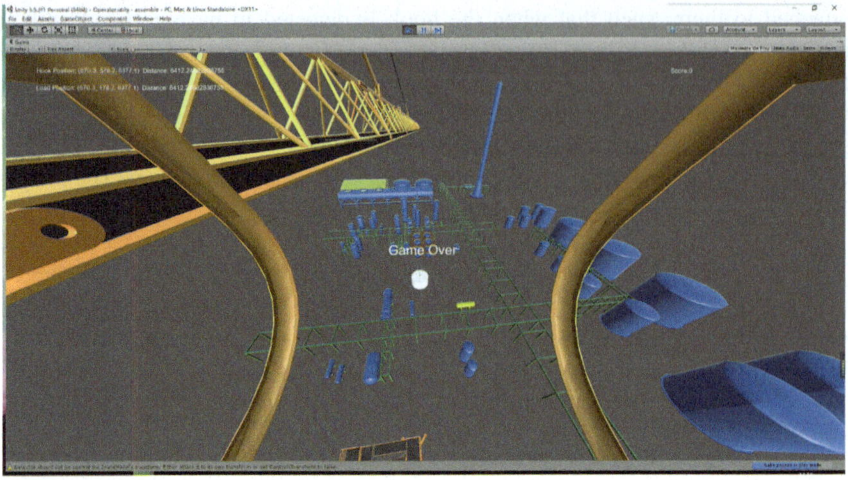

Fig. 16 Tower crane operation simulation interface during "game over"

3.4 *Immersive Visualization*

Finally, the simulation provides a first-person perspective of being a signalman at a construction site as shown in Fig. 17. trainees learn to instruct tower crane motions through hand gestures, approximating how signalmen communicate with crane operators from the ground on site. Apart from allowing novice trainees to experience the job scope of signalmen, this part of the simulation can also be used as a training tool. Although this simulation system is unable to completely replicate the experience of communicating with an actual crane operator, it suffices as a training tool with its reduced cost, reduced safety risk and increased availability.

Comparable with the other parts of the simulation, the model is sorted into two-parent groups: top and bottom. The bottom part of the tower crane model is fixed while the top is further divided into trolley and hook groups, providing three degrees of freedom. For this segment of the simulation, the camera controller *LMHeadMountedRig* is detached from the top group, allowing trainees' field of view to be fixed near the ground and load. For the rotation of the top part of the tower crane that involves components above the slewing unit, there are two scripts, Topoperation1 and Topoperation2, used to control the clockwise and anticlockwise motion of the rotation, respectively. In each of the scripts, three detectors: *Finger Direction Detector*, *Extended Finger Detector*, and *Detector Logic Gate* used for controlling operation conditions are included. The script below includes the order for both activation and deactivation of three hand gestures detectors.

Example of three hand gestures detectors script

```
using UnityEngine;
using System.Collections;
```

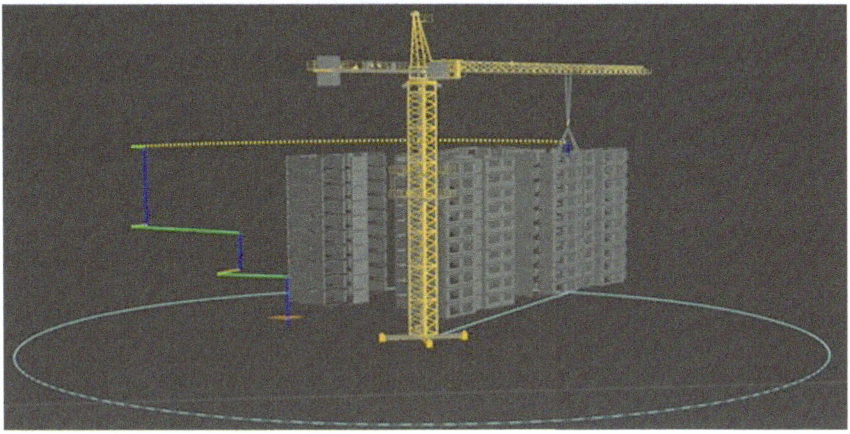

Fig. 17 Signalman's perspective in simulation

```
public class topoperation1 : MonoBehaviour {
    private bool _isActive = false;
    private float step = 0.001f;
    private int maxStepNum = 10000;
    public bool ActivateOnEnable = true;
    public bool DeactivateOnDisable = true;
    public float velocity;
    void Start () { }
    public bool IsActive
    {
        get
        {
            return _isActive;
        }
        set
        {
            _isActive = value;
        }
    }
    public virtual void Activate(bool rotation = true)
    {
        IsActive = true;
    }
    public virtual void Deactivate(bool rotation = false)
    {
        IsActive = false;
    }
```

```
    void Update () {
        if (IsActive == true)
        {
            for (int i = 0; i < maxStepNum; ++i)
            {
                transform.Rotate(0,step*velocity,0)
            }
        }
    }
    private void OnDisable()
    {
        if (DeactivateOnDisable)
            Deactivate(false);
    }
    private void OnEnable()
    {
        if (ActivateOnEnable)
            Activate(false);
    }
}
```

The first detector is the *Finger Direction Detector*, used to detect the direction of a chosen finger, using the x, y, z-axis. The second is the *Extended Finger Detector*, used to detect if every finger of each hand is extended. When the hand detector detects that trainees' fingers are extended such that they meet the conditions for a preset condition, the function of the detector will be activated. The final detector is the *Detector Logic Gate*, a switch for several detectors combined. The *Detector Logic Gate* works such that the switch condition will only change when the logic gate of all detectors is active. Hence, so long as one detector is inactive, the switch will be inactive as well. For this simulation, the *Detector Logic Gate* is a combination of both the *Finger Direction Detector* and the *Extended Finger Detector*.

The motions of the trolley and hook are controlled by a *Palm Direction Detector* used to activate the various scripts with the appropriate hand gestures. The trolley and hook have a limited range of motion that is predetermined by the script, constricted by the height of the trolley and the length of the jib. The script *trolleyoperation* controls the trolley by moving it forwards along the jib and the script *trolleyoperationback* controls the trolley by moving it backwards. *Hookoperation1* controls the hook by hoisting it up when activated whereas *hookoperation2* controls the hook by releasing it.

The communication method between signalmen and crane operators via complex body movements are simplified to simple hand gestures for this signalmen simulation, to provide a more natural and realistic experience. The simulation only aims for trainees to complete the foundational functions with the corresponding hand gestures shown in Fig. 18 and described below:

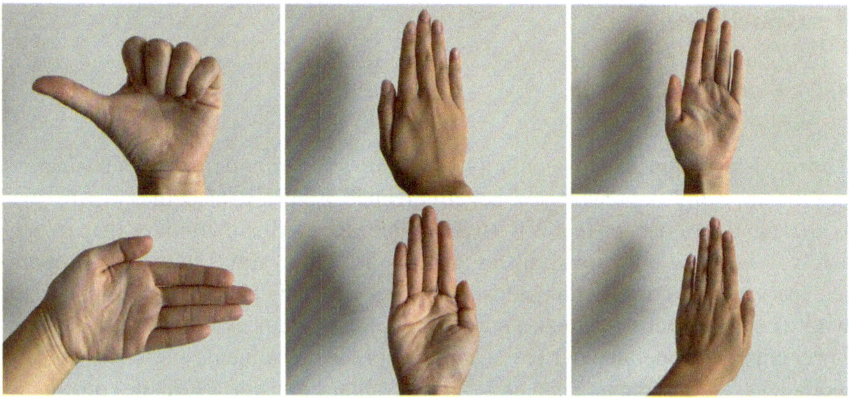

Fig. 18 Signalman's perspective in simulation

1. Rotation of the tower crane via trainees' left hand. To rotate the top part of the tower crane clockwise, the left hand should be clenched, before extending the thumb and positioning the hand such that the thumb points towards the direction $(-1, 0, 0)$, left with respect to the camera. To rotate anticlockwise, extend all five fingers of the left hand and position the hand such that the left thumb points towards the direction $(1, 0, 0)$, right with respect to the camera.
2. The motion of the trolley via trainees' right palm. To activate the script *trolley-operation*, which moves the trolley forward, direct the right palm towards $(0, 0, -1)$, forward relative to the camera. As for activating the script *trolleyoperationback* that moves the trolley backwards, direct the right palm towards $(0, 0, 1)$, backwards with respect to the camera.
3. Control of hook via trainees' left palm. To activate *hookoperation1* which hoists the hook up, direct the left palm towards $(0, 1, 0)$, upwards relative to the camera. To activate *hookoperation2* which moves the hook down, direct the left palm towards $(0, -1, 0)$ which is downwards relative to the camera.

The signalman's simulation consists of collision settings for the load and environmental objects, the position indication system for both the hook and load, as well as a scoring system resembling the operator's simulation. The objects in the simulation have box colliders attached to them, such that when the hook collides with the load, the load will become attached to the hook and will follow the movements of the hook. If the load collides with environmental objects, 20 marks will be deducted from the initial 100 marks, and when there have been 5 collisions, the system will end and the text "Game Over" will appear on the trainees' screen to indicate the failure of the operation, identical to the operator's simulation.

4 Conclusion

4.1 Contribution

The rapid development and popularity of VR allow for the anticipation of the increase in efficiency and unprecedented benefits reaped from VR technology. This work is the proof of concept that VR is a suitable approach for training that requires safety. With this simulation, trainees are provided with various immersive scenarios to experience the construction industry without the safety hazards. Furthermore, this work is suitable for educational purposes, assisting trainees in understanding how tower cranes are operated, what crane operators and signalmen's job scopes are. This work demonstrates the large potential and advantage of using VR in the construction environment which can be extended to other fields.

4.2 Recommendations for Further Studies

Future studies are required to address the following limitations. First, the current work is only used for tower cranes, limiting its merits to only trainees of a single type of vehicle. By adapting the simulation to be used on other crane models and even heavy machines and vehicles from various industries, it will benefit a larger circle of people. Secondly, the current work is a simulation suitable for providing a safe environment for training tower crane operators, but it cannot be used to replace the need for trainees to operate the actual tower crane. With further developments in technology, be it Leap Motion or hand tracking, the simulation will be able to simulate even more closely to real-life conditions, replacing tower crane training completely and possibly even replacing operating tower cranes. This will help to improve the trainee's experience of using the simulation and increase the adequacy of using this work as a training program.

References

1. Alsop, T.: Virtual reality (VR)—statistics and facts (2021). https://www.statista.com/topics/2532/virtual-reality-vr/. Last accessed 28 February 2021
2. Anderson, J., Brody, W., Kriz, C., Wang, Y., Chui, C.K., Cai, Y., Viswanathan, R., Raghavan, R.: daVinci: a vascular catheterization and interventional radiology-based training and patient pretreatment planning simulator (1996)
3. Cai, Y., Chia, N.K.H., Thalmann, D., Kee, N., Zheng, J., Thalmann, N.: Design and development of a virtual dolphinarium for children with autism. In: IEEE Transactions on Neural Systems and Rehabilitation Engineering : A Publication of the IEEE Engineering in Medicine and Biology Society, pp. 208–217 (2013). https://doi.org/10.1109/TNSRE.2013.2240700
4. Cai, Y., Chiew, R., Nay, Z., Indhumathi, C., Huang, L.: Design and development of VR learning environments for children with ASD. Interact. Learn. Environ. **25**, 1–12 (2017). https://doi.org/10.1080/10494820.2017.1282877

5. Cai, Y., Chui, C.K., Ye, X., Fan, Z., Anderson, J.: Tactile VR for hand-eye coordination in simulated PTCA. Comput. Biol. Med. **36**, 167–80 (2006). https://doi.org/10.1016/j.compbiomed.2004.10.002
6. Cai, Y., Goei, S.L.: Simulations, Serious Games and Their Applications (2014). https://doi.org/10.1007/978-981-4560-32-0
7. Chiang, P., Zheng, J., Yu, Y., Mak, K., Chui, C.K., Cai, Y.: A VR simulator for intracardiac intervention. Comput. Graph. Appl. IEEE **33**, 44–57 (2013). https://doi.org/10.1109/MCG.2012.47
8. Condon, M.: Global construction sector to grow 35 softer pandemic impact (2020). https://www.icis.com/explore/resources/news/2020/09/11/10551509/global-construction-sector-to-grow-35-to-2030-on-urbanisation-softer-pandemic-impact/. Last accessed 05 May 2021
9. Facebook to acquire oculus (2014). https://about.fb.com/news/2014/03/facebook-to-acquire-oculus/. Last accessed 16 December 2021
10. Global construction industry set to grow by 5.2 globaldata (2021). https://www.globaldata.com/global-construction-industry-set-grow-5-2-2021-according-globaldata/. Last accessed 05 May 2021
11. Learn about tower cranes—their uses, basic components and specifications (2009). https://www.brighthub.com/education/homework-tips/articles/44062/. Last accessed 05 May 2021
12. Lu, A., Chan, S., Cai, Y., Huang, L., Nay, Z., Goei, S.L.: Learning through VR gaming with virtual pink dolphins for children with ASD. Interact. Learn. Environ. **26**, 1–12 (2017). https://doi.org/10.1080/10494820.2017.1399149
13. Martinelli, K.: Crane safety hazards and control measures (2018). https://www.highspeedtraining.co.uk/hub/crane-safety-hazards-control-measures/. Last accessed 05 May 2021
14. Meisels, M.: 2021 engineering and construction industry outlook (2020). https://www2.deloitte.com/us/en/pages/energy-and-resources/articles/engineering-and-construction-industry-trends.html. Last accessed 05 May 2021
15. Microsoft announces global expansion for HoloLens (2016). https://news.microsoft.com/en-au/2016/10/12/microsoft-announces-global-expansion-for-hololens/#sm.0000hacx3x7r6denyg81m2pmrthzn. Last accessed 16 December 2021
16. (NCCCO), N.C.F.T.C.O.C.O.: Load chart manual (2013). https://www.nccco.org/docs/default-source/reference-materials-2014/lift-director-tower-crane-load-charts-102913a-web.pdf?sfvrsn=2&sfvrsn=2. Last accessed 05 May 2021
17. News, C.N.: Crane operator hand signal guide (2016). https://cranenetworknews.com/crane-operator-hand-signal-guide/. Last accessed 05 May 2021
18. Rosenfeld, J.: What osha safety data tells us about tower crane accidents (2021). https://www.rosenfeldinjurylawyers.com/news/what-osha safety-data-tells-us-about-tower-crane-accidents/. Last accessed 05 May 2021
19. Rr820—tower crane incidents worldwide (2010). https://www.hse.gov.uk/research/rrhtm/rr820.htm. Last accessed 05 May 2021
20. Sutherland, I.E.: The ultimate display. In: Proceedings of IFIP Congress (1965)
21. Top race diecast tower crane construction vehicles. https://www.amazon.com/Top-Race-Diecast-Construction-Vehicles/dp/B07YSX5DQB. Last accessed 05 May 2021
22. Tower crane code of practice 2017 (2017). https://www.worksafe.qld.gov.au/__data/assets/pdf_file/0021/20991/tower-crane-code-of-practice-2017.pdf. Last accessed 05 May 2021
23. ultraleap: Leaprts. https://di4564baj7skl.cloudfront.net/documentation/v2/unity/unity/Unity.LeapRTS.html. Last accessed 05 May 2021
24. Unity (game engine). https://en.wikipedia.org/wiki/Unity_(game_engine). Last accessed 05 May 2021
25. What are tower cranes, and how to buy tower cranes? (2021). https://camamach.com/blog/what-are-tower-cranes-and-how-to-buy-tower-cranes/. Last accessed 05 May 2021
26. Workplace safety and health (operation of cranes) regulations 2011 (2011). https://sso.agc.gov.sg/SL/WSHA2006-S515-2011. Last accessed 05 May 2021

Mixed Reality Applications in Tertiary Veterinary Education: A Systematic Review

Xuanhui Xu, David Kilroy, Arun Kumar, Muhammad Zahid Iqbal, Eleni Mangina, and Abraham G. Campbell

Abstract Virtual Reality (VR), Augmented Reality (AR), and Mixed Reality (MR) provide a novel way to support medical education, helping medical students learn human anatomy and surgery. Veterinary students face additional learning challenges, as they need to deal with many different species. The same techniques may potentially address the problems associated with increased class sizes. A systematic review was carried out for existing articles up to 30/06/2021 in Embase, PubMed, Scopus, ProQuest, Cochrane Reviews, CNKI and Xplore. Searches from Google Scholar and other resources are also acceptable. PRISMA guidelines were adhered to when reporting the results of this study. The electronic searches generated a total of 425 studies. One additional record was identified through other sources. After applying inclusion criteria and exclusion criteria, a total of 22 studies (887 participants) were identified for inclusion in the review. The systematic review reported the current state of VR/AR/MR applications in veterinary education. The simulator was generally used to support the training process with mainly haptic feedback and the learning outcomes could be further enhanced if adapted to the vet curriculum. There are some initial explorations on applying VR and AR in several subjects and the preliminary results showed that the MR techniques had great potential to be utilised in the daily teaching routine. However, further investigation is still needed.

Keywords Mixed reality · Virtual reality · Augmented reality · Simulator · Veterinary · Education · Systematic review

---

X. Xu (✉) · M. Z. Iqbal · E. Mangina · A. G. Campbell
School of Computer Science, University College Dublin, Dublin, Ireland
e-mail: xuanhui.xu@ucdconnect.ie

D. Kilroy · A. Kumar
School of Veterinary Medicine, University College Dublin, Dublin, Ireland

Y. Cai et al. (eds.), *Mixed Reality for Education*, Gaming Media and Social Effects,
https://doi.org/10.1007/978-981-99-4958-8_10

1 Introduction

Medical students encounter many challenges when studying human anatomy and learning to perform detailed complex surgeries. The students' basic teaching materials are mainly paper-based notes, slides, and 2D videos, which might cause misunderstandings due to the dimension difference and the loss of binocular parallax to the actual object. Thankfully, the applications of Virtual Reality (VR), Augmented Reality (AR), and Mixed Reality (MR) techniques stimulate learning motivation and improve the learning outcomes of medical students [1, 17, 21, 38].

However, veterinary students face additional learning challenges. Compared with the human field, they need to deal with multiple species that have huge variations in size and anatomy, from small rodents to large horses. The veterinarians are not only facing challenges in theoretical knowledge acquisition but also in conducting operations in abnormal patient positions based on the species and surgery types. One example is that laparoscopic procedures are typically performed with animal patients standing instead of lying down [8, 13]. So bespoke veterinary simulators would be needed instead of those used in human medicine. This is currently not possible due to the costs and the limitations of facilities used for veterinary training [3]. Moreover, in radiography diagnostic training, veterinary requires large animal, such as horses and cows, that is further limited in the most of institutions, whereas the students can role play the projections as patients in medical education. VR and AR have the potential to offer a unique new tool to solve this problem since the virtual environments bring the the potential to the classroom by merging different VR and AR techniques in veterinary face-to-face and remote education [27, 37, 39].

This study focuses on a systematic review to evaluate the effectiveness and efficiency of applying MR in veterinary education. To this end, the systematic review will answer the following questions:

- What kind of VR or AR techniques have been applied to support veterinary teaching?
- What modules or surgeries apply these techniques?
- Is there any study proving or trying to prove the effectiveness and efficiency of applying these techniques?
- To what extent do these techniques support the training process when compared with other methods?

1.1 Protocol

A systematic review was carried out up to 30/06/2021. PRISMA guidelines were adhered to when reporting the results of this study [25]. Methods of analysis and inclusion criteria were specified in advance and documented in a protocol, which was not published or registered.

1.2 Search Strategy

The literature search and initial screening were conducted by the author; abstract screening was conducted independently by four researchers with different opinions being rectified by discussion; full paper screening and data extraction were conducted by the author. Databases searched were Embase, PubMed, Scopus, ProQuest, Cochrane Reviews, CNKI, Xplore on title, abstract, and keywords. Searches from Google Scholar or other resources were also acceptable. The terms used for searching were: (augmented OR virtual) AND reality AND (veterinary OR vet) AND (training OR education OR study OR trial).

1.3 Selection Criteria

The selection criteria of this review are shown as follows:

- Inclusive criteria: all studies examining the general adult population or healthy adults and people who are or were veterinary-related (novel or experts) were included.
- Exclusive criteria: studies in which individuals were selected with extreme motion sickness, other diagnosed illness or disability and studies in which individuals were not veterinary-related were excluded.
- Year limitation: no year publication limits were set.
- Language: English and Chinese text publications were included as one author is a native Chinese speaker, which provided a unique opportunity to expand the search.

The search was last updated on 03/06/2021. The titles and abstracts database searches were screened to identify potentially relevant records for full-text screening. The titles and abstracts of all remaining records were screened for eligibility to identify records for full-text screening. All records identified for full-text screening were examined to identify records for inclusion in the review. All data potentially relevant to the review were then extracted from the studies selected for final inclusion and collated in a spreadsheet as follows: Details of publication; Participant characteristics; Sample size; Setting; Intervention; Study design; Data type; Result.

A meta-analysis was not undertaken due to the considerable heterogeneity among the studies included in this review. Therefore, a descriptive approach to data synthesis was adapted, whereby summaries of included studies will be presented. Included studies will be presented in line with the outcomes identified from active interventions that involve VR or AR, specifically, changes in training or outcomes related to the trainer or trainee's experience (satisfaction, motivation); population characteristics (study design, study outcome measures); and methodology (pilot studies, new approaches).

Data were extracted using a standardised template to capture information relating to PICO: Population (grade); Intervention (characteristics of the VR and AR tools); Comparator (traditional/other teaching methods); Outcomes (assessment score, time, subjective feeling).

1.4 Quality Assessment

The study's quality information was collected using RoB 2.0 Tool [31] for assessing the risk of bias of the study. The risk-of-bias plots were created by using the robvis tool [24].

2 Results

2.1 Study Selection

The electronic searches generated a total of 425 studies. One additional record was identified through other sources. After removing duplicates, 238 remained. Of these, 172 studies were excluded because the papers did not meet the criteria after reviewing the abstract. The full text of the remaining 66 papers was reviewed. Among those, 43 papers were discarded due to the reasons given in the flow chart (Fig. 1). After applying the inclusion criteria and exclusion criteria, 22 studies were identified for inclusion in the review (Table 1). No unpublished relevant studies were obtained. Twenty-one studies were written in English. One study was written in Japanese [29], the content of which was translated by the online translator.

2.2 Study Characteristics

2.2.1 Methods

Among the included studies, seven were randomised control trials (RCTs) [2, 4, 6, 14, 19, 22, 23]; one was a case–control study [18]; three were pilot studies [30, 34, 35]; five were cross-sectional studies [3, 11, 16, 29, 32]; two were evaluation studies [12, 20], and the remaining four were new methodologies without any evaluation [7, 9, 28, 33].

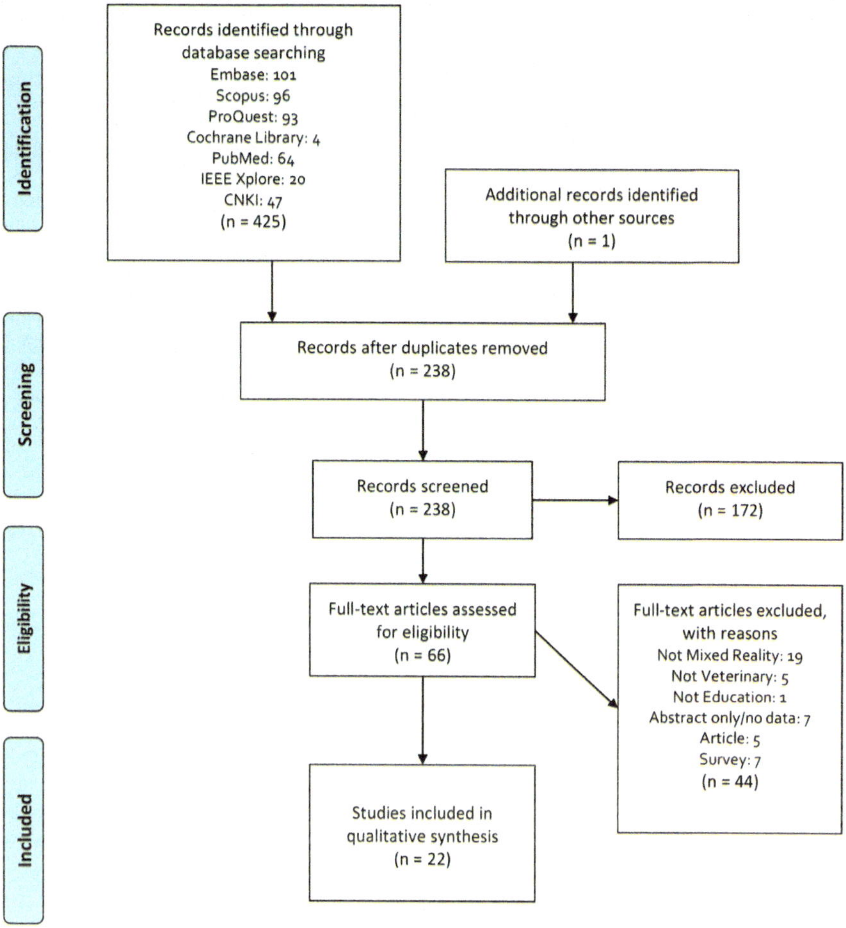

Fig. 1 PRISMA flowchart

2.2.2 Participants

The included studies involved 887 participants. The main inclusion criteria resulted in a population of veterinary students (first-year to master students), pre-veterinary students, veterinary lecturers, and veterinarians.

2.2.3 Intervention

The interventions applied in these included studies are classified into three different categories: non-immersive and semi-immersive simulators, AR devices, and immersive VR devices. Simulators included PHANToMs haptic simulator (n = 6) [2–4, 11, 16, 28], ProMIS simulator (n = 3) [6, 12, 20], Lap-Sim simulator (n = 2) [6, 12], VR

Table 1 General data of included studies

Author (s)	Sample size (n)	Intervention	Study design	Key findings
Baillie et al. [4]	16	PHANToMs	Randomised control study	The study group were significantly better at finding the uterus ($p < 0.01$). The feedback from students was positive and indicated that students found the simulator easy to use
Forrest et al. [11]	26	PHANToMs	Cross sectional study	The veterinarians achieved almost two perfectly identifiable stiffness levels and veterinary students achieved one correctly identified level. The difference between the two groups was significant ($p < 0.001$). Veterinarians used a greater mean maximum force (2.0 N) compared to students (1.6 N) ($p < 0.05$)
Lee et al. [18]	60	AR intravenous (IV) injection simulator & VR IV simulator	Case control study	The result showed that the injection simulator enhanced performance compared to conventional approaches
Lee et al. [19]	40	AR IV simulator	Randomised control study	The study group were more proficient at IV injection technique using real dogs than the control group ($P < 0.01$). The students agreed that they learned the IV injection technique through the AR simulator
Fransson et al. [12]	25	ProMIS & LapSim	Evaluation study	All metrics and basic laparoscopic skills (BLS) score were significantly correlated with experience (except for the hand dominance metric) in AR, whereas none of them showed that in VR
Seo et al. [30]	11	HTC VIVE	Pilot study	The pilot study showed that participants most enjoyed interacting with anatomical content within the VR program. Participants spent less time assembling bones in the VR, and instead spent a longer time tuning the orientation of each VR bone in the 3D space

(continued)

Table 1 (continued)

Author (s)	Sample size (n)	Intervention	Study design	Key findings
Lencioni et al. [20]	26	ProMIS	Evaluation study	For horizontal group, scores for 3/5 tasks were significantly worse when tested in the vertical plane. For vertical group, scores for 2/5 tasks were significantly worse when tested in the horizontal plane. the performance of the vertical group was slightly better than the horizontal group
Chen et al. [6]	33	LapSim & ProMIS	Randomised control study	Post-training BLS results were improved in both training groups ($p < 0.001$). Based on motion metrics analysis, participants completed tasks in a shorter time ($P = 0.0187$), and with better economy of movement ($P = 0.0457$) after training
Christ et al. [7]	N/A	HTC One M8 & Samsung Tablet Galaxy 10	N/A	Showed a workflow of converting medical data to virtual model and applied it into AR application
Xu et al. [34]	9	HTC VIVE	Pilot study	Showed a workflow of converting medical data to virtual model and applied it into VR application. Feedback from students indicated that VR HMD had been better for canine anatomy training assessments than paper-based materials
DeBose [9]	N/A	HTC VIVE & Oculus Rift CV1	N/A	Feedback from the students indicated the VR program was very or extremely effective for learning anatomical features, the labelling was effective, and the anatomical locations from the VR dog model could be transferred to the live dog
Hunt et al. [14]	44	Voxkin VR headset	Randomised control study	Groups spent similar time preparing to perform surgery. Study group watched VR videos for a median of 90 min. Groups did not differ in surgical performance scores or time ($p > 0.05$)

(continued)

Table 1 (continued)

Author (s)	Sample size (n)	Intervention	Study design	Key findings
Xu et al. [35]	7	Samsung Galaxy S7 & HTC VIVE Focus	Pilot study	Communicability. Mobile VR had a better feeling of control and a more detailed model, and is more immersive than AR ($p < 0.05$)
Wilkie et al. [33]	N/A	Hololens	N/A	A MR application to guide practitioners in the femoral nerve block procedure was developed. The anes the siologist found the visualization of the leg would not affect continuous visual contact with the patient
McCool et al. [23]	12	EndoVR	Randomised control study	No significant differences were found between training methods. The intervention was less stressful for participants than the live dog laboratory (LDL) ($p = 0.02$). Feedback from participants found that the VR was a useful and acceptable alternative to the LDL for training of early endoscopy skills
Satoshi et al. [29]	142	Smartphone VR	Cross sectional study	Feedback from students indicated the 360° video was good in learning canine tracheal intubation. There was no significant difference in the evaluation of VR teaching materials by students who wished to practice using live animals or alternative models
Vega-Garzón et al. [32]	48	Acarovet (smartphone VR) & Entomovet (smartphone AR)	Cross sectional study	The grade value was higher in the class after implementation. The SUS total score was 80.05. Feedback from students considered it a useful tool

(continued)

Table 1 (continued)

Author (s)	Sample size (n)	Intervention	Study design	Key findings
Little et al. [22]	74	iPad	Randomised control study	Both groups improved in cardiac knowledge. No significant difference in post-test improvement between the two groups (p = 0.9). A positive correlation was found between spatial awareness scores and post-test improvement regardless of cohort group (p = 0.03). Participants preferred learning with AR compared to traditional methods

intravenous injection simulator (n = 1) [18], and EndoVR simulator (n = 1) [23]. AR devices included AR intravenous injection simulator (n = 2) [18, 19], mobile AR devices (n = 4) [7, 22, 32, 35], and AR head-mounted headset (n = 1) [33]. VR devices included VR head-mounted headset (n = 4) [9, 30, 34, 35], smartphone VR headset (n = 3) [14, 29, 32].

2.2.4 Outcomes

Figure 2 shows a trend of all the interventions applied in veterinary education. In 2005, Baillie et al. first used PHANToMs haptic simulator to train veterinary students to find the bovine uterus by two studies [2, 3]. They then explored and improved this PHANToMs teaching system, while the latest article was published in 2011. This haptic simulator was replaced in 2016 by more advanced human medical operation simulators such as ProMIS AR Simulator (©CAE Inc. 2021) or LapSim VR Simulator (©Surgical Science 2018) to support vet students' learning process. After 2017, the simulator was not on-trend, and the last article on the simulator was published in 2020.

The first researcher who tried to apply AR in veterinary education was Lee et al. In 2012, this team developed and improved an AR intravenous (IV) injection simulator [18, 19]. The AR trend started in 2018; researchers used AR techniques such as Hololens and Smartphone AR, which in recent years have displaced simulators.

VR had just been utilised in veterinary education in 2017, becoming a rising star in this area. VR techniques such as Smartphone VR (Card-broad VR, Acarovet VR) and immersive VR head-mounted display (HMD) with controllers (HTC VIVE, HTC VIVE Focus, Oculus Rift CV 1) attracted researchers' attention as novel tools to support veterinary students' knowledge acquisition and progress. Most notably, in the past two years (up to mid-2021), over half of the included articles used VR as an intervention.

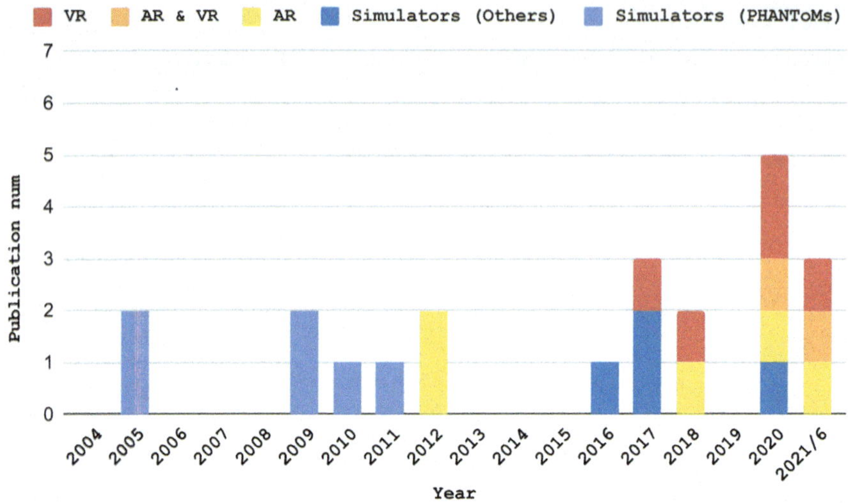

Fig. 2 Number included studies published in the given years classified by intervention

In general, since 2005, researchers have used haptic simulators to teach veterinary students and have still used them in recent years. Although the haptic simulator used in those studies was updated over time, its popularity has decreased. Since 2017, VR and AR have become the new trend towards the future. These MR techniques were increasingly applied as interventions, which indicates that MR has great potential in supporting veterinary education.

Simulator (Non-immersive and Semi-immersive haptic)

Simulators are widely used in human medical operations and education [1, 17, 21]. In medical publications, these haptic simulators are also known as VR/AR/MR techniques, focusing on haptic feedback and hands-on experience. They generally use a conventional 2D monitor to display the virtual content to the user. The simulated environment on the 2D display does not align with reality, losing the similarity to the actual case. In this review, those non-immersive and semi-immersive haptic simulators are classified as simulators, not as recent VR/AR/MR techniques.

The first veterinary education usage of this technique was in 2005. Baillie et al. [2] used PHANToMs haptic simulator to create a bovine rectal palpation simulator to train veterinary students. They conducted an RCT to evaluate this simulator. Compared with the group which was trained by the traditional method ($n = 8$), the study group ($n = 8$) was significantly better at finding the uterus ($p < 0.001$) in the on-farm exam. In the same year, attempts were made to integrate the simulator into the undergraduate veterinary curriculum. The students (session one $n = 69$, session two $n = 43$) found it useful for learning to locate the internal anatomical structures [3], and their later research ($n = 184$) supported this outcome [16]. In 2010, Baillie et al. [4] improved their system to an automated version, allowing the student to learn bovine rectal palpation without an additional instructor. The researchers conducted a similar RCT to evaluate this improved system. The study group (n = 8) was significantly better at finding the uterus in the on-farm exam (proportions of successes, $p < 0.01$) than the control group (n = 8). The questionnaire's feedback was mostly positive, which indicated that students found the intervention enjoyable and useful [2–4, 16]. Eventually, before 2011, they again used PHANToMs to compare the ability of veterinarians and students to identify the stiffness of virtual surfaces through palpation [11]. The difference between the veterinarian group (n = 14) and veterinary student group (n = 12) in mean information transfer was significant (0.97 bits, almost two perfectly identifiable stiffness levels, versus 0.58 bits, one correctly identified level; $p < 0.001$). The result indicated that the simulator could be a summative assessment tool, despite neither group identifying more than two levels of stiffness reliably.

After Baillie et al. explored the application of PHANToMs in veterinary bovine rectal palpation training, the simulator was not investigated in other skills or disciplines until Fransson et al. [12] started. Those simulators first appeared in 2002 and were widely applied in clinical medicine education [12, 21]. In contrast, this simulation training for veterinary education was limited. To determine the construct and concurrent validity of instrument motion metrics for laparoscopic skills assessment in those simulators for veterinary students, an evaluation study (veterinary student, n = 14; veterinarians, n = 11) was conducted and showed that the AR simulator (ProMIS) performed better than the VR simulator (LapSim) [12]. The results indicated the potential usefulness of this AR simulator in assessing veterinary students' laparoscopic skills and associated learning outcomes. A further investigation about the VR LapSim simulator was conducted in 2017.

Using ProMIS AR simulator as an assessment tool, the study found that after training with LapSim simulator, the basic laparoscopic skills (BLS) were significantly improved in the group (n = 17) trained with the standard BLS programme, and the group (n = 16) trained with a variety of basic skills in a canine abdominal model and on the cholecystectomy module ($p < 0.001$). There was no significant difference found between the two curricula, suggesting that students could gain the essential skills during the process of learning from veterinary anatomy models and surgeries [6].

Later that year, these researchers explored what effect the simulator's orientation had on veterinary students' BLS performance using the ProMIS AR simulator. Unlike in the human field, veterinarians are challenged by multiple species with immense variations in body size and anatomy and the most common laparoscopic procedures are performed with animal patients standing [8, 13]. This study found that the BLS acquired during training in the one plane generally did not transfer to the other plane; in the meantime, the performance of the vertical orientation group (n = 14) was slightly better than the horizontal orientation group (n = 12). This result suggested that surgical training for veterinary students should modify the existing training models to maximise the learning experience [20].

After Fransson et al. investigated the use of BLS trainers in veterinary education, there was a three-year hiatus in research on simulators in veterinary education, until McCool et al. [23] evaluated a human virtual-reality endoscopy simulator for teaching these skills to veterinarians. The researchers compared the two groups of veterinarians with limited or no endoscopy experience trained with the EndoVR simulator (©CAE Inc. 2021, n = 6) and a live dog laboratory (LDL, n = 6). No significant difference was found between the two groups, and the participants found the simulator less stressful ($p = 0.02$), which indicated the potential of using these simulators as an alternative or as a precursor to working with live animals.

Augmented Reality

The first attempt at applying AR techniques into veterinary teaching was in 2012. Lee et al. [19] developed an AR IV injection simulator to train pre-veterinary and veterinary students who lacked experience of IV injection. They created a silicone model using a silicone-casting technique based on computerised tomography (CT) Digital Imaging and Communications in Medicine (DICOM) data of a canine forelimb and used it as a base for the system. An electric syringe that could provide haptic feedback was assembled as a controller for the system. The display of the system was a conventional monitor linked to the rear camera. By attaching different AR markers to the model and the syringe individually, the system could track the position of two objects in the real world and reflect it in the virtual space. In late 2012, they improved their system by adding a fur wrapper to the model and replacing the AR marker tracking with gyro sensors [18]. This case–control study compared the AR IV injection simulator with the traditional teaching method and a VR simulator version. This simulator version acts similarly to the previously-described models: a haptic device combined with virtual content delivered by a conventional 2D display.

In subsequent years, the AR feature appeared on all kinds of mobile devices, while augmented user experience no longer relied on bulky computers or lab environments. In 2018, Christ et al. [7] developed a workflow for creating a veterinary education tool by using a smartphone AR (HTC One M8 and Samsung Tablet Galaxy). With AR markers, the mobile devices were able to generate a 3D virtual model generated from actual CT DICOM medical data with reality presence [7, 32]. A more advanced model, Xu et al. [35] used ARCore[1] to implement the application, letting the learning process run without additional printed AR markers (Fig. 3). The pilot study result showed that AR was superior regarding the portability and communication aspects in teaching. On iPad, the IVALA[2] application provides similar AR teaching content for veterinary students, the content of which was assessed by Little et al. [22]. Although the group using the tablet AR performed better than the group using traditional methods, there were no significant differences between the two groups in their knowledge of cardiac anatomy

In 2016, Microsoft announced an AR HMD—Hololens (© Microsoft 2021). Its introduction enhanced the ability of researchers to explore AR within education. Many studies investigated the benefits that Hololens could bring to medical education [38], while few focused on veterinary education. Four years after the launch of Hololens, Wilkie et al. [33] proposed a workflow to apply Hololens for

[1] https://developers.google.com/ar.

[2] https://www.ivalalearn.com/.

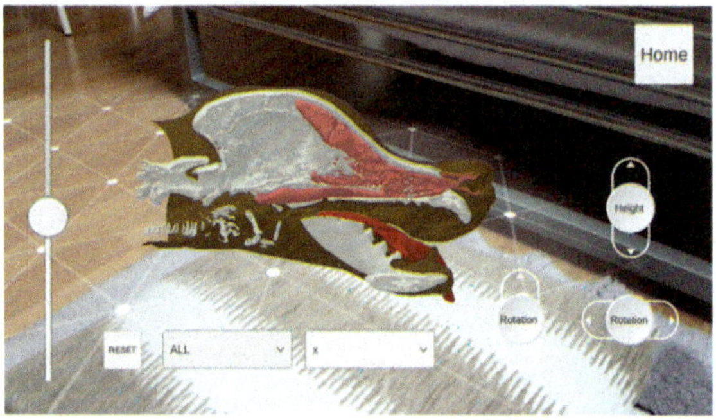

Fig. 3 Research Lab in mobile AR environment [35]. The system would be aware of the spatial information of the surroundings and generate the virtual model based on this information without referring to an AR marker

canine femoral nerve block visualisation. Instead of using CT or Magnetic resonance imaging DICOM data to generate the virtual model, the researchers used a photogrammetry scan. This method maintained the details of the visible part of the cadaver, but the information about internal structures would be lost. An expert was asked to assess the application and found that the hologram generated by the HMD could help the veterinarian identify the target lesion.

Virtual Reality

As shown in the timeline Fig. 2, VR HMD was applied to veterinary learning since 2017, a much shorter time frame than the haptic simulator or AR. Compared to the haptic simulators with non-immersive or semi-immersive displays, VR HMD provides a fully immersive virtual environment and more flexible control to the learners. In 2017, Seo et al. [30] developed an immersive VR teaching tool using Unity 3D Engine (© Unity Technologies 2021) to teach veterinary students canine skeletal systems. After donning the HTC VIVE VR headset (© HTC Corporation 2011–2021), the user could interact with virtual bone parts using controllers. The researchers conducted a pilot study to provisionally evaluate the VR system compared to the traditional teaching material (bone box). The results showed that the ability to directly interact with virtual anatomical contents was more enjoyable. The participant spent much less time assembling the bones in VR due to the benefit of the antigravity field and snapping guidance. However, they spent more time adjusting the coordination of the VR bones as the controllers were probably not compatible with real hands.

In 2018, Xu et al. [34] implemented a canine head teaching and summative assessment tool, in which the virtual model was converted from CT DICOM data (Fig. 4). The pilot study received similar feedback to Seo et al. [30]'s: the interactive 3D visual made the test in VR easier than the traditional multiple choices questions (MCQs). These two provisional research studies indicated the potential benefits that the VR initiative could bring to veterinary education.

Using a similar workflow, DeBose [9] had developed a VR program that was integrated into the first-year students' physical examination labs in 2019. The program was developed by Unreal Engine 4 (© Epic Games, Inc. 2004–2021) and running on HTC VIVE and Oculus Rift CV1 (© Facebook Technologies, LLC.) with the

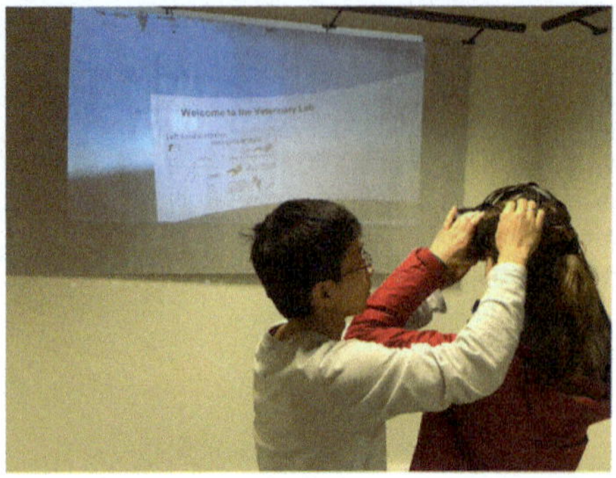

Fig. 4 Pilot study environment [34]. The HTC VIVE VR headset provided the user a fully immersive environment to manipulate the virtual object generated from actual medical data. The lecturer can monitor student's performance via an external display

virtual model generated from medical DICOM data. Experimental results and feedback are needed to prove the efficiency and effectiveness of integrating this system into veterinary students' teaching routines.

VR 360°videos were also raised as a new teaching medium in medical and veterinary education. By using the 360°camera to capture the operation process, the module could repeatedly show the same clips to multiple students. Moreover, most students found it helpful during the learning process [29]. Hunt et al. [14] added 360°VR video into students' preparation as an intervention for the first canine sterlisation surgery. They conducted an RCT to see how participants would use the resources and compared the surgical performance between the two groups. Besides the traditional methods such as text and videos, the study group students also had access to a 2D and a 3D VR surgical video. The 2D VR video was different from a normal video as the user could adjust the view angle by spinning their smartphone. The 3D VR video shared the same logic, while the user needed to insert the smartphone into a headset to play it. Students from the study group spent 11% time on 3D VR video and 22% time on 2D VR video on average (a median of 90 min). Most students reported that VR training was helpful; however, there was no significant difference between groups in the time taken and the total rubric scores.

The 360°video may provide a more immersive experience for the student to observe the teaching material than the traditional 2D videos. These HMD devices are low cost (cardboard or plastic) and portable, making it possible to extend the application scale to the class. However, without controllers, the learning process would not be as interactive as the standalone VR workstation. Vega-Garzón et al. [32] developed a smartphone VR teaching app by using Google VR SDK[3]. The app allowed the user a limited interaction with the teaching materials, which increased the engagement of the learning process. To achieve an interactive experience with controllers while maintaining the low cost and portable benefits, some researchers proposed using the mobile VR HMD such as VIVE Focus (© HTC Corporation 2011–2021) and

[3] https://developers.google.com/vr.

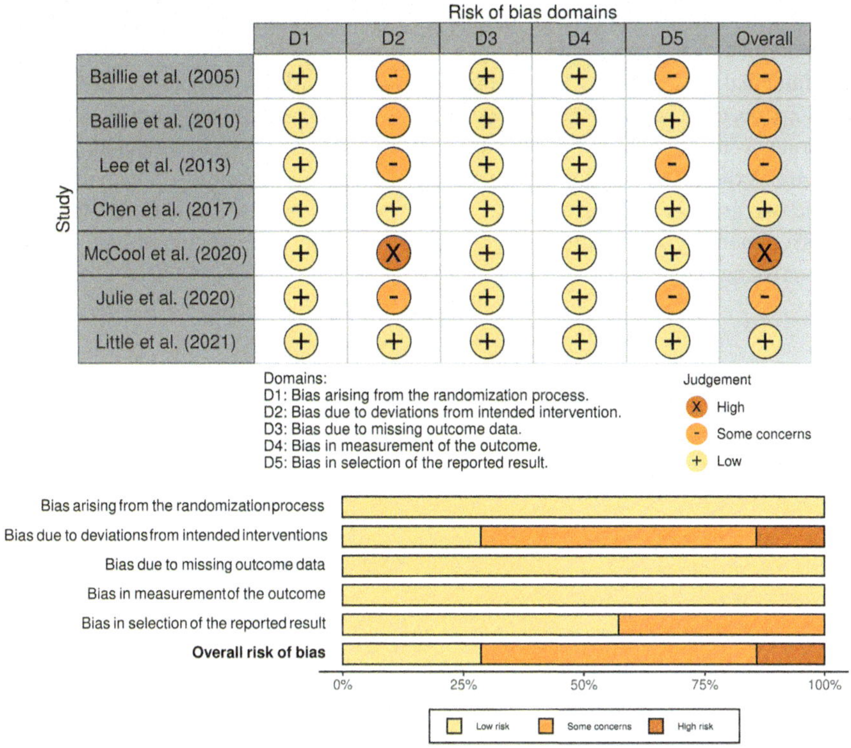

Fig. 5 Risk of Bias analysis

Oculus Quest(© Facebook Technologies, LLC.) [9, 35, 36]. Compared with smartphone AR, the mobile VR HMD provided a more immersive environment and more detailed teaching content for the students [35].

2.3 Risk of Bias Within Studies

To reducc the bias of language, the systematic review screened English studies and Chinese studies; however, there were no Chinese articles included in the review. The quality information from seven randomised control studies was collected using RoB 2.0 Tool [31] for assessing the risk of bias. The analysis process was conducted by the author, using the Excel tool to implement ROB2.[4] All RCT studies' risk of bias was shown in Fig. 5.

Two articles had a low risk of bias, four English articles had some bias concerns, and one had a high risk of bias. All papers clarified that the allocation sequence was

[4] https://www.riskofbias.info/welcome/rob-2-0-tool/current-version-of-rob-2.

random, all outcome data were included without intentional neglect, and the outcome measurement was appropriate. The primary concern was bias due to deviations from the intended intervention. For instance, researchers did not mention whether the participants or the data accessors were blinded to the random assignment, and information was lacking about baseline analysis between groups. McCool et al. [23] had a high risk of bias in this domain due to the different backgrounds of the participants of the two groups, and there was no appropriate analysis used to estimate the effect of assignment to intervention. Some did not mention a pre-specified analysis plan for analysing the result.

3 Discussion

3.1 Summary of Evidence

3.1.1 What Kind of VR or AR Techniques Have Been Applied to Support Veterinary Teaching

As the statement shown in Sect. 2.2.4, different forms of VR/AR/MR techniques had been used in veterinary education and training, generally in three classifications: non-immersive and semi-immersive simulators, VR HMDs, and AR devices.

The simulators used in the included studies are PHANToMs haptic simulator, LapSim VR simulator, ProMIS AR simulator, and EndoVR VR simulator. LapSim and ProMIS were widely used in acquisition of laparoscopic suturing skills [10].

The AR techniques used in the included studies can be classified into three categories: Magic-mirror-alike AR, mobile AR, and AR HMD. The magic mirror concept has generally been used in anatomy education [5]. The system provides intuitive visualisation by using the camera to capture the real surroundings and adding virtual objects to the image of the real scene. The tracking technique can be achieved by using AR markers or gyro sensors. Similar to this form of AR, mobile AR or smartphone AR can add virtual content onto the real scene but with high portability and scalability. Those platforms provide the user with a virtual object to manipulate and they maintain the connection between the user and the real world. AR HMD (Hololens) offers a binocular parallax for the user, which enhances the 3D vision and presence of the virtual object. It also provides a hands-free experience for the user to interact with the virtual and the real world simultaneously.

The VR HMD techniques used the reviewed studies can be categorised into workstation-based VR (interactive), 360°video VR (passive), and mobile VR (interactive). The workstation-based VR such as HTC VIVE and Oculus Rift normally needs a high-capacity PC to drive the dual vision of the HMD so that it can handle complex scenes and detailed models. The virtual model generated from the actual medical data will be the most detailed and it allows users to interact with it in this form of VR. However, workstation-based VR is costly and hard to move around to fit a full class teaching scenario. The 360°video VR is normally seen in a smartphone VR. The producer uses a 360 camera to capture the operation and play it on the smartphone installed into a cardboard (or plastic) helmet. The 360°VR provides an immersive feeling on top of the actual operating room environment footage, where students can observe the surgery at a close distance without any limitations. Thanks to the low price of the helmet, this technique can be easily adapted to the classroom routine. Its disadvantages are three degrees of freedom (3-DOF: pitch, yaw, and roll) and passive learning. To maximise the benefit from the VR technique while maintaining its portability and scalability, researchers started to apply mobile VR HMD (HTC VIVE Focus, Oculus Quest) to the learning routine. The mobile VR HMD does not require a workstation to power it up but supports a 6-DOF (x, y, z, pitch, yaw, and roll) full interactive user experience. These advantages may mean that this becomes a new trend for future research.

3.1.2 What Modules or Surgeries Apply These Techniques

The modules that apply AR/VR/MR techniques are shown in the table (Table 2). As the simulators are built especially for surgical skills, the modules selected for interventions mainly relate to those skills. The PHANToMs haptic simulator was used only in bovine rectal palpation. Similarly, laparoscopic trainers such as LapSim and ProMIS were only applied to evaluate whether the existing devices could benefit veterinary students rather than medical students. Lastly, McCool et al. [23] used the EndoVR simulator to train students in endoscopic skills instead of using an LDL. In contrast to the human medical field, veterinarians are challenged by huge variations in body size and anatomy among species. In this case, the existing simulators could not be directly integrated into the veterinary curriculum but needed essential adjustments and necessary evaluations. Among the included studies, one example was the BLS acquired from a laparoscopic trainer along only one directional plane generally did not transfer to other planes [20]. The student had to be trained with the vertical-placed simulator instead of the standard horizontal placement to maximise the learning effectiveness and perform the surgery with the patient standing.

When it comes to AR techniques, the modules appear to have more options and variations. In one specific instance, Lee et al. [18] implemented the IV injection simulator to virtually train students in injection skills. A different module from practical training, the anatomy modules (such as canine head anatomy and general veterinary anatomy, shown in Table 2) that mainly require observation and understanding tend to use mobile AR (Table AR, and smartphone AR) techniques. Only one study investigated the AR HMD (Hololens) application in canine femoral nerve block visualisation [32], indicating that there is a research gap in evaluating the AR HMD application in veterinary education compared to human medicine.

Almost all modules applying VR as an intervention are canine-related as the dog is the first species studied by veterinary students. Anatomy modules normally use workstation based VR HMD to create an interactive and immersive learning environment.

Table 2 Modules applying non-immersive simulators, AR devices, and VR devices

Category	Modules	Interventions	Researchers
Simulator	Bovine rectal palpation	PHANToMs haptic simulator	Baillie et al. [2, 3], Kinnison et al. [16], Parkes et al. [28], Baillie et al. [4], Forrest et al. [11]
	BLS[a]	LapSim & ProMIS	Fransson et al. [12], Lencioni et al. [20], Chen et al. [6]
	Endoscopic	EndoVR	McCool et al. [23]
AR	IV[b] injection	IV[b] injection simulator	Lee et al. [18, 19]
	Canine head anatomy	HTC One M8 & Samsung Tablet Galaxy & Samsung Galaxy S7	Christ et al. [7], Xu et al. [35]
	Canine femoral nerve block	Hololens	Wilkie et al. [33]
	General veterinary anatomy	iPad	Little et al. [22]
	Veterinary acarology and entomology	Entomovet smartphone AR	Vega-Garzón et al. [32]
VR	Canine skeletal system	HTC VIVE	Seo et al. [30]
	Canine head anatomy	HTCVIVE & HTCVIVE Focus	Xu et al. [34, 35]
	Canine anatomy	HTC VIVE & Oculus Rift CV1	DeBose [9]
	Canine sterilisation surgery	Smartphone VR	Hunt et al. [14]
	Canine tracheal intubation	Smartphone VR	Satoshi et al. [29]
	Veterinary acarology and entomology	Acarovet smartphone VR	Vega-Garzón et al. [32]

[a]Basic laparoscopic skill, [b]Intravenous

Instinctually, we would assume the operation and surgery course uses more interactive VR techniques rather than the passive VR like 360°VR video, because they are more practical compared to the theoretical anatomy course. However, the surgery courses in the included studies (canine sterilisation surgery, and canine tracheal intubation) were both using 360°VR videos. The reasons for this might be that workstation-based VR is more resource-intensive to simulate the operation process when compared to the 360°videos at that stage, which suggests a research gap of comparing the application of both interactive and passive experience VR in veterinary training.

Modules such as canine head anatomy and veterinary acarology and entomology applied both VR and AR technologies in the teaching routine. VR and AR both have advantages and disadvantages; knowing how to maximise their strong points will optimise teaching efficiency and learning effectiveness.

3.1.3 Is There Any Study Proving or Trying to Prove the Benefit of Applying These Techniques

Several researchers from the included articles conducted studies to evaluate the effectiveness and efficiency of VR and AR in veterinary teaching. Baillie et al. [2] proved that the PHANToMs haptic simulator improved student's ability to find the uterus on real cows on-farm. After training with the simulator, students in the study group achieved 18/32 successful identifications verified by ultrasound, significantly greater than those in the control group who achieved 1/32 identifications ($p < 0.001$). Five years later, Baillie et al. [4] improved the PHANToMs haptic simulator to an automated version and then conducted the same experiment with the new system. Students in the study group achieved 12/32 successful identifications verified by ultrasound, significantly greater than those in the control group who achieved 2/32 identifications ($p < 0.001$). The results of those two studies showed that the simulators could improve the training effectiveness. For training in BLS, researchers Fransson et al. [12] found that the abilities of students were significantly improved after training with the laparoscopic simulator ($p < 0.001$), regardless of the applying training curriculum [6]. If the simulator were adapted to the curriculum based on the difference between human medicine, this learning outcome would be enhanced [20].

The AR IV injection simulator implemented by Lee et al. [18] was proven helpful to assist in learning injection technique. The study group trained using the AR simulator was more proficient at IV injection technique using real dogs than the control group ($P < 0.01$). When compared with the traditional simulator and the traditional teaching materials (multimedia contents, tangible prop, and a syringe), the AR technique helped the learner to reach a significantly better performance ($p < 0.01$), proving it could enhance performance compared to conventional approaches [19]. However, there is not enough evidence proving the same for the other AR techniques (smartphone AR and tablet AR). The tablet AR helped students perform better in the exam compared with the traditional teaching method, but without demonstrating a significant difference. More experiments need to be conducted to produce more data, so this can be identified as a research gap.

Similar to the AR technique, VR has been applied to veterinary education in recent years. Few studies investigated the benefits VR could bring to training effectiveness and efficiency as most of the studies described novel methodologies and systems that could potentially achieve that. Only one study evaluated the 360°VR videos, the result of which showed that the 360°video might help with learning the operation process, but no significant difference was found between the intervention group and the control group [14]. Further investigations are needed to determine the usefulness of applying 360°VR videos in the related area. VR techniques such as work-station-based VR and mobile VR need more evaluations based on the current pilot studies, which can also be identified as a research gap in this systematic review.

3.1.4 To What Extent Do These Techniques Support the Training Process When Compared with Other Methods

Several researchers used these techniques as an additional training method to the standard curriculum, shown in Table 3.

The study group that was trained using the various interventions performed better in surgery skill or knowledge gain than the control group ($P < 0.01$) as discussed above [4, 18], while no significant difference was found between groups using 360°VR videos as an intervention [14].

Compared with traditional teaching materials, the AR technique helped the learner to reach a significantly better performance ($p < 0.01$) [2, 19]. Although no significant difference was found between groups in some studies, students tended to gain more knowledge and report more positive feedback using the intervention system [14, 30]. The VR simulation could also be compatible with LDL that was normally limited by resources, proving that VR could serve as a reasonable alternative to LDL for teaching endoscopy skills.

Table 3 Control method and intervention method in each study category

Category	Category control	Intervention	Researcher
Simulation	No additional training	PHANToMs	Baillie et al. [4]
	Traditional training method[a]	PHANToMs	Baillie et al. [2]
AR	No additional training	AR IV[b] injection simulator	Lee et al. [18]
	Traditional training method[a]	iPad AR	Little et al. [22]
	Control 1: multimedia contents, tangible prop, and a syringe. Control 2: VR IV[b] simulator	AR IV[b] injection simulator	Lee et al. [19]
VR	No additional training	Voxkin VR headset	Hunt et al. [14]
	Traditional bone box method	HTC VIVE	Seo et al. [30]
	LDL[c]	EndoVR	McCool et al. [23]

[a] Traditional textbook, slides, standard 2D video resources [b] Intravenous [c] live dog laboratory

3.2 Limitations

The main limitations of this systematic review are the following three points:

- The search strategy included English and Chinese articles to reduce language bias. Also, one Japanese article was included and translated by an online translator. However, to avoid language bias, more languages (such as Spanish and German) are preferred to be added to the search strategy.
- According to the risk of bias analysis table (Fig. 5), over half of the included articles has some bias concerns, mainly due to deviations from the intended intervention.
- This systematic review did not include articles or studies that provided only abstracts as it was difficult to judge whether the study met all the inclusive criteria. However, this fact became one of the limitations in this review as it did not cover all the articles, including grey publications and clinical trial protocols.
- This paper is based on PRISMA 2009 statement and checklist [25]. After work was started and the majority of work had been completed for the review, the authors became aware of a new update [26]. The authors do believe the new protocol is a step forward and the inclusion of an interpretation section would help reviews such as this. Overall though it was felt after examining the new guidance, it would not substantially change the review as such the existing work was retained, and the PRISMA 2009 checklist was followed.

3.3 Conclusions

This systemic review has explored the MR applications in tertiary veterinary education. Throughout this review, a snapshot is provided of current VR and AR techniques and their application in supporting veterinary teaching and enhancing specific modules within the degree programme. With this detailed review of this area, even with the relatively limited number of studies, it can be argued strongly that VR and AR techniques have proven effective, but further research is needed. They are not a panacea, though, and through this review, we can compare how they support the training process when compared to more traditional methods. The field is nascent in nature, and for some areas of training such as surgery, passive techniques such as 360°video are mainly used, while in anatomy training, more immersive interaction techniques already appear common. By its very nature, surgery requires a much larger investment in terms of software development to allow it to be fully interactive, e.g. Modelling of Soft Tissues [15]. It is difficult to gauge how long this transition to more interactive techniques will take, but with more inexpensive hardware and more advanced simulations becoming available both to create and display these more complex scenes, it will be a important future area of research.

Future research can evaluate the currently proposed VR system to prove whether those techniques can increase the training efficiency and effectiveness. More experiments need to be conducted to produce more data in smartphone MR and MR HMDs applications. Novel methods applying such techniques to expand the veterinary learning opportunities, including anatomy and radiology, are needed.

Finally, the review demonstrates the historical trend to move from non-immersive simulators into AR and VR spaces. As with human medicine, there is a far greater number of people who have the interest and the ability to train as veterinarians, but numerous bottlenecks exist that prevent people from pursuing their passion within this field. The introduction of MR could prove to be a revolution in training for tertiary education in both human and animal medicine and create a world where both humans and our animal companions live long and happier lives.

References

1. Alaker, M., Wynn, G.R., Arulampalam, T.: Virtual reality training in laparoscopic surgery: a systematic review & meta-analysis. Int. J. Surg. **29**, 85–94 (2016)
2. Baillie, S., Crossan, A., Brewster, S.A., Mellor, D., Reid, S.: Validation of a bovine rectal palpation simulator for training veterinary students. Studies Health Technol. Inf. **111**, 33–36 (2005)
3. Baillie, S., Mellor, D.J., Brewster, S.A., Reid, S.W.: Integrating a bovine rectal palpation simulator into an undergraduate veterinary curriculum. J. Vet. Med. Educ. **32**(1), 79–85 (2005)
4. Baillie, S., Crossan, A., Brewster, S.A., May, S.A., Mellor, D.J.: Evaluating an automated haptic simulator designed for veterinary students to learn bovine rectal palpation. Simul. Healthcare **5**(5), 261–266 (2010)
5. Blum, T., Kleeberger, V., Bichlmeier, C., Navab, N.: Mirracle: an augmented reality magic mirror system for anatomy education. In: 2012 IEEE Virtual Reality Workshops (VRW), IEEE, pp. 115–116 (2012)
6. Chen, C.Y., Ragle, C.A., Lencioni, R., Fransson, B.A.: Comparison of 2 training programs for basic laparoscopic skills and simulated surgery performance in veterinary students. Vet. Surg. **46**(8), 1187–1197 (2017)
7. Christ, R., Guevar, J., Poyade, M., Rea, P.M.: Proof of concept of a workflow methodology for the creation of basic canine head anatomy veterinary education tool using augmented reality. PLoS ONE **13**(4), e0195866 (2018)
8. De Bont, M.P., Wilderjans, H., Simon, O.: Standing laparoscopic ovariectomy technique with intraabdominal dissection for removal of large pathologic ovaries in mares. Vet. Surg. **39**(6), 737–741 (2010)
9. DeBose, K.: Virtual anatomy: expanding veterinary student learning. J. Med. Library Assoc. JMLA **108**(4), 647 (2020)
10. Dehabadi, M., Fernando, B., Berlingieri, P.: The use of simulation in the acquisition of laparoscopic suturing skills. Int. J. Surg. **12**(4), 258–268 (2014)
11. Forrest, N., Baillie, S., Kalita, P., Tan, H.Z.: A comparative study of haptic stiffness identification by veterinarians and students. IEEE Trans. Haptics **4**(2), 78–87 (2010)
12. Fransson, B.A., Chen, C.Y., Noyes, J.A., Ragle, C.A.: Instrument motion metrics for laparoscopic skills assessment in virtual reality and augmented reality. Vet. Surg. **45**(S1), O5–O13 (2016)
13. Goodin, J.T., Rodgerson, D.H., Gomez, J.H.: Standing hand-assisted laparoscopic ovariectomy in 65 mares. Vet. Surg. **40**(1), 90–92 (2011)

14. Hunt, J.A., Heydenburg, M., Anderson, S.L., Thompson, R.R.: Does virtual reality training improve veterinary students' first canine surgical performance? Veterinary Record **186**(17), 562–562 (2020)
15. Jayasudha, K., Kabadi, M.G.: Soft tissues deformation and removal simulation modelling for virtual surgery. Int. J. Intell. Sustain. Comput. **1**(1), 83–100 (2020)
16. Kinnison, T., Forrest, N.D., Frean, S.P., Baillie, S.: Teaching bovine abdominal anatomy: use of a haptic simulator. Anat. Sci. Educ. **2**(6), 280–285 (2009)
17. Larsen, C.R., Oestergaard, J., Ottesen, B.S., Soerensen, J.L.: The efficacy of virtual reality simulation training in laparoscopy: a systematic review of randomized trials. Acta Obstet. Gynecol. Scand. **91**(9), 1015–1028 (2012)
18. Lee, J., Kim, W., Seo, A., Jun, J., Lee, S., Kim, J.I., Eom, K., Pyeon, M., Lee, H.: An intravenous injection simulator using augmented reality for veterinary education and its evaluation. In: Proceedings of the 11th ACM SIGGRAPH International Conference on Virtual-Reality Continuum and its Applications in Industry, pp. 31–34 (2012)
19. Lee, S., Lee, J., Lee, A., Park, N., Song, S., Seo, A., Lee, H., Kim, J.I., Eom, K.: Augmented reality intravenous injection simulator based 3d medical imaging for veterinary medicine. Veterinary J. **196**(2), 197–202 (2013)
20. Lencioni, R.D., Ragle, C.A., Kinser, M.L., Coffey, T., Fransson, B.A.: Effect of simulator orientation during skills training on performance of basic laparoscopic tasks by veterinary students. J. Am. Vet. Med. Assoc. **251**(10), 1196–1201 (2017)
21. Li, L., Yu, F., Shi, D., Shi, J., Tian, Z., Yang, J., Wang, X., Jiang, Q.: Application of virtual reality technology in clinical medicine. Amer. J. Transl. Res. **9**(9), 3867 (2017)
22. Little, W.B., Dezdrobitu, C., Conan, A., Artemiou, E.: Is augmented reality the new way for teaching and learning veterinary cardiac anatomy? Med. Sci. Educ. **31**(2), 723–732 (2021)
23. McCool, K.E., Bissett, S.A., Hill, T.L., Degernes, L.A., Hawkins, E.C.: Evaluation of a human virtual-reality endoscopy trainer for teaching early endoscopy skills to veterinarians. J. Vet. Med. Educ. **47**(1), 106–116 (2020)
24. McGuinness, L.A., Higgins, J.P.: Risk-of-bias visualization (robvis): an r package and shiny web app for visualizing risk-of-bias assessments. Res. Synthesis Methods **12**(1), 55–61 (2021)
25. Moher, D., Liberati, A., Tetzlaff, J., Altman, D.G., Group, P. et al.: Preferred reporting items for systematic reviews and meta-analyses: the prisma statement. PLoS Med. **6**(7), e1000097 (2009)
26. Page, M.J., McKenzie, J.E., Bossuyt, P.M., Boutron, I., Hoffmann, T.C., Mulrow, C.D., Shamseer, L., Tetzlaff, J.M., Moher, D.: Updating guidance for reporting systematic reviews: development of the prisma 2020 statement. J. Clin. Epidemiol. **134**, 103–112 (2021)
27. Pan, X., Zheng, M., Xu, X., Campbell, A.G.: Knowing your student: targeted teaching decision support through asymmetric mixed reality collaborative learning. IEEE Access **9**, 164742–164751 (2021)
28. Parkes, R., Forrest, N., Baillie, S.: A mixed reality simulator for feline abdominal palpation training in veterinary medicine. In: MMVR, pp. 244–246 (2009)
29. Satoshi, T., Yoshiharu, F., Yukihiro, F., Takuma, A., Miyoko, S., Akikazu, I., Eiichi, K., Yoko, F.: Development of educational virtual reality teaching materials in small animal practice for veterinary students. J. Jpn. Vet. Med. Assoc. **74**, 249–254 (2021)
30. Seo, J.H., Smith, B.M., Cook, M., Malone, E., Pine, M., Leal, S., Bai, Z., Suh, J.: Anatomy builder vr: applying a constructive learning method in the virtual reality canine skeletal system. In: International Conference on Applied Human Factors and Ergonomics, Springer, pp. 245–252 (2017)
31. Sterne, J.A., Savović, J., Page, M.J., Elbers, R.G., Blencowe, N.S., Boutron, I., Cates, C.J., Cheng, H.Y., Corbett, M.S., Eldridge, S.M. et al.: Rob 2: a revised tool for assessing risk of bias in randomised trials. bmj 366 (2019)
32. Vega-Garzón, J.C., Robayo-Sánchez, L.N., Cruz-Maldonado, O.A., Cortés-Vecino, J.A.: Visualization technologies for learning and teaching veterinary acarology and entomology. J. Veterinary Med. Edu., e20200034 (2021)

33. Wilkie, N., McSorley, G., Creighton, C., Sanderson, D., Muirhead, T., Bressan, N.: Mixed reality for veterinary medicine: Case study of a canine femoral nerve block. In: 2020 42nd Annual International Conference of the IEEE Engineering in Medicine & Biology Society (EMBC), IEEE, pp. 6074–6077 (2020)
34. Xu, X., Mangina, E., Kilroy, D., Kumar, A., Campbell, A.G.: Delaying when all dogs to go to heaven: virtual reality canine anatomy education pilot study. In: 2018 IEEE Games, Entertainment, Media Conference (GEM), pp. 1–9. IEEE (2018)
35. Xu, X., Kilroy, D., Mangina, E., Campbell, A.G.: Work-in-progress—adapting a virtual reality anatomy teaching tool for mobility: pilot study. In: 2020 6th International Conference of the Immersive Learning Research Network (iLRN), pp. 328–331. IEEE (2020)
36. Xu, X., Mangina, E., Kilroy, D., Curran, K.M., Healy, J.J., Campbell, A.G.: Doctoral colloquium—a snapshot of the future: Virtual and augmented reality training for radiology. In: 2020 6th International Conference of the Immersive Learning Research Network (iLRN), pp. 407–410. IEEE (2020)
37. Xu, X., Pan, X., Campbell, A.G.: Arls: an asymmetrical remote learning system for sharing anatomy between an hmd and a light field display. In: 26th ACM Symposium on Virtual Reality Software and Technology, pp. 1–2 (2020)
38. Xu, X., Mangina, E., Campbell, A.: Hmd-based virtual and augmented reality in medical education: aA systematic review. front. Virtual Real. **2**, 692103 (2021). https://doi.org/10.3389/frvir
39. Zheng, M., Pan, X., Xu, X., Campbell, A.G.: Metal: Explorations into sharing 3d educational content across augmented reality headsets and light field displays. In: 2021 7th International Conference of the Immersive Learning Research Network (iLRN), pp. 1–6. IEEE (2021)

Exploring Gestural Agency in Collaborative and Embodied Rates of Change Simulations

James Planey and Robb Lindgren

Abstract This chapter describes the theories, design, and analysis applied to a pair of collaborative mixed-reality simulations addressing different scientific content but linked by the crosscutting concept of "rates of change". Body tracking technologies allowed users to interact with these simulations using one of two conventions: through gestures chosen by participants, or a prescribed gesture representing the slope of a line using one's arms. The two simulations explore the relationship between inputs and outputs in the context of population ecology and climate change systems. Participant pairs were randomly assigned to one of 4 treatment groups for gestural convention (learner-defined or prescribed) and simulation order (population ecology or climate change first). Knowledge and engagement surveys found a significant overall increase in individual knowledge, but no significant difference across the four treatments. This motivated deeper analysis of the dialog and actions of the participant pairs as they engaged with the system, where it was found that the order of simulation exposure produced a shift in the ratio between collaborative directives and simulation behavior dialog. In addition, the gestures recorded in the learner-defined gesture group were also leveraged in un-expected ways, such as in facilitating synchronization between partners.

1 Introduction

The role of technology in facilitating the use of gesture within the context of STEM learning simulations has become an area of increasing interest in recent years [1, 8, 10, 15, 22]. The body can act as a critical link to a learner's understanding and integration of STEM ideas, especially those which heavily leverage formalized representations that can be barriers to engagement [21]. In addition, using the body as a means

J. Planey (✉) · R. Lindgren
University of Illinois Urbana-Champaign, Champaign, IL, USA
e-mail: planey@illinois.edu

R. Lindgren
e-mail: robblind@illinois.edu

Y. Cai et al. (eds.), *Mixed Reality for Education*, Gaming Media and Social Effects,
https://doi.org/10.1007/978-981-99-4958-8_11

of externalizing one's thinking and promoting reflection has been shown to be a productive way to enhance learning [4, 6, 7]. These interventions are based in theories that view learning and cognition as fundamentally embodied, inseparable from our movements and systems for perceiving and acting within the physical world [9, 26].

Despite significant promise, however, there remain challenges to developing robust and impactful embodied STEM learning environments. First, there are no clear heuristics for designers to select specific embodied actions such that they effectively map onto target learning concepts. We know that not all forms of embodiment are equal in terms of their potential for generating new learning [9]. Interventions involving embodied acts that are 'congruent' with target concepts will be most effective [25], but it is not always obvious what constitutes congruency, particularly for more abstract constructs such as the crosscutting concepts (e.g., patterns, rates of change) that bridge multiple STEM domains [23]. Second, even when learning through embodied interaction is achieved, the staying power of that learning and its transferability to new domains is largely unexplored. Few demonstrations of embodied learning show effects that go beyond the specific context of initial learning even though embodiment theorists frequently argue that embodied knowledge is intuitive and core (e.g., [13]. However, recent research by Lindgren et al. [17] has found that engaging with crosscutting concepts in multiple science domains is facilitated by an embodied interaction scheme leading to significant increases in knowledge retention as well as knowledge transfer. Third, the research literature on embodied learning is mixed when it comes to the degree to which embodied actions should be explicitly prescribed versus learner-generated. Cueing specific actions can support metaphors that have been shown to be productive for learning [16], but inventing one's own representations can make learning more personal and potent [24]. Finally, studies have emerged recently that have examined embodied learning in more collaborative contexts such as classrooms (e.g., Enyedy et al. [5]), but there is still much to be explored pertaining to how embodied interactions can be synergistic in multi-learner environments. For example, how should the learning environment assist in the distribution of attention when there may be multiple visualizations and physical enactments occurring simultaneously? To begin to address some of these issues, the study described here aims to answer the following questions:

RQ1. What is the role of embodied interactions (e.g., gestures) when engaging collaboratively with abstract and crosscutting concepts (magnitude and scale, rates of change, etc.) in the context of specific STEM content learning?

RQ2. What is the effect of specific embodied interaction schemes (prescribed or personalized) on collaborative engagement and problem-solving?

These questions will be addressed in the context of a full-body simulation platform developed as part of the ELASTIC^{3}S (Embodied Learning Augmented through Simulation Theaters for Interacting with Cross-Cutting Concepts in Science) project. These embodied simulations will allow two users to explore learning in multiple science domains employing the "rates of change" crosscutting concept.

2 Methods

2.1 Simulation Design

To allow for the exploration of the proposed research questions, a series of collaborative gesture-based simulations were developed utilizing a mixed-reality simulation theater (three projection screens spanning the participants' visual field). The simulations below build off of previous research conducted as part of the ELASTIC^{3}S project, establishing the utility of the body tracking systems and the potential to impact learning with individual participants [17]. The simulations developed here apply the findings from the previous ELASTIC^{3}S work to collaborative multi-user interactions. The target domains identified for simulation development were population ecology and climate change. While sharing many of the same core concepts around rates of change, both domains occur in distinctly separate scales across space and time. Both of these systems have complex rates related interactions which were simplified for use in this study to 3 elements. For population ecology the goal was a simulated ecosystem exploring the relationship between wolves, their prey (sheep) and the sheep's food source (grass). For climate change, the aim was to illustrate the relationship between a CO_2 sink (trees), a CO_2 source (industrial output via factories), and resulting mean global temperature change due to atmospheric CO_2.

The design of these simulations was guided by 3 core goals, informed by the work of Lindgren & Johnson-Glenberg [14]: (1) body-sensing should be unobtrusive and flexible enough to allow personalized gestures for each user, (2) the simulations should extend interface conventions across learning topics to facilitate conceptual connections, and (3) the simulations should promote collaborative engagement between the participants.

Unobtrusive, yet robust body tracking (Goal 1) was achieved via the use of a Microsoft Kinect depth sensor, coupled with custom gesture recognition software that allows for the use of pre-programmed or personalized gestures [11]. This motion tracking system requires no additional equipment to be worn by the user, allowing for natural body interactions with the simulations and a high degree of gesture recognition accuracy.

Goal 2 was addressed via extensive planning and iterative design work to ensure that the core modes of information access were consistent across the domains of the simulations. Figure 1 shows the use of these conventions across the two simulations utilized in this study. The participant representation and rate limit graphs (see areas (a) in Fig. 1) show the connection between the participants' movement and the more formal representation of the rate of change as a time vs. quantity graph, linking embodied action to more abstract data representations. The rate limit icons (areas (b) in Fig. 1) provide a more immediate visual representation of the limitations of the simulated systems, informing participants if they have reached the maximum allowable quantities of the target simulation elements. The dynamic models (areas (c) in Fig. 1) were designed to provide an active representation of the relationship between the inputs and outputs for each simulation.

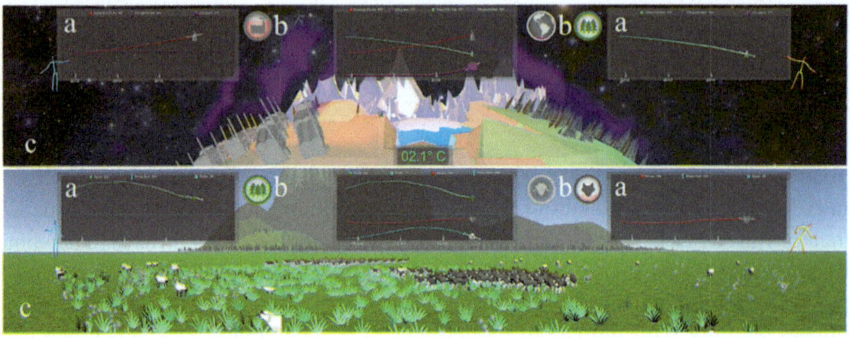

Fig. 1 Sample screen layout of the climate change (top) and population dynamics (bottom) rates of change simulations. Shared interface elements include participant representation and position graphs (a top and bottom), rate limit icons (b top and bottom), and dynamic model placement (c top and bottom)

Finally, to address Goal 3, the facilitation of collaborative discourse between the participants was achieved through the design of a series tasks pertaining to the manipulation of rates embedded within the context of the simulations. Each task within one simulation domain increases in complexity, requiring communication between the participants in order to achieve the target input, output, or dynamic equilibrium needed to reach the task objective. Within the domain of climate change (top of Fig. 1) the tasks have participants embody the rate of industrial activity releasing carbon dioxide into the atmosphere (left side of screen) or the number of trees absorbing carbon dioxide via photosynthesis (right side of screen), with the resulting climate effects observed in the center of the simulation (ice melt, sea level rise, and average global temperature change).

For the domain of population dynamics (bottom of Fig. 1) participants control either the amount of food presented as grass (left) or amount of predation presented as wolves (right) on a sheep population (center). For both content domains, each participant has access to graphed rate representations of their input, as well as the combined resulting effect on the system. Tasks were structured to promote dynamic interaction within the systems, for example, asking participants to maintain a sheep population of 500 and a wolf population of 200 in the population dynamics simulations requires the participant controlling the grass to overproduce to counteract the predation effects in order to maintain the target population (see Table 1 for all the tasks given to participants the simulations).

The tasks were also designed to promote reflection between them during a single simulation, and also comparison across both simulations. For example, task 1 in the context of maintaining a sheep population can be achieved simply through the addition of food (plants) until a steady-state is achieved with the natural death rate of the sheep. In contrast, increasing the amount of factories to begin increasing the level of CO_2 towards the target will not eventually hit a steady-state, and additional coordination with a partner is needed to reach this goal.

Table 1 Simulations task progression presented to participants

	Population simulation	Climate simulation
Task 1	Reach and maintain a sheep population of 500	Reach and hold a CO_2 level of 500 parts per million (ppm)
Task 2	Maintain a sheep population of 500 while adding at least 200 wolves	Maintain a CO_2 level of 500 ppm while adding at least 200 factories
Task 3	Maintain a population of 700 sheep while keeping the amount of plants and wolves in constant flux	Maintain 700 ppm of CO_2 while keeping the amount of trees and factories in constant flux

2.2 *Experimental Design*

With the ELASTIC^{3}S simulations addressing all of the stated design goals, a 2 × 2 factorial design was constructed entailing both individual and collaborative pre and post assessments (see Fig. 2 for full experimental sequence). Each data collection session consisted of an initial pre-assessment focusing on rates of change and domain-specific (climate change or population dynamics) content. This was followed by a video and audio recorded dyad assessment where both participants worked together on a multi-part short answer dynamic equilibrium word problem (modifying flow rates in a large aquarium and controlling rates of charge and discharge in a battery).

For the simulation itself, participant pairs were placed into one of four potential treatments based on the type of gestural convention (*Personal* or *Prescribed*) and the order of the simulations (*Population* or *Climate* first). As discussed previously, the degree to which an embodied actor makes meaningful metaphoric connections to the system, and the relationship of these connections and their own gestural agency, is an area of developing research. Alternating the gestural input convention allows for the exploration of the costs and benefits of increased gestural agency afforded to the participants. At the same time, alternating the order of the simulation that is first introduced to the participants could provide insight into relationships between the gesture conventions and the nature of the content (and the underlying simulated system).

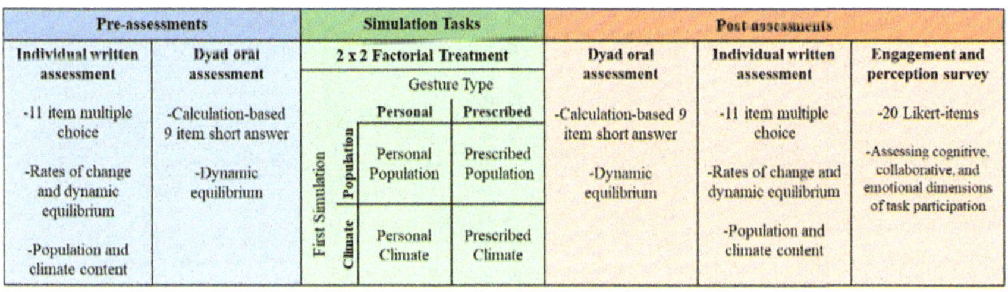

Fig. 2 Experimental individual assessment, collaborative assessment, and gesture/sim treatment sequence

2.2.1 Simulation Order

While both simulations attempt to maintain consistency in the location and representation of simulation data, there are some specific differences in the simulations inherent to the target content they explore. The population simulation displays both the inputs (grass and wolves) and the result (sheep) as discrete units on screen and is a smaller scale phenomenon when compared to the climate simulation where factory and tree levels contribute to a more abstract and diffuse representation of CO_2. The potential here is that as the introduction to the session, a particular combination of gesture convention and simulation may set the stage for subsequent interactions (facilitating or inhibiting the development of a collaborative dynamic) as well as transfer of content relationships to the next simulation and potentially influencing performance on the knowledge assessments.

2.2.2 Gesture Convention

For the personal gesture group, participants were prompted to think of two gestures: one representing a decrease and another representing an increase. Participants were told that the gestures should be repeatable actions that would remain comfortable for the user when repeated for an extended period of time. The resulting gestures were then recorded as template motions into the simulation individually for both participants at the start of the session. These gestures were then repeated at varying rates to increase or decrease the rate of change in the participants' role in the simulation during task completion. For the prescribed gesture group, participants were instructed to spread their arms wide and alter the angle of their arm span to directly control the slope of the rate of their system component. The facilitator ensured all tasks were completed and provided structured clarifying questions based on the performance of the participants, but otherwise did not interact with the participants directly (see Fig. 3 for exemplar images of each gesture convention in use in the simulation theater).

Simulation sessions were audio and video recorded, and software log data (task completion milestones and body position data) were also collected. The sessions

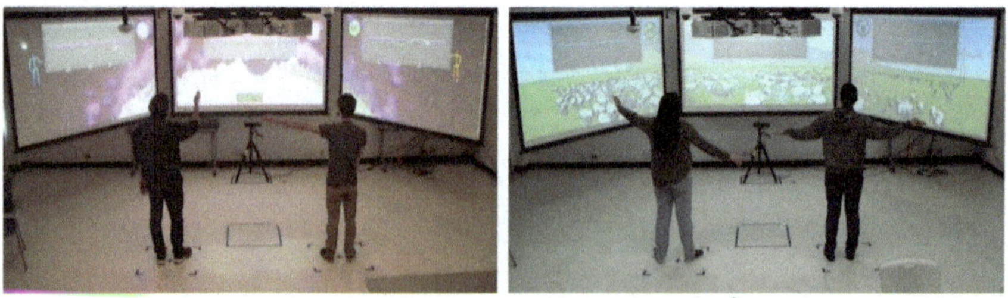

a b

Fig. 3 Sample video captures of both gestural input conventions in use (personalized gesture (**a**), prescribed slope gesture (**b**)) during data collection concluded with post-assessment iterations of both the dyad and individual assessments, as well as a 20 item Likert-scale engagement and perception survey where they were asked to rate items such as the quality of their collaboration, synergy with their partner, their interpretation of the simulations, and general interest in the subject material.

2.3 *Participants*

70 undergraduate students were recruited from a non-STEM pre-service education course at a large midwestern US university to participate in the study (16 male, 54 female). All participants completed consent forms and demographic surveys prior to their participation in the study. Participants received a nominal incentive for their involvement. The demographic breakdown of all participants was 70% White, 21% Asian, 4.5% Black, and 4.5% Multiracial. The mean age of participants was 19.8. Due to the need for two participants for each session, scheduling was opportunistic within predefined blocks and participants could sign up together or be placed with another participant during the desired session time. As a result, 35 total sessions were completed (9 *Prescribed Gesture—Population First*, 9 *Personal Gesture—Population First*, 9 *Prescribed Gesture—Climate First*, and 8 *Personal Gesture—Climate First*). Of the 35 sessions, 15 signed up together, with the remaining 20 pairs assigned semi-randomly based on schedule compatibility. At the conclusion of the sessions, 63 of the 70 participants completed the post-session engagement and perception survey.

3 Results and Discussion

3.1 *Characterizing Personalized Gesture*

The recorded gestures from the personalized gesture convention group were video recorded and summarized for the frequency of gesture occurrence within four body regions (left/right arms, left/right legs). Gesture recording was prompted in the same sequence for each session, with participants first asked to select and record a gesture representing increase. When recording the gesture, 70% of participants used their right arm with 30% utilizing their left arm, and no gestures occurring below the waist. This distribution is likely partially due to right-handed bias, but participants were not instructed on a specific hand to utilize. Increase gestures consisted largely of lifting or raising of the hands, moving hands in sweeping arcs, or lifting and extending an arm at shoulder height. Participants were then subsequently asked to record a gesture that represented a decrease, with a higher variety of gesture distributions noted (18% right arm, 27% left arm, 37% left leg, 18% right leg). Commonly observed gestures were often a novel movement located far from the body area used to record the increase (such as lifting, kicking, or stomping the left leg for decrease as an opposition to raising the right hand for increase), or a mirrored version of the increase gesture (lifting left arm instead of right).

3.2 Learning Metrics and Engagement

Individual learning gain performance based on pre and post assessment scores analyzed via a Welch's two sample t-test shows a significant gain in assessment scores when comparing pre-test ($M = 0.50$ $SD = 0.17$) and post-test ($M = 0.63$, $SD = 0.16$) scores after interacting with the simulation ($t(137.59) = -4.26$, $p < 0.001$). Additional means comparisons found no significant differences ($p < 0.05$) for gender, race, ethnicity, or year in school of the participants. As the simulations were the primary method of instruction (with the facilitator providing context and support), this significant difference highlights the value of engaging with the population and climate simulations while learning about general rates of change concepts such as dynamic equilibrium.

In order to analyze the potential effects on learning gains from the treatments, a normalized change score was calculated for each individual. Normalized change is a ratio measure of gain compared to maximum possible gain, as well as loss compared to maximum possible loss, resulting in a normalized score for each participant ranging from −1 (maximum loss) to 1 (maximum gain) [18]. Each individual's normalized change was calculated using their individual pre and post scores and a resulting whole group average normalized change score of 0.22 was calculated. A 2 × 2 factorial ANOVA comparing normalized change scores across the four treatment groups (Fig. 4) found no significant main effect for simulation order, $F(1,66) = 1.55$, $p = 0.22$, as well as no main effect of gestural convention, $F(1,66) = 0.44$, $p = 0.51$. As the simulation tasks were collaborative in nature, an analysis of covariance for a potential friendship status covariate (whether the two participants signed up together or were randomly paired) also found no significant effects $F(1,62) = 0.004$, $p = 0.98$. In addition, a factorial ANOVA of the average post session engagement and perception survey ratings found no significant difference in average participant self-report ratings when compared across the treatment groups (simulation order, $F(1,31) = 0.02$, $p = 0.88$, or gesture convention, $F(1,31) = 0.31$, $p = 0.58$.

3.3 Collaborative Simulation Interactions

3.3.1 Coding Development

Following the above learning gains analysis, the research team was motivated to take a deeper dive into the collaborative discourse while participants were interacting with the simulation system. As a foundation for the development of an emergent

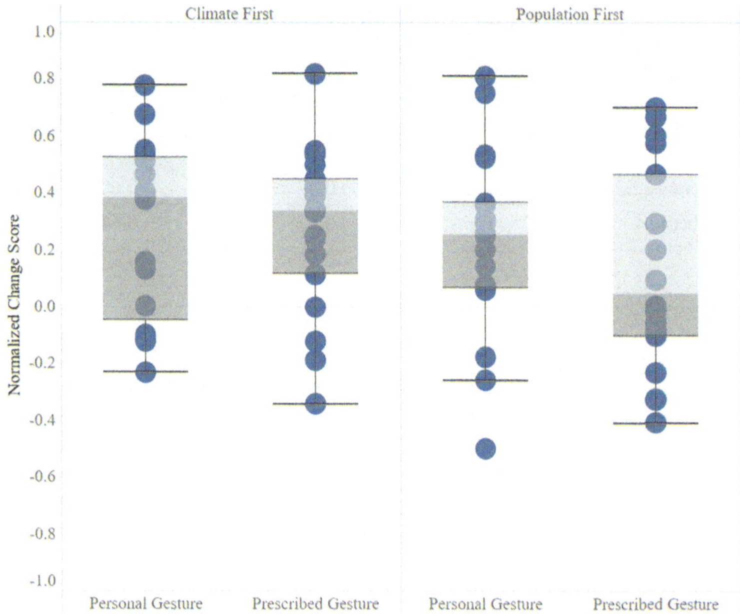

Fig. 4 Quartile plot of simulation order and gesture type treatment factors for normalized change score

conceptual coding scheme with which to analyze interactions, two participant pairs were selected for their individual assessment score similarity, one from the prescribed gesture treatment and one from the personalized treatment (both interacting with the population simulation fist). Pairs were identified based on proximity of the group means for pre and post-tests. Pairs 31 (personal gesture) and 42 (prescribed gesture) were both comprised of participants who's individual pre and post-test scores fell consistently close to within one standard deviation of the group means for both assessments, with participant 42B only having a post-test score slightly above the one SD range (Table 2). Both pairs completed the simulation with a friend as indicated on their registration form, and both showed similar levels of friendly discourse and comfort interacting as they completed the simulation tasks. In pair 31 (personal gesture), participant 31A chose a rapid raising of their left arm over their head to represent an increase in rate, and a slow raising of their right arm to represent a decrease. Participant 31B chose to raise their right arm above their head for an increase gesture and performed a small circular motion at hip-level with their left hand for their decrease gesture. When asked to reflect on their gesture selection by the facilitator, both noted that they wanted gestures that felt sufficiently distinct from each other to reduce confusion while working in the simulation.

The audio and video recording from the pair's simulation task portion of the experiment was then reviewed with the focus on identifying key discourse themes that could be used to code the remainder of the videos. As a result, two major themes emerged

Table 2 Sample pair metrics for pre-post assessment performance

Collaborative pair	Participant code	Pre-test score (M = 0.50, SD = 0.17)	Post-test score (M = 0.63 SD = 0.16)
31 (personal)	A	0.56	0.70
	B	0.47	0.60
42 (prescribed)	A	0.67	0.64
	B	0.55	0.82

for coding the remainder of the sessions: statements around directives coordinating each other's actions in the simulation, and discussion around the specific mechanism or behavior of the simulated system. Tables 3 and 4 show transcript samples of the exemplar pairs (31 and 42) with the resulting coded dialog. This coding scheme was then applied to 16 sessions from the data corpus for further interpretation (4 from each treatment group).

All four of these samples provide a glimpse into how pairs typically engaged with the simulations as they addressed the tasks. In Table 3 both pairs 31 and 42 spend considerable time determining each other's roles within the simulation. While the representation and effect of each portion of the simulation is introduced by the facilitator before they engage with the tasks, it was often observed that once the simulation was activated both participants would need to re-establish their roles with each other as they began to gesture.

In Table 4 we see both pairs 31 and 42 shift frequently between attention to each other and their representations within the simulated system as they coordinate. Here we see some of the potential benefits to the prescribed gestures as commands such as "let's level off first" relate immediately to both the embodied action (leveling outstretched arms) as well as the representation of rate within the simulation.

3.3.2 Coded Interaction Analysis

To examine the subset of the data corpus coded for these interactions, the resulting code counts for each session were converted into frequencies for visual analysis. The goal with this review was to establish evidence for larger shifts in interaction across the treatment groups which could motivate subsequent, targeted analysis. Globally across all 16 sessions coded interactions were 44.2% coordination and 58.8% model behavior. While in this novel simulation environment and interaction scheme no statements can be made around what distribution constitutes higher-quality engagement or learning with the simulation, locating shifts in the proportions of these two codes can help direct future investigations into what might be driving some of the observed interaction diversity, or shed light on interactions between the gestures and the simulations that are not accessible via the pre-post knowledge assessments. Separating out the sessions by simulation order (8 population first, 8 climate first) and gestural

Table 3 Sample model behavior dialog code occurrences while pairs 31 and 42 engage in climate change simulation. Coded occurrences are in parentheses as (M)odel behavior, or (C)oordination

Model Behavior Samples Facilitator asks participants to compare the rate of change model exhibited in the climate change simulation to the population change model they completed previously at two points during their tasks
Pair 31—Personal Gesture [Facilitator prompts for differences after initial experimentation with the system] A: So, I think I'm the one that makes it increase? (M) [performs increase gesture]. Hmmm…what are we noticing B? B: I don't know…[gesturing increase] A: So, make a factory [gestures increase]. (C) Isn't it the same as the sheep one though? (M) It's like an inverse relationship between the two? (M) B: [gestures decrease] [Facilitator prompts for additional reflection] A: I feel like we didn't have to do as much together (M), you [participant B] were like, "go wolves, go wolves!" [rapidly performs participant B's decrease gesture] B: …for that one. [population simulation] A: But for this one… [points to participant B's screen] B: For this one I just add trees. (M) [performs increase gesture] A: Exactly, and I have the negative one this time (M)…and I'm like, "go!"...[gestures increase while commenting about negative global impact of high CO_2]
Pair 42—Prescribed Gesture [Facilitator prompts for differences after initial experimentation with the system] A: We are both doing rate this time rather than population, right? (M) B: Yeah… A: Like, we need to change? [Both gesture to reach initial goal quantity] A: The last simulation was population, and the slope was population change over time [gestures change in CO_2 simulation slope line]. (M) B: [laughing while varying positive and negative slope gestures] A: This time we are looking at a rate instead of the actual number…(M) [begins to adjust slope gesture] [Facilitator prompts for additional reflection] B: It's more proportional. (M) A: CO_2 is the difference between our rates. (M)

convention (8 personal and 8 prescribed) shows a notable shift when the order of the simulations is considered (see Fig. 5). The distribution of the collaborative directives and model behavior statements shifts closer to an even distribution for the population simulation first group, while moving further away from the 44.2% coordination and 58.8% model behavior means for the climate simulation first group. This shift may in part be due to the previously discussed conceptual and representational complexity of the climate simulation, resulting in a need for more extensive discourse around the function and relationships of the model. It is notable here that this shift in behavior when engaging with the climate simulation first would have gone unnoticed when examined solely through the lens of the pre-post knowledge assessments. This has potential significance when examining how participants transfer their gesture, cross-cutting concept knowledge, and collaborative dynamic across one simulation to the

Table 4 Sample collaborative coordination dialog code occurrences while pairs 31 and 42 engage in climate change simulation. Coded occurrences are in parenthesis as (M)odel behavior, or (C)oordination

Coordination Samples Facilitator asks participants to maintain a CO_2 value of 700 ppm while simultaneously adjusting both their input and output to the system
Pair 31—Personal Gesture [CO_2 level is below goal] B: We need more factories. (C) A: [begins increase gesture] More factories? B, you are supposed to be moving! (C) B: [begins increase gestures] [700 ppm goal reached] A: I'm just gonna do a one [gestures increase once] and one [gestures decrease once] (C)...look at that! Look at that wave B! [referring to the up and down slope of their rate line produced by the contrasting gestures] B: [continues to gesture increase] [Departure from 700 ppm goal] A: Oop! [quickly performs multiple decrease gestures]...we're good! Way to save, B [sarcasm]! B: How? A: What do you mean how? I'm just swimming over here! [performs contrasting gestures wildly for B] B: [begins increase gesture] [Net CO_2 levels off long enough to reach goal]
Pair 42—Prescribed Gesture [CO_2 level is at goal] A: We need to keep the difference at 0... (M) [CO_2 level drops below goal] B: Wait, come back! (C) [Both participants begin using slope gestures] A: Let's get it back in there... [the goal range] (C) [Both continuing to gesture] A: What are you doing?! B: Oscillating...[laughter] A: Ok, we need to communicate...let's level off first. (C) [levels arms] B: Ok... [levels arms] [CO_2 level returns to goal] B: Ok, so what do we want to do? Do you want me to start decreasing while you start increasing? (C) A: No, we have to do the same thing (M)...so just tell me what you are going to do and I'll copy. (C) B: Alright...go...down. (C) [Both align decreasing slope gestures, while both quantities decrease the net CO_2 level stays constant]

next. The result here suggests that an approach to the simulations around more extensive discourse of the model behavior established during the climate simulation was carried over to the population simulation.

This also highlights a potential amplifying effect of both the novelty of the embodied input schemes along with the increased conceptual difficulty of the climate simulations. With no significant difference in learning gains between the

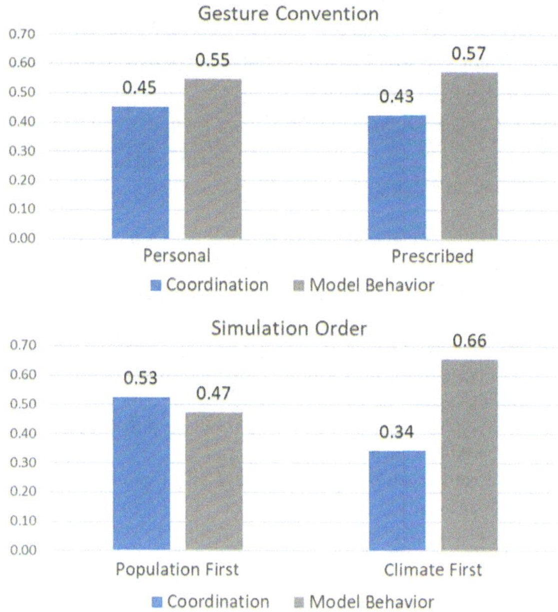

Fig. 5 Frequency of coordination and model behavior codes across sim order or gestural convention

treatments, the order of simulation exposure could be manipulated to emphasize conceptual discussions around the model (climate first) or the collaborative processes (population first) depending on the goals of the implementation.

3.3.3 Gesture as a Unit of Coordination

One additional dimension of these embodied interactions that warrants further investigation is the influence of the "unit" of a gestural input (or if a gesture used to control a simulation can be considered a discrete repeated action or part of a continuous motion). This is supported in part by dialog with groups using personalized gestures to engage with the simulations in which the performance of the gesture itself acts as a visual representation of the rate of change and being leveraged by the groups to coordinate their actions (see Table 2 pair 31's coordination sample and Table 5).

In Table 5, the coordination and references to being "in sync" were specifically directed at the rate with which their actions were taken and not at their representations within the simulation. These instances of gesture reference are not captured directly in the coding scheme utilized here, however re-review of the 8 sessions leveraging personalized gestures found that 3 pairs used this technique to some degree during their interaction with the simulations, with a 4th pair leveraging gestures that produced sounds (clapping and stomping) adding a level of audio feedback to their inputs not found in any other session and possibly allowing them to coordinate in yet another medium outside of the simulations themselves.

Table 5 Sample Interaction with personal gestures used to coordinate inputs to the simulation

Facilitator asks participants to maintain a population of 700 sheep while both participant inputs are in a state of constant flux
Pair 7—Climate Simulation (personal gesture)
[Facilitator asks participants to begin to try and reach the goal]
A: I need to go up... [repeats increase gesture]
B: Yeah...
A: I need to be…[continues to perform increase gesture at a steady rate]
B: [Turns to observe A's gesture] You need to be increasing when I am increasing [shows increase gesture], and decreasing when I am decreasing [shows decrease gesture]
A: Yeah...yeah
B: Ok, so decrease
[Both begin performing gesture repetitions at the same rate]
B: Shoot, no...so increase
[Both swap to their increase gestures but keep their repetition rate in sync]
B: Ok, we both have to be moving…
[Participants reach goal and facilitator asks them to reflect on why this task was more difficult than the previous goal]
B: Because we had to stay moving...we had to keep increasing and decreasing [alternates hands up and down]
A: Yeah, we had to be in sync

4 Conclusion

At the highest level of analysis, we found that our embodied simulations were well-received and produced significant individual learning gains. The significant individual pre-post performance increase shows the overall value of the simulations, tasks, and facilitation as a productive engagement with critical crosscutting concepts knowledge and potential transfer effects between simulations. Initial design concerns around the potential shifts in perception between the two gesture input conventions (possibly one being perceived as physically more difficult or requiring more cognitive load to enact) are not supported by the self-report survey data, suggesting that the input convention played a neutral role from a standpoint of engagement and perception of the tasks. This suggests that perhaps there is less urgency to choose-sides in the prescription versus user-generated embodiment debate than previously assumed. The lack of significant differences in learning gain or participant engagement across the treatment groups also established the necessity for a more focused analysis of the nuanced interactions contained in the paired simulation task video data. The increase in model behavior dialog (potentially due to challenges interpreting participants' roles in the simulation) without a corresponding negative effect on learning assessments shows the durability of complex simulated systems controlled via gesture. This opens up new dimensions to consider in the design of not only individual embodied learning simulations, but how multiple simulations of varying complexity could be ordered to facilitate a specific kind of discourse (collaboration or model centric).

The flexibility of the gesture systems not only supported learning and engagement with the simulation, but also allowed for novel uses of the personalized gesture beyond what was originally designed (using gestural units to coordinate, or even gestures that include additional modality such as sound). Previous work on how facilitators leveraged the simulation space in this system with the single-participant ELASTIC^{3}S simulation may help to inform how to approach some of these "off the screen" interactions that have not been captured sufficiently by the current analysis methods [12]. Improvements in how we track and monitor attention when engaging with mixed reality simulation theaters such as the system utilized here could lead to valuable insights into how attention is shared across the simulation representations (graphs and visual elements) and the participants (gesturing and collaborating). There is also now motivation to innovate the gesture input system further to move beyond the current "repeat a gesture" framework and potentially leverage some of the advantages of a continuously variable gesture utilized here but with the agency of a user-defined motion.

Ultimately, more thorough analysis of all participant dialog for emergent themes and interactions will be necessary to fully characterize the benefits and challenges to dynamic collaborative and embodied simulations like the ones described here. Recently, several authors have begun to highlight the need for revised approaches to documenting collaborative interactions when engaging with novel technology platforms. Coding schemes and interaction frameworks such as those outlined in [2, 3, 25] have the potential to highlight even more subtle changes in collaborative interaction and inform future development of collaborative mixed-reality science learning simulations.

Acknowledgements This research was supported by a grant from the National Science Foundation (IIS-1441563). Additional support from the National Center for Supercomputing Applications and the College of Education at the University of Illinois. A preliminary version of this work was published in the proceedings of the 2020 International Conference of the Learning Sciences (virtual).

References

1. Abrahamson, D., Trninic, D., Gutiérrez, J.F., Huth, J., Lee, R.G.: Hooks and shifts: a dialectical study of mediated discovery. Tech. Know. Learn. **16**, 55–85 (2011). https://doi.org/10.1007/s10758-011-9177-y
2. Chen, Y., Andrews, C.D., Hmelo-Silver, C.E., D'Angelo, C.: Coding schemes as lenses on collaborative learning. Inf. Learn. Sci., 121:1–18 (2019). 10/ghjtht
3. Csanadi, A., Eagan, B., Kollar, I., Shaffer, D.W., Fischer, F.: When coding-and-counting is not enough: using epistemic network analysis (ENA) to analyze verbal data in CSCL research. Intern. J. Comput. Support Collab. Learn. **13**, 419–438 (2018). 10/gfr4xk
4. DeSutter, D., Stieff, M.: Teaching students to think spatially through embodied actions: design principles for learning environments in science, technology, engineering, and mathematics. Cogn. Res. Princ. Implic., 2 (2017). https://doi.org/10.1186/s41235-016-0039-y
5. Enyedy, N., Danish, J.A., Delacruz, G., Kumar, M.: Learning physics through play in an augmented reality environment. Comput. Supported Learn. **7**, 347–378 (2012). 10/f37fgz
6. Gallagher, S., Lindgren, R.: Enactive metaphors: learning through full-body engagement. Educ. Psychol. Rev. **27**, 391–404 (2015)

7. Glenberg, A.M.: Embodiment for education. In: Handbook of Cognitive Science, pp. 355–372. Elsevier (2008)
8. Han, I., Black, J.B.: Incorporating haptic feedback in simulation for learning physics. Comput. Educ. **57**, 2281–2290 (2011). 10/b6dcsg
9. Johnson-Glenberg, M.C., Birchfield, D.A., Tolentino, L., Koziupa, T.: Collaborative embodied learning in mixed reality motion-capture environments: two science studies. J. Educ. Psychol. **106**, 86–104 (2014). https://doi.org/10.1037/a0034008
10. Johnson-Glenberg, M.C., Megowan-Romanowicz, C.: Embodied science and mixed reality: How gesture and motion capture affect physics education. Cognit. Res. Principles Implications **2**, 24 (2017). 10/gd8d2t
11. Junokas, M.J., Kohlburn, G., Kumar, S., Lane, B., Fu, W.T., Lindgren, R.: Using one-shot machine learning to implement real-time multimodal learning analytics. CEUR Workshop Proc. **1828**, 89–93 (2017)
12. Kang, J., Lindgren, R., Planey, J.: Exploring emergent features of student interaction within an embodied science learning simulation. Multimodal Technol. Interaction **2**, 39 (2018). https://doi.org/10.3390/mti2030039
13. Lakoff, G., Núñez, R.E.: Where mathematics comes from: how the embodied mind brings mathematics into being. Basic Books, New York, NY, US (2000)
14. Lindgren, R., Johnson-Glenberg, M.: Emboldened by embodiment: six precepts for research on embodied learning and mixed reality. Educ. Res. **42**, 445–452 (2013)
15. Lindgren, R., Tscholl, M., Wang, S., Johnson, E.: Enhancing learning and engagement through embodied interaction within a mixed reality simulation. Comput. Educ. **95**, 174–187 (2016)
16. Lindgren, R.: Getting into the cue: embracing technology-facilitated body movements as a starting point for learning. In: Learning Technologies and the Body, pp. 51–66. Routledge (2014)
17. Lindgren, R., Morphew, J.W., Kang, J., Planey, J., Mestre, J.P.: Learning and transfer effects of embodied simulations targeting crosscutting concepts in science. J. Educ. Psychol. Advance Online Publication (2021). https://doi.org/10.1037/edu0000697
18. Marx, J.D., Cummings K.: Normalized change. Am. J. Phys. **75**, 87–91 (2006). https://doi.org/10/d9m9fd
19. Nathan, M.J., Walkington, C.: Grounded and embodied mathematical cognition: Promoting mathematical insight and proof using action and language. Cognit. Res. Principles Implications **2**, 9 (2017). https://doi.org/10.1186/s41235-016-0040-5
20. Nathan, M.J.: Rethinking formalisms in formal education. Educ. Psychol. **47**, 125–148 (2012). https://doi.org/10.1080/00461520.2012.667063
21. National Research Council: A Framework for K-12 Science Education: Practices, Crosscutting Concepts, and Core Ideas. National Academies Press (2012)
22. Schwartz, D.L., Martin, T.: Inventing to prepare for future learning: the hidden efficiency of encouraging original student production in statistics instruction. Cognit. Instruction, 22 (1988). 10/dn2d3q
23. Segal, A., Tversky, B., Black, J.: Conceptually congruent actions can promote thought. J. Appl. Res. Memory Cognit. **3**, 124–130 (2014). 10/ggdd5v
24. Shapiro, L.: Embodied Cognition. Routledge (2019)
25. Shehab, S., Mercier, E.: Visualizing Representations of Interaction States during CSCL. In: A Wide Lens: Combining Embodied, Enactive, Extended, and Embedded Learning in Collaborative Settings, pp. 871–872. International Society of the Learning Sciences, Lyon, France (2019)
26. Wilson, M.: Six views of embodied cognition. Psychon. Bull. Rev. **9**, 625–636 (2002). https://doi.org/10.3758/BF03196322

Integration of Wearable Devices in VR-Enhanced Aircraft Maintenance Training

Eugene Ng, Azam Abu Bakr, and Yiyu Cai

Abstract There is a growing demand for using VR for educational purposes due to the realistic and immersive experience that it provides. For hands-on tasks, like aircraft maintenance procedures, several researchers have studied various methods of enhancing this learning experience. While performing the interactive learning of the aircraft hydraulics system for example, it is highly important to incorporate safety elements during the hands-on tasks. This way students have a chance to develop habits for safety hazards control while learning the knowledge of hydraulics systems. This chapter reports wearable devices we designed providing interactive alerts or warnings to avoid hitting in a constrained working area. In this study, custom-made tactile gloves and jackets are developed to improve the learning experience of a VR aircraft hydraulics system learning application available in-house.

1 Introduction

In recent years, there has been a huge development in Virtual Reality (VR) technology. Ever since Facebook bought Oculus in 2014 [1], VR had gained popularity rapidly causing many companies to develop VR products. In 2020, the pandemic situation worsened globally. To reduce the rate of transmission of COVID-19, companies adopted a different working method, and allowed employees to work remotely from home [2]. VR products are once again in the spotlight, with various applications developed including those for education and medical sectors [3–7]. Take education as an example, even though the student might be learning from home, VR allows them to learn if they are being there. Immersive and interactive learning through VR has increasingly been recognized [3, 6, 8] with a potential to increase students' motivation and efficiency [9]. To engage learners in a way that traditional education cannot, serious games should integrate the pedagogical theories and motivational principles [10] as they are effective tools in helping learners to acquire new knowledge and skills [3].

E. Ng · A. A. Bakr · Y. Cai (✉)
Nanyang Technological University, Singapore, Singapore
e-mail: myycai@ntu.edu.sg

Y. Cai et al. (eds.), *Mixed Reality for Education*, Gaming Media and Social Effects,
https://doi.org/10.1007/978-981-99-4958-8_12

Several researchers agree that the traditional pedagogy of learning aircraft maintenance tasks is lacking due to various factors like the cost as well as the availability of real or even functional aircraft. For students to visualize extreme scenarios, it would be difficult and sometimes even dangerous to imitate with real-life equipment [11–14]. The practice of using Augmented Reality (AR) and VR for aircraft maintenance is of good interest to aircraft training. Upon conducting surveys of efficacy amongst volunteers, there seems to be a clear benefit of AR/VR in addition to traditional learning. AR is used during the initial learning phase, where helpful diagrams and 3D models are shown when the AR-capable device scans a target (textbook picture or actual component [15–18]. VR and Mixed Reality (MR) are used for interactive learning scenarios, where VR systems [14, 19–22] are totally virtual while MR systems [15–17] can overlap with real-life equipment. With the inexpensiveness of AR/VR devices and increasing functionality, sequential practical lab lessons or hands-on tasks may be replaced by self-paced or asynchronous tasks for each student instead.

Relevant skill transfer of hands-on tasks will be complete only if the interactions are natural, otherwise applications created with older handheld devices like joysticks and 6-DOF mice, complex-gestures and even speech may not be enough [17, 22–25]. Some of the prior arts focus on several unique concepts of learning like the incorporation of virtual experts or collaboration amongst students during hands-on tasks. Not many works focus on the hand interactions itself. Also, safety is sometimes overlooked during training, for instance in aircraft hydraulics systems training, students have to stay in confined spaces without causing any injuries. Reference [26] discussed boundary awareness using VR enabled training. While visual and auditory cues aid in giving such feedback to users, the sensation of touch is also important allowing direct feedback when interacting. Haptic and tactile technology are being developed to generate this sensation of touch [27]. There are currently five types of haptic feedback: force, vibrotactile, electro-tactile, ultrasound and thermal feedback [28]. Force feedback haptic devices are capable of simulating realistic feedback. VRgluv is able to generate up to 10 pounds of force on each finger [1]. The Aura Interactor Vest is able to provide feedback like a punch, kick, gunshot or explosion [29]. Mikhailovsky [30] in his article talks about force feedback devices, but it is bulkier and heavier. Vibrotactile feedback haptic devices like the Senso gloves can simulate the feeling of touch at a much smaller build [31]. However, they cannot provide realistic feedback as compared to the force feedback devices [30], for example, putting their hands through objects and walls [26].

In this study, we are interested in creating an aircraft maintenance training system (Fig. 1). A virtual hydraulics system is created together with a tactile jacket and a haptic glove to enhance the VR training experience.

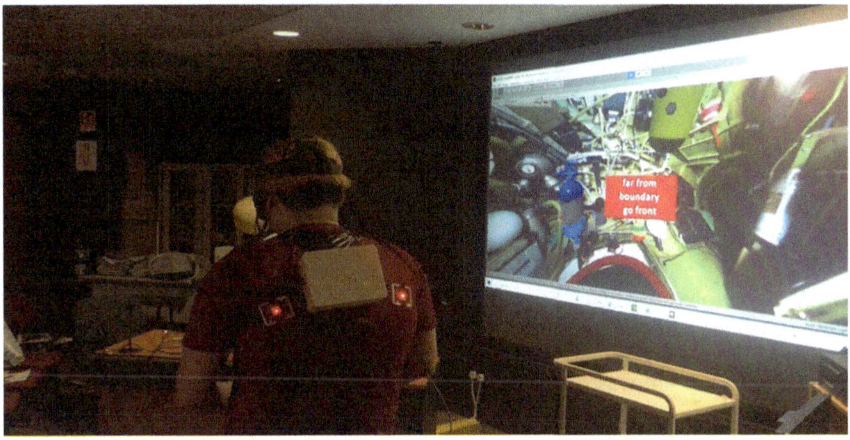

Fig. 1 Vibrotactile jacket and glove for aircraft maintenance learning application

2 Tactile Device Design and Fabrication

This section describes the development of wearable devices. 3D printing is used to prototype the casing of the devices. Specifically, an Arduino Nano, Bluetooth module, battery holder and a boost converter are hosted in each of the cases. The vibrotactile feedback jacket developed in this work (Fig. 2a) consists of the electronic case and four output modules (one for each shoulder, one at the side of the waist). Each output module has one vibration motor providing feedback to the wearer, and one LED indicating the activation of the motor. The total cost of manufacturing this jacket was S$48.35 (Arduino Nano, Boost converter, Bluetooth module, vibration motors LEDs, rechargeable batteries and 3D printing).

Similarly, the vibrotactile feedback glove developed in this work (Fig. 2b) consists of a smaller electronic case and five output modules (one at the back of each finger).

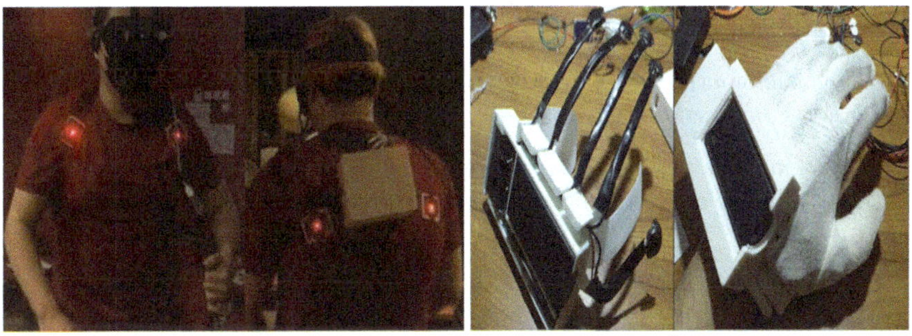

a b

Fig. 2 The design of vibrotactile feedback system: **a** Jacket, and **b** Glove

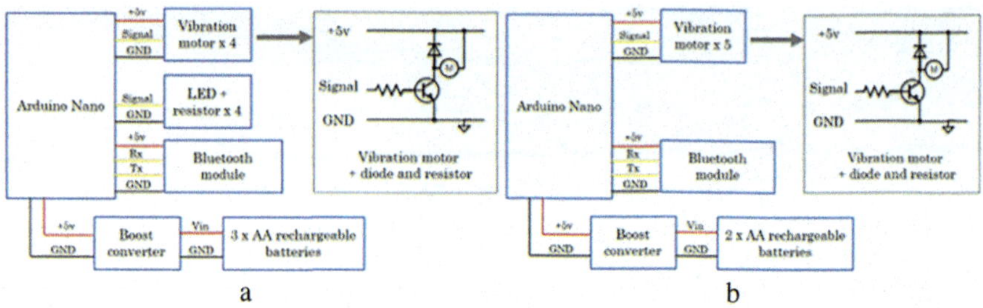

Fig. 3 The circuit design of the vibrotactile feedback system: **a** the Jacket, and **b** the glove

The entire hardware was printed as a whole piece to keep the components securely in place. The cost of all the components (Arduino Nano, Boost converter, Bluetooth module, vibration motors, LEDs, rechargeable batteries, Cloth gloves and 3D printing) is S$37.40 for the glove fabrication. Figure 3 shows the circuit design of the jacket and glove.

3 Development of a VR Aircraft Maintenance Training System

An Oculus Rift CV1 was used for the development of the application and the Leap Motion infrared camera was used for hand-tracking. Unity3D [32] is the game engine used for the development of the training system. The Mixed Reality Tool Kit (MRTK) [33] from Microsoft is used in this project for rapid prototyping. For other needs, custom C# scripts are prepared. The training system has three parts:

1. Introduction: To explore the aircraft, its hydraulics system and its components.
2. Tutorial: To learn one specific hands-on task in hydraulics systems, i.e., accumulator removal using wrenches (Fig. 4).
3. Testing: Based on the guided introduction on the knowledge of hydraulic flow and the hands-on task.

The hands-on task, which is based on the removal of an accumulator, as shown in Fig. 4 is composed of two steps:

1. Unlocking lock nuts using two 7/8 wrenches.
2. Loosening the clamps using a 3/8 wrench.

In step one, the trainee must use one wrench to hold the crossfit and the other to unfasten the locknuts by moving counterclockwise. He/she is successful only if this sequence is followed, otherwise, there is a deduction in marks. The hydraulics system is inside a circular cavity with an opening at the bottom (see the left portion of Fig. 5). Working in such an area makes one prone to injuries. Therefore, there is a need to create a boundary awareness system. Figure 5b is an illustration of the user

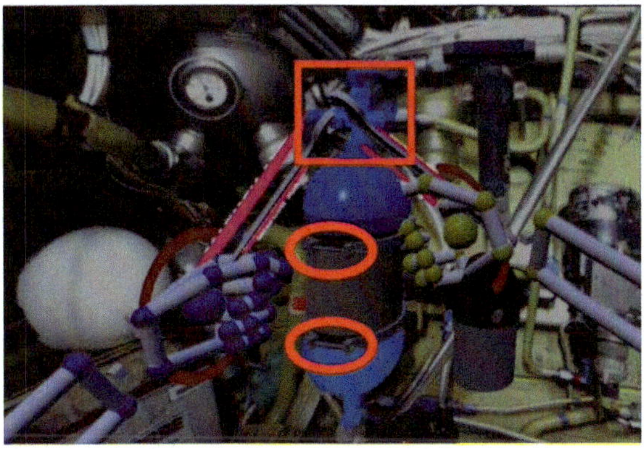

Fig. 4 A snippet of the hands-on task (rectangle: step one, ovals: step two)

as a square. Given the center of the hydraulics system, the algorithm constantly finds which one of the four corners is the closest to the center. Regardless of orientation and distance, there will always be a unique corner that is the closest to the center. Once confirmed, the final orientation of the user is based on the opposite point, e.g., if the back of the user is the closest, i.e., corner 3, the user is heading front. Similarly, if the user is heading right, the left most corner, i.e., corner 2 will be the closest. Based on certain distance thresholds, different warnings starting from visual up to tactile feedback can be given. In the test, moving away from the boundary results in safety test scores being reduced. The three colors in the figure highlight the three different levels of boundary thresholds used in the application. The wearable devices are integrated into the VR application by using a plug-in as shown in Fig. 6 which allows for the devices to be easily integrated into any VR application with minor changes to the original application. The vibrotactile feedback jacket makes use of the Oculus headset's position to provide feedback to the user. This tells whether he/she has moved outside the boundary of the virtual hydraulics space. Figure 7 depicts the working of the associated algorithms for the vibrotactile feedback jacket and glove.

4 Evaluation and Discussion

To evaluate the efficiency of the haptic devices, 39 student participants were invited to try out the developed VR application, along with the devices. They would first try the application without the devices and answer questions about their experience as shown in Fig. 8a. Next, the participants would then wear the devices and try the application again, followed by another set of questions about their experience with the devices as shown in Fig. 8b.

The Mann Whitney U test was used to test the significance of the data. For this test, the null hypothesis is that the two groups (17 with and 22 without devices) are sampled from populations with identical distributions, the alternate hypothesis is

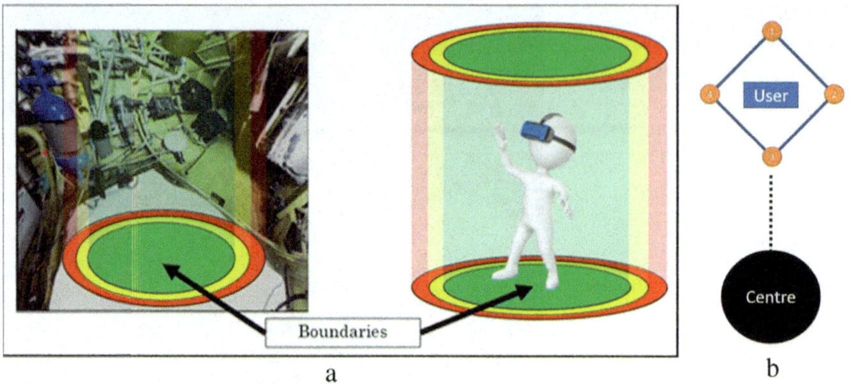

Fig. 5 Boundary safety system: **a** an illustration in the hydraulic training application, and **b** a simple visualization of the boundary safety system

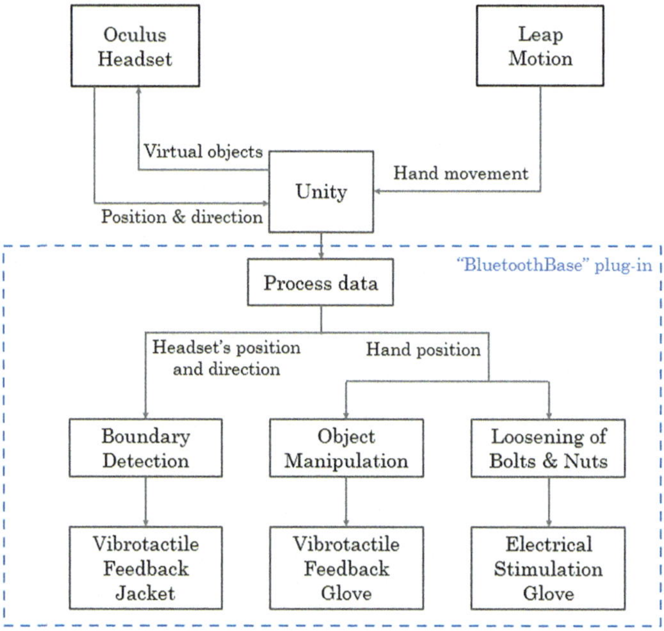

Fig. 6 Working process of wearable devices with the VR application

that the two groups are sampled from populations with different distributions and the significance level (alpha) is set as 5%. The results for the test are shown in Table 1. The p-values for both devices are smaller than alpha, meaning that the null hypothesis is rejected, and that for both devices, the participants felt that there is a difference in their experiences with the wearable devices.

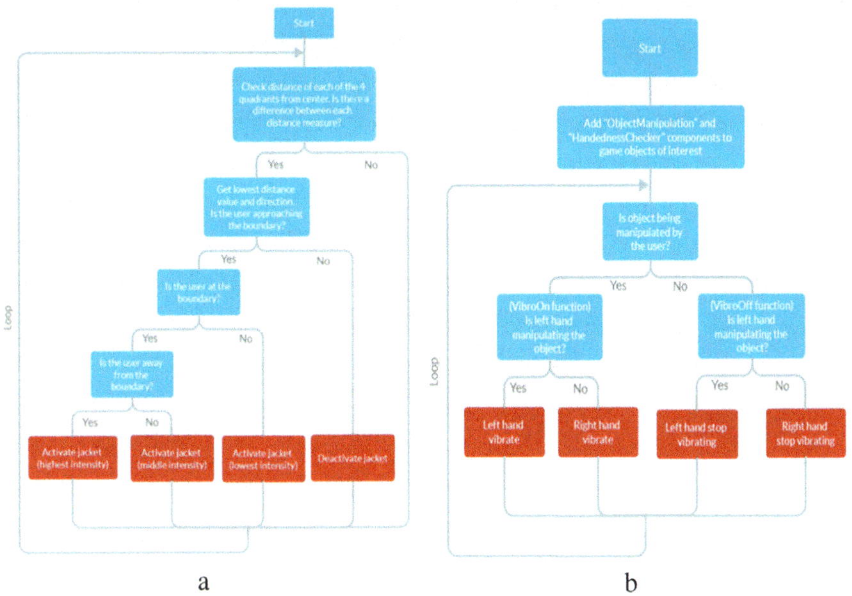

Fig. 7 Illustration of the boundary safety system algorithms: **a** for the vibrotactile feedback jacket, and **b** for the vibrotactile feedback glove

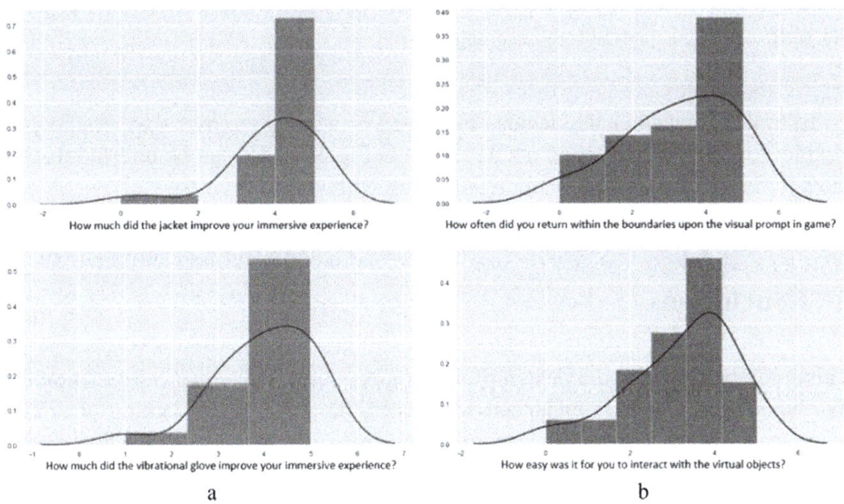

Fig. 8 Plot of data for group: **a** without devices, and **b** with devices

Table 1 Results of Mann–Whitney U-test

Data compared	Test statistics	P-value
Jacket	377.5	0.037456
Glove	266.0	0.010669

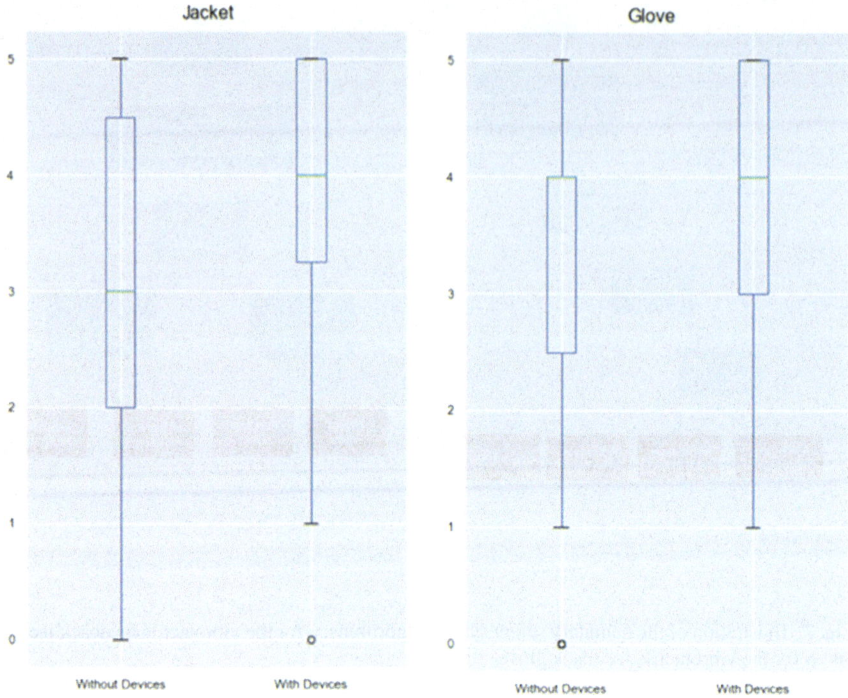

Fig. 9 Box-plots comparison between groups

To evaluate whether the devices improved or worsen the user's experience, the boxplot for the data is plotted out as shown in Fig. 9, and it can be inferred that the plots for both devices have better results than without the devices.

5 Conclusions

This study has presented a design of a VR application with advanced functionality like boundary awareness enhanced with wearable devices for learning of aircraft maintenance tasks. The primary wearable devices are the vibrotactile feedback jacket and glove. The efficacy of these devices has also been proved with the help of a survey conducted amongst 39 participants.

While the VR application and haptic devices developed in this study improves the learning experience, there are some limitations. The Arduino Nano used for the haptic devices is still considerably big compared to the size of the gloves. This can be improved by eliminating its use and using more compact digital circuits instead. The integration of the vibrotactile feedback glove was made easy by using a plug-in which limits its vibration pattern to only work with all 5 motors together, which makes the feedback unrealistic. This can be improved by controlling each motor individually.

References

1. Plunkett, L.: Facebook Buys Oculus Rift For $2 Billion (2014)
2. Osterland, A.: Coronavirus could be catalyst to reinvigorate virtual reality headsets (2020)
3. Cai, Y., van Joolingen, W., Walker, Z.: VR, simulations and serious games for education (2019). https://doi.org/10.1007/978-981-13-2844-2
4. Chiang, P., Zheng, J., Mak, K.H., et al.: Progressive surface reconstruction for heart mapping procedure. Comput. Aided Des. **44**, 289–299 (2012). https://doi.org/10.1016/J.CAD.2011.11.004
5. Chiang, P., Zheng, J., Yu, Y., et al.: A VR simulator for intracardiac intervention. IEEE Comput. Graph. Appl. **33**, 44–57 (2013). https://doi.org/10.1109/MCG.2012.47
6. Chen, G., Moreland, J., Ratko, D., et al.: Virtual reality development for engineering applications. In: ASME 2010 World Conference on Innovative Virtual Reality, WINVR 2010, pp. 127–135 (2010). https://doi.org/10.1115/WINVR2010-3757
7. Nguyen, V.T., Jung, K., Dang, T.: Creating virtual reality and augmented reality development in classroom: Is it a hype? In: Proceedings of 2019 IEEE International Conference on Artificial Intelligence and Virtual Reality, AIVR 2019, pp. 212–217 (2019). https://doi.org/10.1109/AIVR46125.2019.00045
8. Cai, Y.: 3D immersive and interactive learning (2013). https://link.springer.com/book/10.1007/978-981-4021-90-6
9. Yan, Y., Zhang, L., Chen, M.: AGRMTS: a virtual aircraft maintenance training system using gesture recognition based on PSO-BPNN model. Comput Animat. Virtual Worlds **33**, e2031 (2022). https://doi.org/10.1002/CAV.2031
10. Cai, Y., Indhumathi, C., Chen, W.Y., Zheng, J.M.: VR Bio X games. Lecture Notes in Computer Science (including subseries Lecture Notes in Artificial Intelligence and Lecture Notes in Bioinformatics) 5080 LNCS, pp. 278–287 (2008). https://doi.org/10.1007/978-3-540-69744-2_21/COVER
11. Lee, H., Woo, D., Yu, S.: Virtual reality metaverse system supplementing remote education methods: based on aircraft maintenance simulation. Appl. Sci. **12**, 2667 (2022). https://doi.org/10.3390/APP12052667
12. Mustapha, S., Chong, C.A., Mohammed, M.N.: Review on the usage of mixed reality and augmented reality assisted learning tool in aircraft maintenance. In: Proceeding of 2021 IEEE 9th Conference on System, Process and Control, ICSPC 2021, pp. 168–173(2021). https://doi.org/10.1109/ICSPC53359.2021.9689118
13. Rupasinghe, T., Kurz, M., Washburn, C., Gramopadhye, A.: Virtual reality training integrated curriculum: an aircraft maintenance technology (amt) education perspective. Int. J. Eng. Educ. **27**, 778–788 (2011)
14. Sadasivan, S., Vembar, D., Washburn, C., Gramopadhye, A.K.: Evaluation of interaction devices for projector based virtual reality aircraft inspection training environments. Lecture Notes in Computer Science (including subseries Lecture Notes in Artificial Intelligence and Lecture Notes in Bioinformatics) 4563 LNCS, pp. 533–542 (2007). https://doi.org/10.1007/978-3-540-73335-5_58
15. Christian, J., Krieger, H., Holzinger, A., Behringer, R.: Virtual and mixed reality interfaces for e-training: examples of applications in light aircraft maintenance. Lecture Notes in Computer Science (including subseries Lecture Notes in Artificial Intelligence and Lecture Notes in Bioinformatics) 4556 LNCS, pp. 520–529 (2007). https://doi.org/10.1007/978-3-540-73283-9_58/COVER

16. Wang, W., Qi, Y., Wang, Q.X.: An augmented reality application framework for complex equipment collaborative maintenance. Lecture Notes in Computer Science (including subseries Lecture Notes in Artificial Intelligence and Lecture Notes in Bioinformatics) 6874 LNCS, pp. 154–161 (2011). https://doi.org/10.1007/978-3-642-23734-8_25/COVER
17. Webel, S., Bockholt, U., Engelke, T., et al.: An augmented reality training platform for assembly and maintenance skills. Robot. Auton. Syst. **61**, 398–403 (2013). https://doi.org/10.1016/J.ROBOT.2012.09.013
18. Xie, Y., Zhang, Y., Cai, Y.: Virtual reality engine disassembly simulation with natural hand-based interaction. 121–128 (2019). https://doi.org/10.1007/978-981-13-2844-2_11
19. Jamain, N.S., Kasirun, Z.M.: Potential of using virtual environment for aircraft maintenance learning system in making tacit knowledge explicit. Commun. Comput. Inf. Sci. (CCIS) **179**, 506–516 (2011). https://doi.org/10.1007/978-3-642-22170-5_43
20. Sadasivan, S., Vembar, D., Stringfellow, P., et al.: Aircraft maintenance technology education: Integrating asynchronous technology and virtual reality. In: ASEE Annual Conference and Exposition, Conference Proceedings (2006). https://doi.org/10.18260/1-2-1427
21. Shao, X., Wei, X., Liu, S.: Research on aircraft virtual assembly technology based on gesture recognition. In: Proceedings of 2019 IEEE 1st International Conference on Civil Aviation Safety and Information Technology, ICCASIT, pp. 147–150 (2019). https://doi.org/10.1109/ICCASIT48058.2019.8973210
22. Ze, L., Lijie, C., Bo, R.: Study on the virtual simulation training system for SCS maintenance. In: Proceedings of 2020 International Conference on Virtual Reality and Intelligent Systems, ICVRIS, pp. 143–146 (2020). https://doi.org/10.1109/ICVRIS51417.2020.00039
23. Siyaev, A., Jo, G.S.: Towards aircraft maintenance metaverse using speech interactions with virtual objects in mixed reality. Sensors **21**, 2066 (2021). https://doi.org/10.3390/S21062066
24. Siyaev, A., Jo, G.S.: Neuro-symbolic speech understanding in aircraft maintenance metaverse. IEEE Access **9**, 154484–154499 (2021). https://doi.org/10.1109/ACCESS.2021.3128616
25. Suárez-Warden, F., Yocelin, C., Eduardo, G.M.: Assessment of communicative learning via augmented reality versus traditional method for aeronautical transportation. Int. Conf. Transparent Opt. Netw. (2011). https://doi.org/10.1109/ICTON.2011.5970860
26. George, C., Tamunjoh, P., Hussmann, H.: Invisible boundaries for VR: auditory and haptic signals as indicators for real world boundaries. IEEE Trans. Visual Comput. Graphics **26**, 3414–3422 (2020). https://doi.org/10.1109/TVCG.2020.3023607
27. Zhu, L., Cao, Q., Cai, Y.: Development of augmented reality serious games with a vibrotactile feedback jacket. Virtual. Rity. Intell. Hardw. **2**, 454–470 (2020). https://doi.org/10.1016/J.VRIH.2020.05.005
28. Haptics in product design—What is Haptic feedback & Haptic technology. https://engineeringproductdesign.com/knowledge-base/haptic-feedback-and-technology/. Accessed 21 Aug 2022
29. Interactor Virtual Reality. https://www.larryshultz.com/Best-Virtual-Reality-VR-Video-Game-Wear-Invention-by-Larry-Shultz/
30. Mikhailovsky, A.: What is Haptic Feedback (Haptics)? (2017). https://teslasuit.io/blog/haptic_feedback/
31. Senso Devices. https://senso.me/. Accessed 21 Aug 2022
32. Unity Real-Time Development Platform | 3D, 2D VR & AR Engine. https://unity.com/. Accessed 17 May 2022
33. MRTK-Unity Developer Documentation—Mixed Reality Toolkit | Microsoft Docs. https://docs.microsoft.com/en-us/windows/mixed-reality/mrtk-unity/?view=mrtkunity-2021-05. Accessed 1 June 2022

Case Study of AR Digital Literacy Intervention for Students Diagnosed with ADHD

Georgia Psyrra, Eleni Mangina, and Rita Treacy

Abstract The present chapter constitutes an Augmented Reality (AR) literacy survey case study for primary school students diagnosed with Attention Deficit Hyperactivity Disorder (ADHD). It aims to evaluate AR, digital and conventional literacy programmes, examining whether the interventions had a significant impact on participants' literacy skills, and determining whether their use prevails over the standard learning methods used in school. The study is based on data collected during the intervention of the "*ADHD AUGMENTED*" (AHA) pilot project which aimed to support students diagnosed with ADHD by developing an AR-enhanced literacy application. The intervention concerned the use of either the "*WordsWorthLearning*" (WWL) online literacy programme or the WWL programme enhanced with AR, while the control group received the standard primary school educational intervention. A pre-post assessment methodology was applied for the evaluation of students' performance. The results showed that students assigned to the intervention groups scored significantly better in the post intervention assessments and demonstrated higher learning gain than those treated with conventional learning methods. Additionally, the analysis revealed that both web-based and AR literacy programmes can be equally valuable to students' literacy skills.

G. Psyrra (✉) · E. Mangina
School of Computer Science, University College Dublin, Dublin, Ireland
e-mail: georgia.psyrra@ucdconnect.ie

E. Mangina
e-mail: eleni.mangina@ucd.ie

R. Treacy
WordsWorth Learning Ltd, Dublin, Ireland
e-mail: rita@wordsworthlearning.com

Y. Cai et al. (eds.), *Mixed Reality for Education*, Gaming Media and Social Effects,
https://doi.org/10.1007/978-981-99-4958-8_13

1 Introduction

Attention Deficit Hyperactivity Disorder (ADHD) is one of the most common mental health diagnoses in childhood and it is estimated to affect about 5% of school-aged children [1]. It is a neurodevelopmental disorder, characterized by persistent inattention, hyperactivity, and impulsive behaviour [2]. In addition, ADHD is directly related with low frustration tolerance [3]. Research has shown that children with ADHD have a higher prevalence of language problems than those without ADHD, thus contributing to markedly poorer academic functioning [4]. Considering the difficulties that students with ADHD encounter in literacy, this chapter presents a study that aims to assess the impact of web- based technologies on the reading and spelling skills of primary school students with ADHD, by emphasizing on Augmented Reality (AR) digital literacy interventions. Participants recruited for this study were in 3rd, 4th or 5th Class at primary school in Ireland (aged 8–12 years) during the 2018/19 academic year.

1.1 Literacy in Students with ADHD

The symptoms of ADHD are known to impact academic performance, however they are a heterogeneous population showing considerable variation in degree of symptoms and presence of comorbidities [5]. The association between literacy problems and ADHD at the diagnostic level appeared to be part of a broader pattern of high and pervasive levels of inattention [6]. Students with ADHD experience problems with reading, spelling and writing tasks [7]. One-half to two- thirds of children with ADHD have one or more co-existing learning difficulties in reading and/or writing [8].

Learning and attention problems are on a continuum, are inter-related and usually co-exist [9]. Additionally, reading is a complex and slowly learned skill requiring integration of visual, linguistic, cognitive and attentional systems. Working memory is related to processing speed (PS) retardation and can affect reading fluency and comprehension [10]. Research has shown significant relationships between slow processing speed (PS) and reading disabilities and that PS deficits are strongly associated with ADHD [11]. A comparative study of literacy and numeracy levels in students diagnosed with ADHD and non-ADHD conducted in Western Australia, investigating three cross-sectional birth cohorts in School 3rd class (students aged 7–8), 5th class (age 9–10) and 7th class (age 11–12 years), revealed that the percentage of students with ADHD who score below the State benchmark in 3rd class is higher than those without ADHD, and that this proportion increases during elementary school. In relation to spelling, the study showed even higher prevalence rates of students without ADHD than in reading, demonstrating that more than a third of children with ADHD did not reach the State benchmark throughout their primary schooling [12].

1.2 Benefits of AR Technology

AR was first introduced in 1990 by Tom Caudell who designed a head-mounted digital display to facilitate the wiring process in aircrafts [13]. It is a technology used to overlay digital information within real environments, thus combining the two worlds into one experience [14].

Initial evidence on the leverage of AR technology by school-age children have shown that AR can help reduce ADHD-related symptoms such as hyperactivity, inattention and impulsivity [15]. Studies have indicated that AR can support children with disabilities to understand concepts faster and better [16], and that attention skills can be improved during action with AR and multimedia game training [17–19]. A recent study aimed at training selective and focused attention in children diagnosed with ADHD found that AR serious games can enhance children's concentration and reduce frustration during problem-solving activities [20]. Furthermore, a meta-analysis study showed that AR can support people with disabilities, including those with ADHD, in acquiring daily functional skills [21].

AR is used in various areas such as museums, libraries and medical centres to improve the user experience and it can potentially be applied to all senses [22]. It is known for improving students' perception of interaction with the real world by increasing the sense of reality [22, 23]. As technology has become a crucial toolkit in education, educators have begun to utilize AR in their lesson plans.

Various studies have been developed to document the positive impact of AR technology on students' language learning. Previous research have shown that the use of AR/VR technologies can positively effect students' literacy in terms of vocabulary, speaking, writing and cultural language skills [24, 25]. In addition to these effects, AR has been found to benefit students' performance in in reading, comprehension, pronunciation and phonetics [25]. The technology can be implemented in language learning tasks in various formats such as AR web-based literacy programmes, books and games. AR books have been found to be valuable in enhancing students' reading comprehension and story recall tasks [26]. Similarly, AR games can positively affect students' cognitive learning outcomes and comprehension [27].

Furthermore, AR solutions have been found to can effectively impact literacy performance of student's with special needs. A recent study of literacy for students with special needs through AR technology implemented an AR module as self-learning strategy for students with special needs [28]. The results revealed that after using the AR module participants' reading scores increased and that students who viewed the 3D cards demonstrated higher comprehension levels. With specific regard to children with ADHD, a study for literacy through AR mobile-based learning assessed the impact of these technologies on students' Chinese reading and word recognition skills [29]. The results demonstrated that student with ADHD and reading disabilities were positively effected by the use of AR.

Overall, these findings suggest that AR solutions can support students with special needs and ADHD on minimizing some of their symptoms and that may effectively applied to literacy skills acquisition. However, the studies that concern the use of AR as a literacy instructional tool for students with special needs and ADHD seem to be limited, thus revealing a notable gap in this research area [30–32].

2 Research Questions

Considering that the use of online and AR-enhance learning tools could be valuable in acquiring literacy, the present study seeks to enrich research by evaluating the use of such technologies by students with ADHD. It concerns a quite large number of participants that can support statistically valid results, with the aim of providing evidence for future research work in the field. The study is implemented using data from the AHA[1] pilot project, which focuses on primary school students aged 8–12 years (at the time of registration), all diagnosed with ADHD.

Focusing on the evaluation of digital AR educational resources, this study aims to explore the impact of both WordsWorthLearning[2] (WWL) literacy programs used as interventions in the AHA project: the web-based application as well as its enriched version with AR content. WWL is an online literacy application that addresses reading and spelling skills [33].

WWL literacy interventions were delivered by mixed learning methods, thus giving students the opportunity to work either in the school or home environment, or collaboratively between both. On the other hand, students who were assigned to the control group received the typical in-class intervention. The above allow the exploration of different types of student groups based on their assigned intervention programme and the learning environment in which it was conducted. Given this opportunity, the study explores the effects of WWL and WWL with AR programmes compared with the school typical teaching methods and aims at providing evidence of the efficacy of the alternative online programmes.

In conclusion, this study aims to evaluate the WWL with AR and WWL online literacy programmes, used as learning resources to enhance reading and spelling literacy skills of students diagnosed with ADHD, by answering the following research questions:

- RQ1: What is the impact of AR emerging technologies utilised for the advancement of literacy skills of the student participants?
- RQ2: What is the contribution to reading and spelling skills of the AR digital tools compared to typical in-class methodologies?

[1] https://aha.ucd.ie

[2] https://wordsworthlearning.com

3 The Aha Project

The study presented in this chapter is based on AHA intervention project[3], funded by the European Commission, which aimed to support students with ADHD diagnosis by developing an educational application designed to improve students' literacy skills. The application integrates the technologies of the WordsWorth-Learning (WWL) web-based literacy programme[4] along with AR solutions and functionalities of the Web Health Application for ADHD Monitoring[5] (WHAAM) [34].

Monitoring interface. The AHA monitoring interface allows teachers and parents to monitor students' engagement and performance during the WWL activities by providing a dashboard with a series of facilities as follows [35]:

- The WWL Account which allows children's progression monitoring through the WWL educational content.
- The Case Data section which allows the monitoring of participants' perfor- mance on literacy assessment tests.
- The Observations section which provides a web-based tool that teachers and parents can use to measure children's off-task behaviour.
- The data analytics visualizations section which includes several charts foras shown in Fig. 1: (a) scores on evaluation questionnaires that are included in the WWL; (b) estimations of off-task behaviour per session; (c) time spent on programmes per level; and (d) time spent on programmes weekly.

Literacy programme. The AHA system was designed to help students with ADHD diagnosis to improve their learning through interaction and exploratory learning and it has been implemented by two different versions of the programme, the WWL and the WWL with AR. These two versions, along with the standard classroom teaching, are the three implemented methods of teaching delivery and identify the types of interventions received by the project participants.

In 2011, the initial paper-based WWL literacy programme was successfully digitalised, so that interested parties could access the programme at their own pace using a computer, tablet or a smartphone [33, 36]. The programme focuses on teaching reading and spelling rules and strategies to enhance vocabulary and understanding. According to the official WWL website [37] and the parent- teacher guide for the AHA pilot "AHA-booklet" [38], the programme consists of seven hierarchical levels that must to be followed in ascending order, as follows (see Fig. 2):

- *Level 1* concerns sound and symbol association. The learning objective of this level is for students to learn to listen and recognize speech sounds (phonemes) with the help of vowel and consonant maps.

[3] European Commission funded Pilot Project: Technologies and Tools for Children and Young People with Attention-Deficit Hyperactivity Disorder (ADHD), under Grant Agreement No. 30-CE-0885096/00–34 (ADHD2016-13) https://aha.ucd.ie/

[4] https://wordsworthlearning.com/staticpage/program

[5] https://www.whaamproject.eu/

Fig. 1 AHA data analytics dashboard, redrawn with permission from [38]

- *Level 2* focuses on literacy skills related to acoustic recognition and sequencing. Students learn to identify the number and order of sounds in meaningless pseudo words.
- *Level 3* aims to develop the ability to breakdown words so that the student will then be able to deal with unfamiliar words.
- *Level 4* targets at facilitating reading and spelling. It introduces 20 spelling rules essential to explain irregularities, such as the 'ch' and 'que' as k rules, which are presented with interactive exercises and short videos.
- *Level 5* introduces a visualization technique to learn how to spell complex and irregular words. Students practice reading and spelling with simple syllable words of increasing complexity and incorporating vocabulary according to a school curriculum.

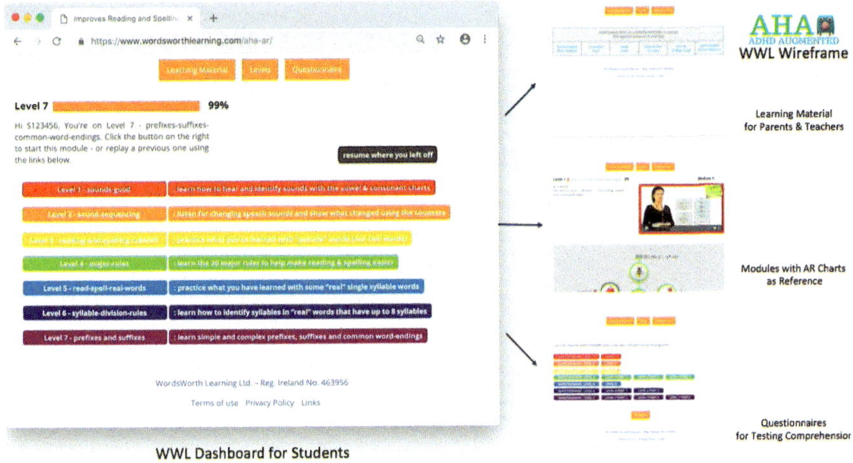

Fig. 2 WWL online dashboard, retrieved from [38]

- *Level 6* aims to enhance students' reading with word analysis as well as spelling with word construction. Students are taught syllable division rules that are used to identify syllables with words of varying complexity.
- *Level 7* is oriented towards teaching students prefixes, suffixes, and common word-endings. It focuses on understanding the rules related to the origin of words as well as recognizing these prefix and suffix 'chunks' as units.

For the needs of AHA project, the WWL literacy programme was enhanced with AR 3D content to facilitate learning with students diagnosed with ADHD. Tutorials and exercises were reinforced with AR technology to include computer 3D graphics of phonemes, rules and prefix/ suffix flashcards.

The AR content built into the WWL literacy programme consists of a marker, the complete repertoire of English vowels and consonants, along with literacy rules and flashcards. Students can click on a specific object and activate their mobile device's camera and by holding it over the marker they can see the AR object displayed. The Vowel and Consonant interactive charts introduce AR elements that bring to life 3D examples of individual letters, digraphs and three and four letter vowel graphemes and their corresponding speech sounds (see Fig. 3). The AR rule flashcards are similar to the AR vowel and consonant charts and they are introduced to enhance a specific lesson and to imprint these complex rules in visual memory. However, their AR functionality is reduced as the animation does not include sound nor purposely do they rotate, in order to minimise confusion related to letter orientation [39, 40]).

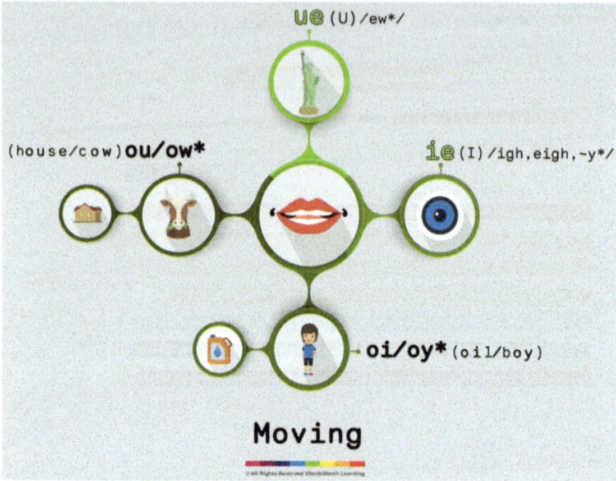

Fig. 3 Example of AR vowels chart, retrieved from [38]

4 Study Design

The inclusion criteria for the AHA project participants [40] were:

- to be in either 3rd, 4th or 5th Class in the primary school system in Ireland (aged from 8 to 12 years) at the time of recruitment (April–June 2018);
- to have a formal diagnosis of ADHD;
- to have an estimated functioning in the average range of Intelligence Quotient (I.Q.);
- to have a broadband connection at home/in school and laptops/tablets with web-browsers to access the learning platform.

Following the recruitment process, a total of 117 pupils were enrolled [36, 39, 40]. The sample size was reasonably large, considering the fact that the target population is demographically narrowed, including students aged 8 to 12 living in Ireland and pre-diagnosed with ADHD.

Assessment and evaluation metrics. Prior to the intervention stage, each child in the AHA programme underwent standard literacy assessments to mea- sure their reading and spelling performance [34]. The assessment consists of the Neale Analysis of Reading Ability (NARA II) and the Graded Word Spelling Test literacy tests which were used later to investigate the change in performance after the intervention. Both tests are able to assess students in the age of AHA participants; NARA II is used to evaluate oral reading in terms of reading accuracy, comprehension and rate [41], while Graded Word Spelling Test used to assess spelling attainment and progress [42].

Case History Data. Prior to the pilot, parents of the participants were re- quested to complete a multidisciplinary (MTD) case history form[6] in order for a wide variety of pertinent clinical information about the students to be collected [40]. Filling in the

[6] Www.wordsworthlearning.com/profiled ©

form consisted of subjective parental reports of the child's developmental, family, medical, educational and social/behavioural history. Items related to demographics, language and literacy reported by the parents were analysed to investigate the distribution of students between the intervention and control groups and to confirm the objectivity of the analysis' results.

Group assignments. In order for the students' activities and performance based on the different teaching intervention they have received to be evaluated, student grouping was needed. After all pre-tests were completed and before the intervention, children were quasi-randomly assigned to either Group 1 (WWL web-based literacy programme enhanced with AR content), Group 2 (web-based literacy programme without AR content) or Group 3 which was the control group (without any WWL intervention during the pilot). Student grouping was mandatory to measure the different effects on student participation based on monitoring the events logs of the web-based programmes (WWL with or with- out AR) and the possible differentiation of student performance between these programmes and the typical education delivery method and programme. Using multidisciplinary data from MDT forms filled in by students' parents, the even distribution of participants into the control and intervention groups was investigated.

Intervention. The intervention occurred during the 2018/19 academic year. Some students completed the online programme before the end of the second semester of the school year (June 2019). Figure 4 presents a description of students', parents' and teachers' expected activities during the involvement in the pilot. Students assigned to both the WWL and WWL-AR intervention groups were asked to use the application (WWL-AR or WWL) according to the group in which they were assigned, ideally both at school and at home with the support of an adult (teacher or parent/guardian depending on the intervention arrangement). Students assigned to the control group only received the typical learning support provided from the educational system, assisted by parental support at home. However they each were provided with full access to the web-based WWL literacy programme after the post-tests were completed [38, 40].

Fig. 4 Distribution of users' expected activities between groups, retrieved from [40]

Students, teachers and parents were given a suggested work schedule. The schedule recommended the involvement of students with the respective web- based literacy programme, ideally on a daily basis during the academic school year. Each scheduled intervention session was designed to last approximately 15 min [34].

Teachers and parents involved in the project were required to submit an off- task rating of students' behaviour during each session. In addition, they could use the monitoring system interface to continuously observe the progress and involvement of children with activities. Before starting the intervention, a technical survey was completed by teachers and parents in order to confirm that the AHA platform was compatible with the hardware and software available at schools and family homes. [40].

Re-assessment. After completing the intervention, each child was retested on the same initial test batteries that were administered prior to intervention, to evaluate their literacy performance over time. Thus, after the re-assessment of students' reading and spelling skills, a performance Change Score of pre and post intervention was calculated.

Participation. Following the exclusion criteria, from the initial dataset of 117 students participated in the AHA project, 116 of them were finally included in the analysis dataset. A students' information, who was one year older after repeating an academic year, was decided not to be included. In conclusion, the present analysis includes 116 AHA participants out of which 104 completed the post assessments.

Outcome measures. In order to measure the students' performance on literacy skills before and after the intervention, all three NARA-II sub-scores of Standard Reading Accuracy, Comprehension and Rate as well as the Standard spelling score of the Graded Word Spelling test were included. A further Change Score (positive or negative) was computed for each participant based on the difference between their 'pre' and 'post' scores.

5 Analysis and Results

The metrics described in the previous section have been analysed for reporting though the utilisation of the "IBM SPSS Statistics Version 24". Continuous variables were descriptively used without any changes. The analysis aimed to detect significant statistical differences on students' performance based on the intervention they received.

The focus of the analysis was to investigate the impact of WWL and WWL with AR on participants' literacy skills as well as whether the intervention programmes prevailed over the typical school programmes. Specifically, in order to detect the significance of the impact of both WWL and WWL with AR programmes on the participants' literacy skills, students' assessment outputs were compared using both the Paired-t-test and the Wilcoxon singed Ranks test. Additionally, in order to investigate possible programmes' prevailing over the scores' change (post minus pre intervention scores) that participants achieved in each group, their outcomes were compared using either the Kruskal–Wallis or the One way Anova test as well as Bonferroni Post-hoc and Pairwise comparisons. The selection process of the applied statistical method has been supported by Kolmogorov Smirnov and Shapiro Wilk distribution tests. Normality distribution assumptions were supported based on the "Sig.2" value with α=0.05.

5.1 *Research Question 1: Did the Intervention Programmes Contribute to the Development of Participants' Literacy Skills?*

This analysis aims to show whether WWL or WWL with AR programmes enhanced the project's participants literacy skills. This is achieved through the comparison of the performance parameters observed between the pre and post intervention assessment. The statistical methods used are the "*paired-t-test*" and the "*Wilcoxon Sign rank test*" depending on the distribution of pre and post intervention scores.

Impact of WWL+AR. The literacy scores of WWL+AR group participants were mostly normally distributed in both pre and after intervention tests, thus the paired-t-test was used to measure the impact of the programme to the participants' literacy skills. The only exception is the spelling score in the pre intervention test for which the Wilcoxon—Sign rank test was used.

Paired-T-Test

The paired-t-test showed that the difference between the literacy score levels before and after the intervention is statistically significant since the probability of erroneously rejecting the null hypothesis (that is, that there is no correlation between intervention and literacy skills levels) is less than $\alpha = 0.05$, as shown Table 1. Furthermore, from the Mean score, it was found that the difference (i.e. post minus pre) in both Accuracy and Comprehension standard scores is positive, which means that the post assessment levels are higher. The opposite seems to have happened for the Reading Rate score. Table 2 demonstrates in detail the descriptive statistics of students' literacy levels for each assessment before and after the intervention.

Table 1 Paired Samples Test for WWL+AR group (change scores)

Change score				95% CI [c]			Sig. [f]	
Variables	Mean	SD[a]	SEM[b]	Lower	Upper	t[d]	df[e]	(2-tailed)
Accuracy	17.595	8.675	1.426	14.702	20.487	12.338	36	0.000
Comprehension	16.568	6.108	1.004	14.531	18.604	16.499	36	0.000
Reading rate	−7.568	8.884	1.460	−10.530	−4.606	−5.182	36	0.000

[a]Standard Deviation; [b]Standard Error of Mean; [c]95% Confidence Interval of the Difference; [d]The test statistic; [e]Degrees of Freedom; [f]The two-tailed probability

Table 2 Descriptive statistics for the WWL + AR group participants' scores

Assessment measures	Mean	N[a]	SD[b]	SEM[c]
Pre-test accuracy	89.73	37	12.764	2.098
Post-test accuracy	107.32	37	14.138	2.324
Pre-test comprehension	90.51	37	12.339	2.029
Post-test comprehension	107.08	37	11.594	1.906
Pre-test reading rate	98.27	37	13.773	2.264
Post-test reading rate	90.70	37	12.539	2.061

[a] Number of valid observations; [b] Standard Deviation; [c] Standard Error of Mean

Table 3 Wilcoxon signed ranks test for the WWL + AR group participants' spelling change scores

	Spelling change
Z (based on negative ranks)[a]	−3.647
Asymp. Sig. (2-tailed)[b]	0.000

[a]Test statistic value; [b]Approximate p-value based on normal distribution

Wilcoxon Signed Ranks Test

Regarding the Spelling score, the Wilcoxon Signed Ranks Test showed that the difference between the spelling score levels before and after the intervention is statistically significant since the probability of erroneously rejecting the null hypothesis (that is, that there is no correlation between intervention and spelling skills levels) is less than $\alpha = 0.05$, as shown in Table 3.

Furthermore, the Descriptive statistics in Table 4 show that the median, the 25th and the 75th quartile of the distribution of participants spelling scores after the intervention are higher in comparison to those before.

Based on the above analysis it is concluded that the WWL+ AR programme results in reducing the reading rate (words per minute) and enhancing the spelling and the reading accuracy and comprehension scores of students diagnosed with ADHD.

Impact of WWL. The scores of participants assigned to the WWL group, both pre and post intervention, were found to be normally distributed and the paired- t-test was used to measure the impact of the programme on the participants' literacy skills.

Table 4 Wilcoxon test's descriptive statistics output for the WWL + AR group participants' Spelling scores

Spelling score					Percentiles[c]			
Variables	N[a]	Mean	SD[b]	Min	Max	25th	25th	75th
Pre-test	39	87.64	14.507	69	118	73.00	86.00	99.00
Post-test	37	91.62	15.426	69	121	79.00	90.00	104.50

[a] Number of valid observations; [b] Standard Deviation; [c] First, second and third quartile

The only exception is the pre intervention participants' Accuracy scores, which were found not to be normally distributed, thus the Wilcoxon—Sign rank test was used.

Paired-t-test

The paired-t-test showed that the difference between the literacy score levels before and after the intervention is statistically significant since the probability of erroneously rejecting the null hypothesis, that is, that there is no correlation between intervention and literacy skills levels, is less than $\alpha = 0.05$, as shown in Table 5.

Furthermore, from the Mean in the same table, it was found that the difference of post minus pre test scores in both Spelling and Comprehension standard scores is positive which means that the post assessment levels are higher than those of pre intervention. The opposite seems to have happened for the Reading Rate standard score. The following table (see Table 6) demonstrates in detail the descriptive statistics of students' literacy levels for each assessment measure before and after the intervention.

Table 5 Paired samples test for WWL group (change scores)

Change score			95% CI [c]		Sig.[f]		
Variables	Mean	SD [a] SEM [b]	Lower	Upper	t [d]	df [e]	(2-tailed)
Comprehension	15.7500	9.326 1.649	12.3880	19.112	9.55400	31	0.000
Reading rate	−10.406	8.739 1.545	−13.557	−7.255	−6.7360	31	0.000
Spelling	6.31300	6.240 1.103	4.06300	8.5620	5.72300	31	0.000

[a] Standard Deviation; [b] Standard Error of Mean; [c] 95% Confidence Interval of the Difference; [d] The test statistic; [e] Degrees of Freedom; [f] The two-tailed probability

Table 6 Descriptive statistics for WWL group participants' scores

Assessment measures	Mean	N[a]	SD[b]	SEM[c]
Pre-test comprehension	92.53	32	13.548	2.395
Post-test comprehension	108.28	32	11.955	2.113
Pre-test reading rate	102.78	32	12.315	2.177
Post-test reading rate	92.38	32	12.659	2.238
Pre-test spelling	87.72	32	14.308	2.529
Post-test spelling	94.03	32	17.015	3.008

[a] Number of valid observations; [b] Standard Deviation; [c] Standard Error of Mean

Table 7 Wilcoxon signed ranks test for the WWL group participants' accuracy change scores

A	Accuracy change
Z (based on negative ranks)[a]	−4.942
Asymp. Sig. (2-tailed)[b]	0.000

[a] Test statistic value; [b] Approximate p-value based on normal distribution

Table 8 Wilcoxon test's descriptive statistics output for the WWL group participants' Accuracy scores

Accuracy score					Percentiles[c]			
Variables	N[a]	Mean	SD[b]	Min	Max	25th	50th	75th
Pre-test	40	91.82	13.980	69	123	84.00	88.50	104.50
Post-test	32	108.13	13.176	74	130	97.00	110.50	117.750

[a] Number of valid observations; [b] Standard Deviation; [c] First, second and third quartile

Wilcoxon Signed Ranks Test

Regarding the Accuracy standard score, the Wilcoxon Signed Ranks Test showed similar results. The change of the spelling score levels before and after the intervention is statistically significant since the probability of erroneously rejecting the null hypothesis, that is, that there is no correlation between intervention and accuracy levels is less than $\alpha = 0.05$, as shown in Table 7.

Furthermore, the Descriptive statistics in Table 8 show that the median, the 25th and the 75th quartile of the distribution of participants accuracy reading scores after the intervention are higher.

Based on the above analysis it is concluded that the WWL programme results in reducing the reading rate and enhances spelling and reading accuracy and comprehension of students diagnosed with ADHD.

5.2 Research Question 2: Did the Intervention Programs Significantly Affect the Performance of Students' Literacy Compared to the Use of Typical In-Class Methodologies?

This part of the analysis aims to respond to the research question about whether WWL and WWL with AR programmes could significantly enhance the project participants' performance on literacy skills in comparison to the participants who receive the traditional teaching methods. The effect of AHA on the reading and spelling skills of students with ADHD is measured through the comparison of the parameters observed between the pre and post intervention assessment that took place at the end of the project for each group.

The evaluation measures were taken into account to ensure that all participants have been evenly distributed based on pre-test scores across the intervention and control groups. Ensuring that participants in each group have similar pre-test scores can support the hypothesis that possible differences between post and pre test students' scores are exclusively due to the intervention they received.

Pre-test scores comparison between the 3 groups. The tests used to evaluate whether students have the same starting point for their literacy skills for each group are the Kruskal-Wallis and the One way Anova. The null hypothesis at both tests assumes that the samples (groups) are from identical populations. The use of the tests is intended to examine which intervention group (independent variable) prevails over the pre-test scores (dependent variable) [43]. According to normality tests as well as visual graphics it was concluded that the distribution of participants of each group based on pre-test measures it is not normal, thus the Kruskal–Wallis test was chosen.

Kruskal-Wallis Test

The Kruskal-Wallis test showed (see Table 9) that the literacy skill levels of participants for each pre-assessment measure do not significantly differ between the three intervention and control groups since the probability of erroneous rejection of the null hypothesis, that is, that there is no correlation between pre-test score levels of literacy measures and the intervention or control groups, is more than α=0.05.

Despite that there is no statistically significant difference in participants' literacy skills between the groups, the following descriptive table (see Table 10) shows that the control group has a slightly higher scoring average for all literacy measures. This could result in higher post-test scores for the control group. Thus, the next analysis uses the Change Score, which is calculated by subtracting the initial score from the final, after the pilot competition, to compare the effect of the interventions and the traditional learning delivery method to students.

Change of scores between the 3 groups (post minus pre-test scores). This part of the analysis focuses on comparing the Change Scores achieved by the participants in each intervention and control group. The aim is to determine whether the use of technologies used in the interventions (WWL and WWL+AR) prevails in enhancing the development of literacy skills of students with ADHD versus typical school based literacy programmes.

Despite that there is no statistically significant difference between students' pre-test scores depending on the group to which they were assigned, there appears to

Table 9 Kruskal–Wallis test statistics for each pre-assessment measure

Statistic measures[a]	Accuracy	Comprehension	Reading rate	Spelling
Chi-square	1.722	1.940	3.164	1.890
df[b]	2	2	2	2
Asymp. Sig[c]	0.423	0.379	0.206	0.389

[a] Grouping Variable is based on WWL + AR, WWL and Control groups; [b] Degrees of freedom; [c] Statistical significance

Table 10 Descriptive statistics of participants' pre-test scores

Pre-test score						95% CI[c]			
Variables	Group	N	Mean	SD[a]	SEM[b]	Lower	Upper	Min	Max
Accuracy	WWL + AR	39	90.95	13.621	2.181	86.53	95.36	69	121
	WWL	40	91.83	13.980	2.210	87.35	96.30	69	123
	Control	37	94.03	13.088	2.152	89.66	98.39	69	123
Comprehension	WWL + AR	39	91.28	12.532	2.007	87.22	95.34	69	112
	WWL	40	92.08	14.101	2.230	87.57	96.58	69	116
	Control	37	95.32	13.197	2.170	90.92	99.72	69	119
Reading rate	WWL + AR	39	99.13	14.144	2.265	94.54	103.71	69	127
	WWL	40	100.60	13.401	2.119	96.31	104.89	69	124
	Control	37	104.59	13.619	2.239	100.05	109.14	75	129
Spelling	WWL + AR	39	87.64	14.507	2.323	82.94	92.34	69	118
	WWL	40	87.48	14.358	2.270	82.88	92.07	69	124
	Control	37	93.32	18.226	2.996	87.25	99.40	69	131

[a] Standard Deviation; [b] Standard Error of Mean; [c] 95% Confidence Interval for Mean

be a significant difference in their Change Scores across all performance variables. Table 11 shows that the absolute value of Change Scores is higher for students who were assigned to the intervention groups compared to those in the control group, for all the included literacy evaluation measures.

The previous descriptive statistic measures are not enough to conclude to a reliable outcome for the dominance of WWL and WWL+AR programmes over the typical school intervention. In order to extract a documented result, Kruskal-W or one-way ANOVA are the appropriate tests. The normality and homogeneity tests revealed that the participants' Change Scores on comprehension and reading rate assessment measures are normal and homogenous distributed, thus the one-way ANOVA test will be used. On the other hand, for the accuracy and spelling the Kruskal-W test will be used.

The results of both tests (see Tables 12 and 13) reveal that the null hypothesis for all Change Scores is rejected (assumption sig. ¡ $\alpha = 0.05$). Therefore, there is a correlation between the groups regarding all the literacy assessments measures.

So far, the results have shown that the intervention on the groups that the participants were assigned to effected their evolvement on literacy skills. How- ever, the type of correlation (either positive or negative), as well as which specific groups it concerns, has not been yet investigated. The Bonferroni Post-hoc (see Table 14) along with the Pairwise comparison (see Table 15) outputs presented below reveal that there is a significant difference on the Change Scores levels for all the assessment measures between the control group with both intervention groups and that there is

Table 11 Descriptive statistics of participants' Change Scores (post minus pre-test)

Change score					95% CI [c]				
Measures	Group	N	Mean	SD[a]	SEM[b]	Lower	Upper	Min	Max
Accuracy	WWL + AR	37	17.59	8.675	1.426	14.70	20,49	−3	41
	WWL	32	15.78	7.312	1.293	13.14	18,42	4	31
	Control	36	1.86	7.754	1.292	−0.76	4,48	−24	18
Comprehension	WWL + AR	37	16,57	6.108	1.004	14.53	18,60	5	26
	WWL	32	15.75	9.326	1.649	12.39	19,11	−8	32
	Control	36	2.58	8.872	1.479	−0.42	5,59	−17	20
Reading Rate	WWL + AR	37	−9.51	14.858	2.443	−14.47	−4,56	−80	9
	WWL	32	−10.41	8.739	1.545	−13.56	−7,26	−33	3
	Control	36	−1.31	6.790	1.132	−3.60	0.99	−21	14
Spelling	WWL + AR	37	5.00	6.658	1.095	2.78	7.22	−10	18
	WWL	32	6.31	6.240	1.103	4.06	8.56	−2	19
	Control	36	0.47	7.542	1.257	−2.08	3.02	−18	19

[a]Standard Deviation; [b]Standard Error of Mean; [c]95% Confidence Interval for Mean

Table 12 One way Anova output for comprehension and reading rate change scores

Change score measures		Sum of squares	df[a]	Mean square	F[b]	Sig.[c]
Comprehension	Between groups	4371.561	2	2185.780	32.552	0.000
	Within groups	6781.824	101	67.147		
	Total	11,153.385	103			
Reading rate	Between groups	1499.183	2	749.591	11.109	0.000
	Within groups	6814.971	101	67.475		
	Total	8314.154	103			

[a] Degrees of freedom associated with the sources of variance; [b] Mean Square Regression divided by the Mean Square Residual; [c] The p-value associated with this F value

Table 13 Kruskal–Wallis test output for spelling and reading accuracy change scores

Statistic measures[a]	Accuracy	Spelling
Chi-square	50.242	14.927
df[b]	2	2
Asymp. Sig[c]	0.000	0.001

[a] Grouping Variable is based on WWL + AR, WWL and Control groups; [b] Degrees of freedom; [c] Statistical significance

no significant difference between the change scores of participants assigned to the WWL group in comparison with those assigned to the WWL with AR group.

The type of correlation between the intervention with the control groups on students' Change Scores can be detected through the descriptive measures presented above (see Table 11). It is revealed that students who were assigned in the intervention groups improved more on the literacy measures of Reading comprehension, Reading accuracy and Spelling. Additionally, they significantly reduced their Reading rate when compared to those who were assigned to the control group whose change was minor.

Furthermore, the analysis demonstrates that students who were assigned to intervention groups reduced their reading rate (see Tables 1 and 5) but significantly increased their reading comprehension and accuracy as well as their spelling scores

Table 14 Multiple comparisons of comprehension and reading rate change scores between the groups of participants

Dependent variable			Std			95%CI[d]	
	(I)Groups	(J)Groups	MD[a] (I-J)	Error[b]	Sig.[c]	Lower bound	Upper bound
Comprehension	WWL + AR	WWL	0.818	1.978	1.000	−4.00	5.63
	Control	WWL + AR	−14.082*	1.932	0.000	18.79	−9.38
	Control	WWL	−13.264*	2.004	0.000	−18.14	−8.39
Reading rate	WWL + AR	WWL	2.839	1.983	0.466	−1.99	7.67
	Control	WWL + AR	6.339*	1.937	0.004	1.62	11.05
	Control	WWL	9.178*	2.009	0.000	4.29	14.07

[a] Mean Difference; [b] The standard error; [c] Statistical significance; [d] 95% Confidence Interval

Table 15 Multiple comparisons of accuracy and spelling change scores between the groups of participants

Dependent variable	(I)Groups	(J)Groups	Test statistic	Std. error[a]	Sig.[b]	Adj. Sig.[c]
Accuracy	WWL + AR	WWL	−3.862	7.275	0.490	1.000
	Control	WWL + AR	22.002	7.106	0.000	0.000
	Control	WWL	25.864	7.371	0.000	0.000
Spelling	WWL + AR	WWL	22.002	7.233	0.593	1.000
	Control	WWL + AR	22.002	7.065	0.002	0.006
	Control	WWL	25.864	7.329	0.000	0.001

[a] The standard error; [b] Asymptotic significances (2-sided tests) are displayed: The significance level is 0.05; [c] Adjustments for multiple testing using the Bonferroni error correction

Table 16 Descriptive statistics and Paired-t-test output for each normal distributed score variable within control group

Change scores				95% CI[c]				
	Mean	SD[a]	SEM[b]	Lower	Upper	t[d]	df[e]	Sig.[f]
Accuracy	1.600	7.705	1.302	−1.047	4.247	1.229	34	0.228
Comprehension	2.486	8.982	1.518	−0.600	5.571	1.637	34	0.111
Reading rate	−1.229	6.873	1.162	−3.590	1.132	−1.057	34	0.298

[a] Standard Deviation; [b] Standard Error of Mean; [c] 95% Confidence Interval of the Difference; [d] The test statistic; [e] Degrees of freedom; [f] The two-tailed probability

Table 17 Descriptive statistics and Wilcoxon Sign rank test for each non normal distributed score variable within each group

Participants' group						Percentiles[b]		Test stats		
	Scores	Mean	SD[a]	Min	Max	25th	50th	75th	Z[c]	Sig[d]
Control	Post spelling	93.26	18.015	69	131	79.00	93.00	106.00	−0.204	0.838
	init. spelling	93.32	18.226	69	131	79.00	89.00	105.00		
WWL + AR	post spelling	91.62	15.426	69	121	79.00	90.00	104.50	−3.647	0.000
	init. spelling	87.64	14.507	69	118	73.00	86.00	99.00		
WWL	Post accuracy	108.13	13.176	74	130	97.00	110.50	117.75	−4.942	0.000
	init. accuracy	91.82	13.980	69	123	84.00	88.50	104.50		

[a] Standard Deviation; [b] First, second and third quartile; [c] Test statistic value based on positive ranks; [d] Approximate p-value based on the standard normal distribution

(see Tables 1, 5 and 17), while the control group did not demonstrate significant changes (see Tables 16 and 17) on any of the literature measures.

Overall, based the results suggest that the WWL and WWL+AR technologies prevail over the typical literacy school programmes in terms of increasing students' spelling, reading accuracy and comprehension scores and reducing their reading rate.

6 Discussion—Conclusion

AHA project was aiming to enhance literacy skills of primary school students diagnosed with ADHD by using introducing new technologies. The work described in this chapter is limited to the specific target group of the AHA project participants and supports results only for ADHD English-speaking students aged 8–12. In addition, data related to the socio-economic status of participants were not analysed, as the occupation of the parents was not collected based on standardized parameters. The main strength of this work is the utilization of standardized literacy tests (NARA-II and Graded Word Spelling Test), which can identify specific problem areas in individual students and provide guidelines for curriculum.

In line with previous research findings [12], this study confirms in the Irish context, the need for ongoing monitoring of reading, spelling and particularly writing skills and the provision of additional supports to children diagnosed with ADHD throughout their primary school years. The idea that treatments for children with ADHD should focus on improvement in function, with the goal of improved long-term functional outcomes not merely improvement in core ADHD symptoms, is not new [8]. Addressing any literacy deficits certainly meets this aim.

Based on previous research [10, 11], it can be assumed that when students with ADHD reduce their reading rate, they focus more on their task and earn more processing time. The decrease in reading rate accompanied by the increase in reading comprehension is in line with the objectives and results of the project and confirms that the use of AR and web-based technologies can enhance student concentration and comprehension and may support students with reading and attention difficulties [16–20]. Students who were assigned to intervention groups reduced their reading rate but significantly increased their reading comprehension and accuracy as well as their spelling scores, while the control group did not demonstrate significant changes on any of the literature measures. It is assumed that this difference is due to the contribution of both WWL and WWL enhanced with AR literacy programmes. Therefore, in accordance with previous research, AR technology was found to contribute on student's literacy skills [24–29].

Furthermore, the analysis on the effect of the WWL online and AR enhanced intervention programmes in comparison to the typical in-class literacy instruction showed that students belonging to the intervention groups achieved significantly better scores than those who were assigned to the control group (in terms of statistical significance, $p=0.05$). ADHD can be linked with the appearance of executive functioning, language and fine motor difficulties thus educational materials should be sufficient to acknowledge students' difficulties [44, 45]. After examining the impact of a web-based programme and an AR edition on primary school students diagnosed with ADHD it is concluded that both programmes, with neither one prevailing over the other, can significantly support participants in acquiring competent literacy skills. Therefore, this study is in accordance with the idea that multimedia technology could substantially contribute to the educational process [17, 46].

Acknowledgements The work described in this chapter presents the results from data collected during the research project ADHD-AUGMENTED (AHA) which was funded from the European Commission (2017-2019) for Pilot Project: Technologies and Tools for Children and Young People with Attention Deficit Hyperactivity Disorder (ADHD), under Grant Agreement No. 30-CE-0885096/00- 34 (ADHD2016-13). The research analysis has been supported by European Union's Horizon 2020 research and innovation programme under grant agree- ment No 856533, project ARETE.

References

1. Neurodevelopmental Disorders: Attention-Deficit/Hyperactivity Disorder. In: Edition F Diagnostic and statistical manual of mental disorders. Am Psychiatric Assoc 21 (2013)
2. Thapar, A., Cooper, M., Jefferies, R., Stergiakouli, E.: What causes attention deficit hyperactivity disorder? Arch. Dis. Child. **97**(3), 260–265 (2012). https://doi.org/10.1136/archdischild-2011-300482
3. Ocay, A.B., Rustia, R.A., Palaoag, T.D.: Utilizing augmented reality in improving the frustration tolerance of ADHD learners: an experimental study. In: Proceedings of the 2nd International Conference on Digital Technology in Education, pp. 58–63 (2018). https://doi.org/10.1145/3284497.3284499
4. Sciberras, E., Mueller, K.L., Efron, D., Bisset, M., Anderson, V., Schilpzand, E.J., Jongeling, B., Nicholson, J.M.: Language problems in children with ADHD: a community-based study. Pediatrics **133**(5), 793–800 (2014). https://doi.org/10.1542/peds.2013-3355
5. Steinhausen, H.C.: The heterogeneity of causes and courses of attention deficit/hyperactivity disorder. Acta Psychiatr. Scand. **120**(5), 392–399 (2009). https://doi.org/10.1111/j.1600-0447.2009.01446.x
6. Carroll, J.M., Maughan, B., Goodman, R., Meltzer, H.: Literacy difficulties and psychiatric disorders: evidence for comorbidity. J. Child Psychol. Psychiatry (2004). https://doi.org/10.1111/j.1469-7610.2004.00366.x
7. Luman, M., Goos, V., Oosterlaan, J.: Instrumental learning in ADHD in a context of reward: intact learning curves and performance improvement with methylphenidate. J. Abnorm. Child Psychol. **43**(4), 681–691 (2015). https://doi.org/10.1007/s10802-014-9934-1
8. Barbaresi, W.J., Campbell, L., Diekroger, E.A., Froehlich, T.F., Liu, Y.H., O'Malley, E., Pelham, W.E., Power, T.J., Zinner, S.H., Chan, E.: Society for developmental and behavioural pediatrics clinical practice guide-line for the assessment and treatment of children and adolescents with com- plex ADHD. J. Dev. Behav. Pediatr. **41**, S1–S23 (2020). https://doi.org/10.1097/dbp.0000000000000770
9. Mayes, S.D., Calhoun, S.L., Crowell, E.W.: Learning disabilities and ADHD: overlapping spectrum disorders. J. Learn. Disabil. **33**(5), 417–424 (2000). https://doi.org/10.1177/002221940003300502
10. Jacobson, L.A., Ryan, M., Martin, R.B., Ewen, J., Mostofsky, S.H., Denckla, M.B., Mahone, E.M.: Working memory influences processing speed and reading fluency in ADHD. Child Neuropsychol. **17**(3), 209–224 (2011). https://doi.org/10.1080/09297049.2010.532204
11. Kramer, E., Koo, B., Restrepo, A., Koyama, M., Neuhaus, R., Pugh, K., An-dreotti, C., Milham, M.: Diagnostic associations of processing speed in a transdiagnostic, pediatric sample. Sci. Rep. **10**(1), 1–11 (2020)
12. Silva, D., Colvin, L., Glauert, R., Stanley, F., Srinivas Jois, R., Bower, C.: Literacy and numeracy underachievement in boys and girls with ADHD. J. Atten. Disord. **24**(10), 1392–1402 (2020). https://doi.org/10.1177/1087054715596575
13. Vaughan-Nichols, S.J.: Augmented reality: No longer a novelty? Computer **42**(12), 19–22 (2009). https://doi.org/10.1109/MC.2009.380
14. Berryman, D.R.: Augmented reality: a review. Med. Ref. Serv. Q. **31**(2), 212–218 (2012). https://doi.org/10.1080/02763869.2012.670604
15. Vahabzadeh, A., Keshav, N.U., Salisbury, J.P., Sahin, N.T.: Improvement of attention-deficit/hyperactivity disorder symptoms in school-aged children, adolescents, and young adults with autism via a digital smartglasses-based socioe-motional coaching aid: short-term, uncontrolled pilot study. JMIR mental health **5**(2), e9631 (2018). https://doi.org/10.2196/mental.9631
16. Hrishikesh, N., Nair, J.J.: Interactive learning system for the hearing impaired and the vocally challenged. In: 2016 International Conference on Advances in Computing, Communications and Informatics (ICACCI), pp. 1079–1083. IEEE (2016). https://doi.org/10.1109/ICACCI.2016.7732188

17. Franceschini, S., Gori, S., Ruffino, M., Viola, S., Molteni, M., Facoetti, A.: Action video games make dyslexic children read better. Curr. Biol. **23**(6), 462–466 (2013). https://doi.org/10.1016/j.cub.2013.01.044
18. Mohd Yusof, A., Daniel, E.G.S., Low, W.Y., Ab. Aziz, K. (2014). Teachers' perception of mobile edutainment for special needs learners: the Malaysian case. Int. J. Incl. Educ. **18**(12), 1237–1246. https://doi.org/10.1080/13603116.2014.885595
19. Deault, L., Savage, R., Abrami, P.: Inattention and response to the ABRACADABRA web-based literacy intervention. J. Res. Educ. Eff. **2**(3), 250–286 (2009)
20. Avila-Pesantez, D., Rivera, L.A., Vaca-Cardenas, L., Aguayo, S., Zun˜iga, L.: Towards the improvement of ADHD children through augmented reality serious games: preliminary results. In: 2018 IEEE Global Engineering Education Conference (EDUCON), pp. 843–848. IEEE (2018). https://doi.org/10.1109/EDUCON.2018.8363318
21. Baragash, R.S., Al-Samarraie, H., Moody, L., Zaqout, F.: Augmented reality and functional skills acquisition among individuals with special needs: a meta-analysis of group design studies. J. Spec. Educ. Technol. (2020). https://doi.org/10.1177/0162643420910413
22. Carmigniani, J., Furht, B., Anisetti, M., Ceravolo, P., Damiani, E., Ivkovic, M.: Augmented reality technologies, systems and applications. Multimed. Tools Appl. **51**(1), 341–377 (2011). https://doi.org/10.1007/s11042-010-0660-6
23. Milgram, P., Kishino, F.: A taxonomy of mixed reality visual displays. IEICE Trans. Inf. Syst. **77**(12), 1321–1329 (1994)
24. Huang, X., Zou, D., Cheng, G., Xie, H.: A systematic review of AR and VR enhanced language learning. Sustainability **13**(9), 4639 (2021). https://doi.org/10.3390/su13094639
25. Parmaxi, A., Demetriou, A.A.: Augmented reality in language learning: a state-of-the-art review of 2014–2019. J. Comput. Assist. Learn. **36**(6), 861–875 (2020). https://doi.org/10.1111/jcal.12486
26. Billinghurst, M., Duenser, A.: Augmented reality in the classroom. Computer **45**(7), 56–63 (2012). https://doi.org/10.1109/MC.2012.111
27. Schmitz, B., Specht, M., Klemke, R.: An analysis of the educational potential of augmented reality games for learning. In: mLearn, pp. 140–147 (2012)
28. Ahmad, N.A.B.: Learning literacy using augmented reality (LitAR): an application of learning through expository, social and technical-scientific using augmented reality as learning strategy. Int. J. Acad. Res. Bus. Soc. Sci **8**, 1772–1778 (2018). https://doi.org/10.6007/IJARBSS/v8-i11/5353
29. Lin, C.Y., Yu, W.J., Chen, W.J., Huang, C.W., Lin, C.C.: The effect of literacy learning via mobile augmented reality for the students with ADHD and reading disabilities. In: International conference on universal access in human-computer interaction, pp. 103–111. Springer, Cham (2016, July). https://doi.org/10.1007/978-3-319-40238-3-11
30. Quintero, J., Baldiris, S., Rubira, R., Cer´on, J., Velez, G.: Augmented reality in educational inclusion. A systematic review on the last decade. Front. Psychol. 10, 1835 (2019). https://doi.org/10.3389/fpsyg.2019.01835
31. Ak¸cayır, M., Ak¸cayır, G.: Advantages and challenges associated with augmented reality for education: a systematic review of the literature. Educ. Res. Rev. 20, 1–11 (2017). https://doi.org/10.1016/j.edurev.2016.11.002
32. Cihak, D.F., Moore, E.J., Wright, R.E., McMahon, D.D., Gibbons, M.M., Smith, C.: Evaluating augmented reality to complete a chain task for elementary students with autism. J. Spec. Educ. Technol. **31**(2), 99–108 (2016). https://doi.org/10.1177/0162643416651724
33. WordsWorthLearning—The Programme. https://wordsworthlearning.com/staticpage/program
34. Mangina, E., Chiazzese, G., Hasegawa, T.: AHA: ADHD augmented (learning environment). In: 2018 IEEE International Conference on teaching, assessment, and learning for engineering (TALE), pp. 774–777. IEEE (2018). https://doi.org/10.1109/TALE.2018.8615222
35. Chiazzese, G., Mangina, E., Chifari, A., Merlo, G., Treacy, R., Tosto, C.: The AHA project: An evidence-based augmented reality intervention for the improvement of reading and spelling skills in children with ADHD. In: International Conference on Games and Learning Alliance, pp. 436–439. Springer, Cham (2018). https://doi.org/10.1007/978-3-030-11548-7

36. Luna, J., Treacy, R., Hasegawa, T., Campbell, A., Mangina, E.: Words worth learning-augmented literacy content for ADHD students. In: 2018 IEEE Games, Entertainment, Media Conference (GEM), pp. 1–9. IEEE (2018). https://doi.org/10.1109/GEM.2018.8516483
37. WordsWorthLearning: Official website. https://wordsworthlearning.com
38. Eleni Mangina (Ed). AHA: A pilot project for evaluating the effect of augmented reality in the reading and spelling skills for children with ADHD. Dublin (2018). ISBN: 978-1-910963-25-8
39. Tosto, C., Hasegawa, T., Chiazzese, G., Treacy, R., Merlo, G., Chifari, A., & Mang- ina, E. "AHA-ADHD AUGMENTED"-PARTICIPANTS' CHARACTERISTICS. In Proceedings of EDULEARN19 Conference (pp. 5637–5645). ISBN: 978–84–09- 12031–4
40. Tosto, C., Hasegawa, T., Mangina, E., Chifari, A., Treacy, R., Merlo, G., Chi- azzese, G.: Exploring the effect of an augmented reality literacy programme for reading and spelling difficulties for children diagnosed with ADHD. Virtual Reality **25**(3), 879–894 (2021). https://doi.org/10.1007/s10055-020-00485-z
41. Neale, M.: Neale analysis of reading ability (NARA-II). In: 2nd Revised British Edition (1958–1997). London, GL Assessment (1997)
42. Vernon, P.E., Crumpler, M., McCarty, C.T.: Graded word spelling test. Hodder and Stoughton (2006)
43. Knapp B.H.: ANOVA and Kruskal-Wallis test. In: Intermediate Statistics Using SPSS, pp. 107–140. SAGE Publications Thousand Oaks (2018). https://doi.org/10.4135/9781071802625
44. Kaiser, M.L., Schoemaker, M.M., Albaret, J.M., Geuze, R.H.: What is the evidence of impaired motor skills and motor control among children with attention deficit hyperactivity disorder (ADHD)? Systematic review of the literature. Res. Dev. Disabil. **36**, 338–357 (2015). https://doi.org/10.1016/j.ridd.2014.09.023
45. Parigger, E.M.: Language and executive functioning in children with ADHD. Amsterdam Center for Language and Communication (2012). ISBN: 978-90-889-1504-8
46. Balu, A.: Contribution of multimedia technology in education. Int. J. Multidiscip. Educ. Res. **9**(2), 127 (2020)

Development of AR Interactive Components for Positive Behavioral Interventions and Supports

Luciano Seta, Sui Lin Goei, Giuseppe Chiazzese, Marco Arrigo, Mariella Farella, Crispino Tosto, Antonella Chifari, Ana Domínguez, and Eleni Mangina

Abstract This chapter reflects the research conducted for the development of Augmented Reality (AR) interactive components for *Positive Behavioral Interventions and Supports* (PBIS) involving the development of an AR mobile app from the basic principles to the software prototype. The research lays the foundations for introducing AR into a new teaching and learning context based on the teaching of behavior according to the PBIS framework. PBIS is a preventative approach for decreasing problem behavior and improving the instruction of expected behaviours, organization of consequence systems, redesign of environmental settings, and use of evidence-based practices. The main aim of this chapter is to introduce the PBIS–AR mobile app

L. Seta (✉) · G. Chiazzese · M. Arrigo · M. Farella · C. Tosto · A. Chifari
Institute for Educational Technologies, National Research Council, Genoa, Italy
e-mail: luciano.seta@itd.cnr.it

G. Chiazzese
e-mail: giuseppe.chiazzese@itd.cnr.it

M. Arrigo
e-mail: marco.arrigo@itd.cnr.it

M. Farella
e-mail: mariella.farella@itd.cnr.it

C. Tosto
e-mail: crispino.tosto@itd.cnr.it

A. Chifari
e-mail: antonella.chifari@itd.cnr.it

S. L. Goei
Vrije Universiteit (VU), Amsterdam, Netherlands
e-mail: s.l.goei@vu.nl

A. Domínguez
Vicomtech Foundation, Basque Research and Technology Alliance (BRTA), San Sebastián, Spain
e-mail: adominguez@vicomtech.org

E. Mangina
School of Computer Science, University College Dublin, Dublin, Ireland
e-mail: eleni.mangina@ucd.ie

Y. Cai et al. (eds.), *Mixed Reality for Education*, Gaming Media and Social Effects,
https://doi.org/10.1007/978-981-99-4958-8_14

and provide an overview of the development process integrating AR with the requirements of PBIS practice. After a short reference to PBIS, the concept underlying the development of the PBIS–AR app is presented. The concept is based on the definition of a new theoretical paradigm that integrates AR and behavioural learning creating an augmented space that we have called *Augmented Reality Behavioral Learning Space* (AR–BLS). The chapter further introduces the structure of the behavioural lessons, designed to instruct learners according to the school behavioral values, and presents the PBIS–AR app architecture as a component of the *AR PBIS ecosystem*. It illustrates the different components of this ecosystem together with the AR solutions adopted.

1 A Positive Approach

School safety, student behaviour, and academic outcomes are priority areas to be addressed in educational agendas [1]. Schools are responsible for creating safe, positive, and meaningful learning environments for their students and adopting approaches that respect students' academic and behavioural needs to promote student success.

Over the last decades, schoolwide frameworks, like *Schoolwide Positive Behavioral Interventions and Supports* (aka SWPBIS) have emerged to support student behaviour and create safe learning environments [2]. In the United States, this approach has been codified within the law, specifically, the Individuals with Disabilities Education Act (IDEA) that mandates "positive behavioral interventions, strategies, and supports" for children "whose behavior impedes his or her learning or that of others."

Using the words of Edward G. Carr's: "*Our job is to redesign the counterproductive and unfair environmental contexts that so many people, with and without disabilities, have to contend with every day of their lives. Our job is to give people the skills, the coping strategies, and the desire to deal with the frustration that is an inevitable part of life, particularly the lives of people with disabilities. We must give them and their loved ones the support they need to challenge and reconstruct systems that serve bureaucratic needs rather than human needs. All of these goals reflect the great ideas that are at the heart of PBS, ideas so great that they are worthy of scientific study*" [3] (p. 3).

Implementation of PBIS at school level is focused on (a) creating systems that establish a positive social culture, (b) adopting a continuum of evidence-based practices to support student behaviour and promote a safe learning environment, and (c) using data to monitor and adjust implementation [4].

The adjective *positive* refers to both behaviour and support: *positive behaviour*, which can be seen as desirable, adaptive, prosocial behaviour. And *positive behavioural support* as differentiated from nonpositive support, which might involve the use of aversive, humiliating, or stigmatizing interventions [5].

More specifically, the PBIS framework aims at preventing behavioural problems, such as disruptive classroom behavior, in order to support the development of a safe school climate and thus promoting students' social-emotional development and learning outcomes [4, 6]. Interventions delivered in PBIS schools follow a three-tiered system of behaviour support [7].

The first tier, the schoolwide level, includes universal interventions that are provided to the entire school population. The second tier focuses on group interventions to students who do not profit from the universal interventions, and the third tier provides individualized interventions for students who do not profit from the interventions provided in the first and second tier.

Basically, PBIS practices implemented in Europe are based on the following pillars [8]:

- Defining and teaching behavioral expectations based on shared values.
- Systematically reinforcing expected positive students' behavior.
- Addressing challenging student behavior.
- Using behavioral data to identify students who require additional support.

PBIS interventions delivered at the primary-tier level usually include a set of universal interventions designed for all school members and across all school settings (e.g. classroom, hallway, corridor, et cetera) to create and guide a positive social culture. To reach this aim, a restricted number of contextually and culturally relevant behavioural expectations, generally three to five, are selected and taught to all school stakeholders (students, school staff, and community members) across all settings. They are defined, modelled, and practiced as other academic skills. PBIS practice also requires corrective and positive feedback being delivered during the educational intervention. More in detail, PBIS interventions comprises both (1) acknowledgement or rewards of students' adherence to behavioral expectations and (2) systematic responses to problem behaviours [9].

When PBIS is implemented with fidelity, students, educators, and schools experience positive outcomes, including increased prosocial skills [10], enhanced perceptions of school safety [11], reduced problem behaviour [12, 13], improved school climate [11, 12, 14], and increased teacher self-efficacy (e.g., [15]) and well-being (e.g., [16]).

2 Augmenting PBIS

The idea to introduce the Augmented Reality (AR) in the context of the PBIS is grounded on both methodological and pragmatic bases.

From to the methodological point of view, the PBIS is characterised by an openness toward new approaches and interventions, as Robert H. Horner and George Sugai declare: "*We are an organization that is committed to application of scientifically validated practices that can be used to achieve socially valued outcomes. We are open to adopting practices from many venues, but we expect any practice we adopt to have documented proof of effectiveness*" [17] (p. 19).

The AR appears particularly intriguing for implementing new learning practice.

In the last decade, research in education and practitioners' interest has focused on emerging technologies, such as AR, and their potential in teaching and learning processes has been widely explored. In this regard, research has clearly shown that AR solutions can enhance students' academic outcomes compared to traditional learning methods [18]. Research also indicated that AR solutions can have a stronger effect on learners' achievement in terms of content understanding and long-term retention, motivation, engagement, and satisfaction with the learning experience than traditional and other digital media-related educational experiences exert [19, 20].

From the pragmatic point of view, an effective implementation of the PBIS approach at schoolwide level requires that students and school staff and community members identify a set of positive expectations. These expectations have to be teach directly and continuously using local and real behavioral examples in real contexts or settings of the school. The examples must be observable, relevant, and doable [4].

The use of AR seems highly appropriate for this purpose. In fact, this technology permits to enrich the real space around the learner with digital objects. These objects are able to modify the learner's learning experience without transport her/his in an *virtual reality* that has no connection with the real school environment. Using the AR technologies, the learner, by means of her/his mobile device, is able to access to relevant behavioral examples in the real context of her/his school setting, in the form of AR objects always available. This approach could represent a step forward compared to video modelling, traditionally used in this field.

Despite the amount of literature on the role of AR technology in the educational field, no existing research has yet investigated the potential impact of AR solutions on the promotion of students' behavioural outcomes at a school-wide level from a preventative perspective [21].

The AR solution here proposed is the results of the *Augmented Reality Interactive Educational System* (ARETE) project, granted by H2020 EU program. The project tries to overcome this gap by specifically designing an application within the PBIS framework [22].

3 The AR Behavioural Learning Space

Within the scope of the PBIS, an approach aimed at transforming the school in a supporting environment, the integration of the AR technologies can steer the behavioural learning activity in the direction of an engaging experience able to modify permanently values, attitudes, beliefs and emotions of the learners.

This valuable experience goes through the interaction with AR objects embedded in the learning process of the behavioural lessons, using them as a support for the behavioural teaching, practice and reinforce phases, as it will be explained in the

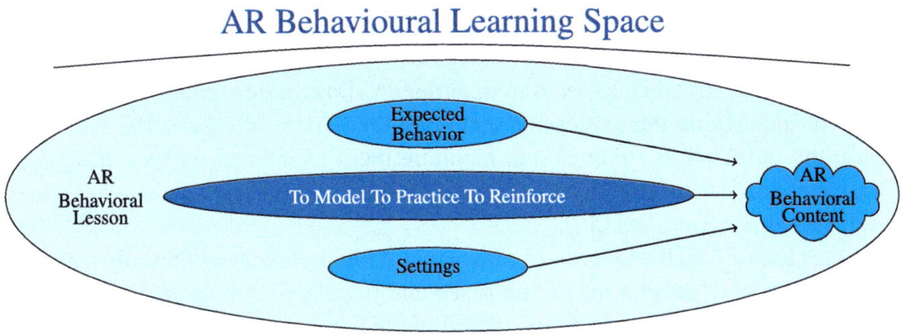

Fig. 1 The AR-BLS

following. The student is facilitated in using AR to gain autonomy in managing expected behaviours in school settings and applying different social skills according to the school values. In our vision, AR technologies can be effectively integrated in PBIS learning phases to allow a new PBIS AR Learning experience.

To achieve this result, it can be useful to think about a new learning space, that is an *ecological environment* in which different *agents*, *systems* and *practices* interact. The main components of this space are shown in the Fig. 1 and briefly illustrated below.

The *AR behavioral lesson* is both the start point and the first environment within the learner's experience occurs. One of the defining characteristics of the PBIS is the *instructional focus*, so that priority is given to directly teaching social behaviors.

At the schoolwide level, a small number of behavioral expectations are shared and a set of related *expected behaviours* is taught directly to all student, in order to establish a common language among students, family and staff members [4]. Another important pillar of the PBIS approach is that strict relation between behavioural values/expectations and the learner's real life contexts. Any behaviour has to be *modelled*, *practiced* and *reinforced* within specific school *settings*, typical school-life spaces and moments, as arrival/dismissal, corridors, breaks/lunch, playground, restroom, and classroom, where the behaviours will be put in act.

The augmented reality is a technology especially advisable to support this learning process. The digital objects are embedded in the real physical space in which the subject is moving and, in this way, the real context is enriched without the feeling of isolation characterising virtual reality (VR). Moreover, the subject engaged with the AR objects can also interact with peers and other subjects on the scene reinforcing the *social learning* process, starting point of PBIS. The school setting is always present to the learner, and the relation between situation/setting and expected behaviour will be fixed and reinforced in the *embodied memories* of the child [23].

Concerning other modelling techniques, such as video modelling or role-playing, the AR permits better control of the level of personal engagement, and the enrichment of the physical space with information able to activate multiple sensory channels. These features can be useful tools both for students with physical/cognitive disabilities and without. As noted by various authors, the learning of behavior management skills improves when an AR-based strategy is applied [24].

For example, for autism spectrum disorder (ASD) children, characterised by difficulty in identifying the expression and emotional states of others, the AR games arouse the participants' interest and facilitate them to interact with each other, so social interaction and emotion recognition and control are the most addressed skillsets in AR-based interventions [25].

In the case of ADHD, Ocay et al. affirm that AR features can potentially entice the ADHD learners in developing an improved emotional strength through the intuitive and interesting learning process by combining real and virtual objects [26].

Moreover, AR enables the presentation of instructions in a step-by-step manner which can be particularly beneficial to individuals with special educational needs (SEN) and there is strong evidence to indicate that using AR is a viable strategy for teaching functional skills to individuals with SEN [24].

Starting from these considerations, the natural PBIS environment has been enriched with specific AR contents. These resource, correlated to specific moments of the behavioural lesson, will help children to engage in a lived experience the expected behaviors in their real school settings.

The *AR Behavioural Learning Space* (AR-BLS) can be defined as a real learning space enriched with AR behavioral contents, described in Sect. 4, that the teacher can create and use along the different phases of teaching, according to the structure of a typical behavioural lesson. In the model phase the teacher demonstrates the expected behavior. The teacher clarifies the difference between expected and unexpected behavior by providing positive examples and a negative example (or non-example). Augmented Reality can be embedded in this process for supporting teachers in the creation of AR examples and non-example behaviours that can explain to the student how to perform the different steps. In the practice and reinforce phases students are involved in multi-user interactive activities aimed at giving them the opportunity to practice and reflect on new learned skills in the variety of settings where the behaviors should be used [22].

4 AR Contents for the Behavioral PBIS Lesson

A relevant feature of the PBIS approach—in line with social learning and modeling theory—is the clear stance that the behavior must be explicitly taught and learned. The design and the supplying of lessons addressed to teach specific positive behaviors is a crucial phase of the PBIS implementation at the school and classroom level.

School staff are responsible for creating a safe learning environment for students in which they can learn expected behaviour. An important tenet within PBIS regards the positive learning of expected behaviour by giving students direct, specific, and positive feedback and/or positive reinforcement on their behaviour as well as ample opportunities to practice the expected behaviour in all settings where these behaviours are required of them.

Schools' full set of expected behaviors for all school settings and their link to the school values are summarized in a *behavioural expectation matrix* which is shared with all students and staff [4].

In the following, a new application, called *AR-PBIS app*, will be described. It has been designed to support school staff and students in teaching and learning expected behaviours within the PBIS framework, using AR learning resources. In order for the PBIS-AR app to reach its full potential in this regard, the contents of the application need to be useful for all schools working with PBIS. As such, an important first step in the development of the PBIS-AR app is to establish an behavioural expectation matrix to guide the creation of AR content within the PBIS-AR application.

Regarding to this, it is important to highlight as schools individually select their core school values and decide on expected behaviours in line with these school values. Moreover, schools differ within countries and across countries in the school settings available and deemed important for teaching students expected behaviour, as well as in how expected behaviour is positively reinforced.

Thus, to improve the PBIS-AR app effectiveness a preliminary research has been carried on to define a behavioral expectation matrix able to cover an ample spectrum of schools and teachers/experts desiderata.

Using a mixed approach, based on the review of the relevant literature, the organization of focus groups involving PBIS-teachers and PBIS-experts, and the analysis of the answers to an online questionnaire, the behavioral expectation matrix for the AR-PBIS app has been defined.

With regards to school values, a large-scale study among 155 US primary schools in 12 US states by Lynass et al. [9] suggested that school values are relatively homogeneous, at least in the US. By far, most schools were found to define the following three school values: *respect*, *responsibility*, and *safety*.

With regards to school settings, tentative input for the behavioural expectation matrix was derived from a recent representative study by Lane et al. [27] describing the development and psychometric properties of a survey to guide schools in selecting schoolwide behavioural expectations for their matrices. A total of seven common and important school settings was established: classroom, corridors, cafeteria, playground, restroom, bus, and arrival/dismissal. Again, these school settings were based on the situation corresponding to the US schooling system. To adapt this result to more universal contexts, three more general settings were considered: *general/all setting*, *classroom setting*, *social skill/all setting*.

Finally, nine expected behaviors for these settings were selected, using focus groups and online questionnaire results. The final list is shown in Table 1.

From a pedagogical point of view, the design of the AR–PBIS app follows the organization of the traditional (non–AR) behavioral lessons as outlined for the ARETE project by the SVU–team in collaboration with three teacher teams of Dutch primary schools and three Dutch PBIS–experts.

The process of lesson development is based on the following preferred practices [28] (pp. 500–501):

Table 1 AR-PBIS app settings and expected behaviors

Setting	Expected behavior
General/all settings	Greet others
	Walk with a goal
Classroom setting	Keep your working space organised
	Store your belongings
	Work independently at your desk
Social skill/all settings	Stand up for others
	Keep my hands and feet to myself
	Help others with questions
	Let others be in peace (and ask others to let me be in peace)

- provide multiple examples and non–examples and variations of the behavioural skill relevant to the specific setting.
- Sequence positive and negative examples that are minimally different to maximize discrimination about when and where the skill should be used. Teach the skills with and across a range of contexts where they will be applied.
- Teach directly and actively by modeling and demonstrating the skill, variations of the skill, and the conditions under which the skill should be used (a critical rule). Teachers point out the critical features of the modeling or demonstration as they are presented ("See how he is moving quietly along the right side of the hallway").
- Provide opportunities for students to practice and to rehearse the skill with assistance (verbal prompts) and feedback and then without assistance to test their knowledge and accurate use of the skill.
- Review skills and routines regularly, for instance by demonstrating the skill or describing the context in which a skill is used.
- Acknowledge and reinforce appropriate displays of the skill frequently.

This activity produced a template for a behaviour lesson that was subsequently used in the lesson design process. The template focuses on the integration of four key elements:

1. *Outcomes*: A clearly articulated description of the behaviours required to promote safe and effective school environments that are endorsed by staff and students.
2. *Practices*: A set of evidence–based interventions and strategies used to teach, supervise and monitor both non–classroom and classroom settings.
3. *Data*: Information used to identify the current status and need for change, and to monitor effects of interventions and guide decisions.
4. *Systems*: The support needed to implement and sustain systems of PBIS.

The template for a scripted behavioral PBIS lesson to teach appropriate behavior is described below in Table 2. This template will also form the basis for the technological affordances of the implementation of a PBIS lesson through the AR application.

Table 2 Behavioural PBIS lesson structure

(Schoolwide) Expected behavior This entails the following specific behaviour in this setting: ………
Lesson goals Subgoals: ………
PBIS value(s) According to the school value ………
Starter/prompt/remind Short and tangible, attract attention by showing/telling (approx. 1 min)
Teach: instruction and explain Includes stating lesson objectives and making them visible
Modeling Teacher models expected behaviour: 3 example behaviour, 1 non-example behaviour.
Practice A set of evidence-based interventions and strategies used to teach, supervise and monitor both non-classroom and classroom settings
Reflection/review Give and solicit the students feedback and deliver consequences as necessary
Acquirement and retention *Post-lesson*: call attention to practice behavioural expectation daily during one week
Evaluation *Post-lesson*: evaluate learned behaviour with students after one week

Different types of AR resources have been planned to support the behavioral learning process throughout some specific moments of this general behavioral lesson. In particular, three different type of AR contents have been designed:

1. The **AR Behavioural Learning Resources** (AR-BLRs), designed to support the learner during the *teaching/modeling* phase, when she/he has to play a specific behavioral routine (an example or non–example behavior) in defined setting. The AR-BLR will be displayed in the position decided by the teacher so that the student can learn in her/his daily learning environment.
2. The **Behavioral Reflection Game in AR**, a gamification learning content provided to the students during the *reflection/review* phase, useful to solicit feedback and deliver consequences.
3. The **Multi–user Interaction Activity**, designed to give students the opportunity to train behavioral activities through augmented reality, using 3D objects and characters, interacting with peers in a mixed environment during the *practice* phase.

The *AR-PBIS app* is the technological interface that permits learners, lead by the teacher, to access all these resources, localised in the school environment, using a mobile device, a smartphone, or an head–mounted device, so transforming the physical space of the school in a "learning space", that is in the *AR-BLS* illustrated above in Sect. 3.

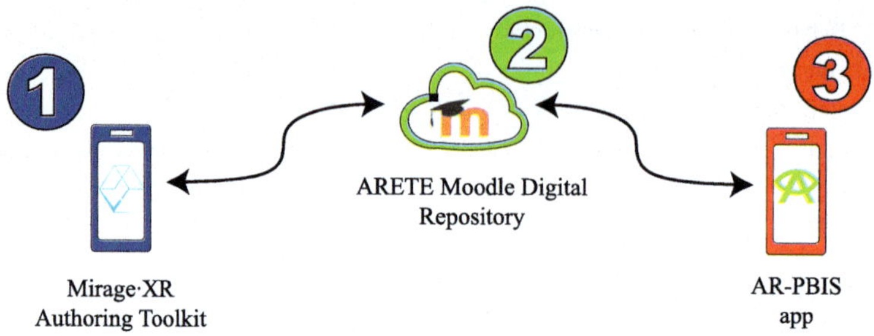

Fig. 2 The AR PBIS ecosystem: outline

5 The PBIS–AR App

The PBIS–AR app is part of a broader (eco)system, designed and realized during the European Union's Horizon 2020 project denominated *ARETE*. This part of the system is aimed for supporting learners involved in the PBIS learning program. Before to describe the application the system will be briefly outlined.

Figures 2 and 3 present how the AR-PBIS app is an integral part of the *ARETE ecosystem for PBIS* and contributes to the creation of an AR–BLS.

The ecosystem consists of three main subsystems that interplay to allow learners more rich learning experiences.

The first subsystem is an authoring component permits teachers to create the AR–BLRs, AR content specific for the PBIS, localised at the exact school place, and related to a well–defined expected behavior. This component is based on the Mirage·XR Open Source core, an in–situ authoring tool for creating content directly in XR, cheap and efficiently. Mirage·XR has received funding from the European Union's Horizon 2020 research and innovation programme through the *XR4ALL* project. The principal users of this authoring tool are the teachers engaged in the design of the PBIS behavioral lessons at school.

The AR–BLRs create using the toolkit will be stored using the second component (in green, label 2, in Figs. 2 and 3) of the AR PBIS ecosystem, the *ARETE Moodle Digital Repository*.

Finally, the components of the AR-PBIS app, the third sub-system, are presented in Fig. 3. Using this app students can access and activate different AR resources designed to support the behavioral lesson. The teacher is always in control of students' activities and the app was designed to support the lessons and not substitute them. As Fig. 3 shows, the student have three modalities of use:

1. In the **learning mode** the app enables learners to view AR-BLRs, created by teachers through the Mirage·XR toolkit (subsystem 1), directly in real school contexts using AR technology. Indeed, such a system is capable of interfacing with the Moodle server (subsystem 2) from which it is able to extract and launch the contextualised animations created through the Mirage·XR toolkit. In this case,

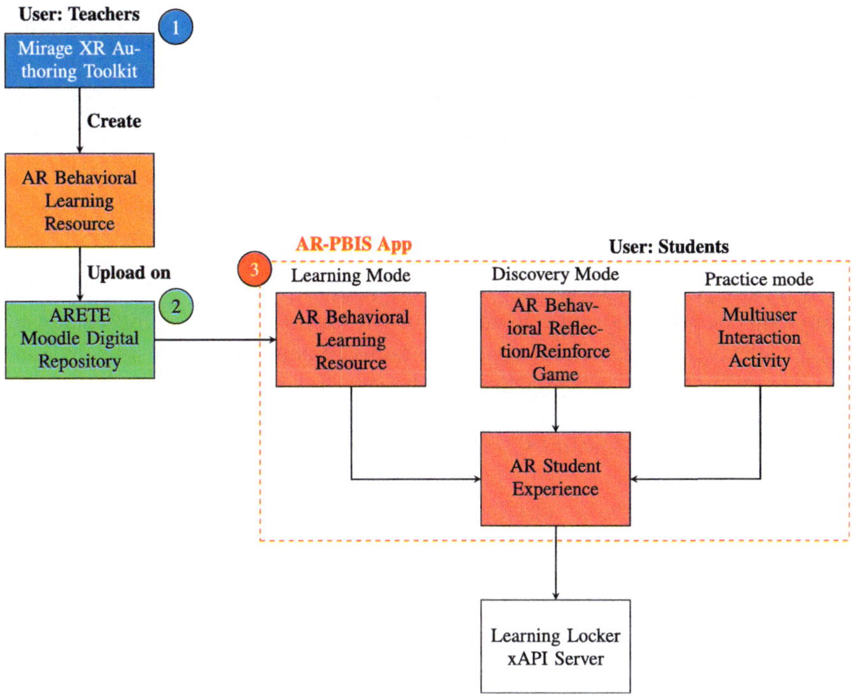

Fig. 3 The AR PBIS ecosystem: main components

AR technology does not use markers, but is based on mapping the environment through, for example, tracking planes and points.

2. Using the app, students can also access the **discovery mode** in which there are reflective games to practise and reinforce expected behaviours through the use of AR markers needed to activate augmented content. The various markers will be placed within the school and students, by framing the markers, can access AR-based learning activities that will allow them to earn points according to a reward system provided by PBIS.
3. Finally, the application provides a multi-user, interactive and collaborative learning section, in the **practice mode**, to practise and reinforce some of the expected behaviours.

To monitor and record students' interaction with the PBIS–AR app, and especially with AR-BLRs, the *ExperienceAPI* (xAPI) standard is used, integrating a cloud–based learning recording repository and PBIS analysis tools. This will provide objective feedback on the learning process.

The combination of these technologies and systems creates an example of a technological and innovative ecosystem designed for creating behavioural lessons in AR.

In the following, these modalities of the AR–PBIS app will be illustrated with some examples of use.

5.1 Learning Mode

During the lesson, the teacher may decide to have the students use the application to access some AR content (AR–BLR) s/he has previously created and stored on the ARETE moodle digital repository.

The teaching phase enriched with the use of the PBIS-AR app basically applies the video-modeling technique showing AR examples of expected (and unexpected) behaviors to users in which different characters (teachers, students, principal) demonstrate how to perform a behavior in a specific setting.

In terms of designing and construction of AR learning resources, the teacher can build a specific AR behavioral content in terms of animations that show example and non-example of a behavior through the selection of virtual characters, actors of the scene and the compositions of the different steps of the animation itself. In other terms, augmented reality 3D characters can be used for showing to children the performance of simple expected behavioral routines that are claimed to be particularly attractive and motivating to young students.

To this aim, a set of virtual characters were designed and developed. The main character, see Fig. 4, is an alien coming from space that is completely unaware of life on Earth, so that it can be considered as a neutral behavioural coach. The alien character was chosen as the main actor to interpret the behavioural rules in the identified scenarios [29]. This choice is supported by the following reasons:

- PBIS practice makes abundant use of visuals, metaphors and social stories to show all stakeholders (students, teachers and school staff, community members) in the school setting what positive behaviour is [30, 31]. The power of these metaphors is that self-management strategies can be learned in a less intrusive way. The main metaphor underlying the design of the PBIS–AR app is that an alien is new to planet Earth and has to explicitly learn all values, procedures and routines.
- The alien is a boundary crossing object for behavioral learning [32]; it has no feelings, expectations, or emotions and can be 'told' and 'learn' what to do and in this way can automate the expected behavior via prompting, teaching, practicing, and reinforcing the new behavior. For non–responders and resisters (i.e. children who need more intensive and layered instruction in behavior management and self regulation), this is a practice-based way to let the alien model the expected behaviour.
- The alien has no gender implications; this because gender specification could have made students' gender salient in the process of identification with the character (i.e. making identification of male students more likely or intense in case of a male character and vice versa) thus contributing to some extent to the outcome of the learning process.
- If the avatar is a human character there can be dissociation with this character as the children—especially non–responders—will not identify with the values, procedures and routines related to this character, and they will not be triggered or motivated to exhibit the expected behaviour.

Fig. 4 Arpro, the alien character

The alien is only one of the characters which can be involved in the creation of examples and non-examples of a behavioural routine in the PBIS-AR application. At school he should indeed have relationships with other characters, such as teachers, the administrator and fellow peers. Secondary characters are being designed and developed following the same process, see Fig. 5.

The AR–BLR basically takes the form of structured animations superimposed to real settings and involves the alien and secondary characters performing examples and non-example behaviours. Starting from the expected behaviors defined in Table 1 and from the example and non-example defined in each behavioral lesson as set of atomic animation has been created an integrated in the Mirage·XR authoring toolkit, so these animations can be used by all users to create new example and non–example of behaviors in their school setting.

Moreover, some basic 3D objects have been added to create more realistic scenarios, such as school desk, chair, backpack, lunchbox, books, pen, rule, etc. All this object, stored on *Sketchfab* under *Creative Commons* license, can be added to the real scene via the authoring toolkit.

Figure 6 illustrates the process of creation of an AR–BLR. This process starts with the definition of example and non–example for the expected behavior, topic of the lesson. Using these examples as a plot, teachers create a storyboard representing a

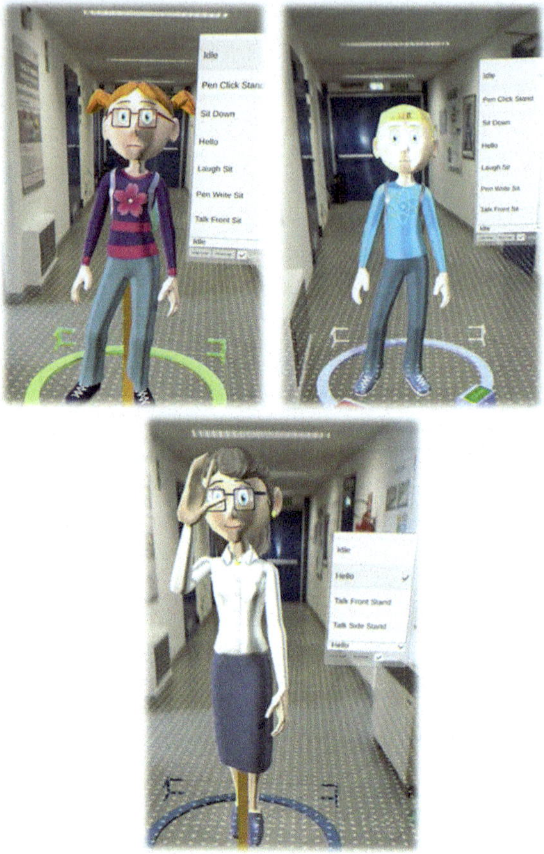

Fig. 5 Examples of secondary characters

specific behavioral routine. The Mirage·XR authoring toolkit allows to put together 3D objects and characters, animate this one, record in–situ the routine, and store it on the ARETE repository.

Students could play the AR–BLR representing the specific behavioral routine using in–situ the AR–PBIS app, in the "Learning mode" section, on her/his mobile device, during the phase of the behavioral lesson devoted to teach and modeling the expected behavior.

5.2 Discovery Mode

An important part of a correct PBIS implementation is the design of positive reinforcement systems for rewarding students' compliance to expected behaviour.

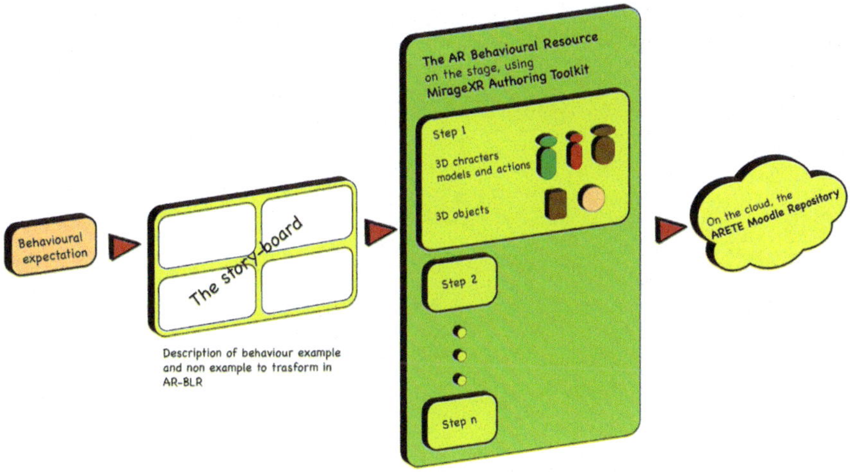

Fig. 6 The creation process of an AR–BLR

According to this consideration, gamification learning contents will be provided to the students. This is the "Discovery mode" section defined in Fig. 3.

Using this feature, students participate in a behavioural reflection game in AR. Through a series of markers located in the various settings, examples and non-examples of behaviours are displayed, as animated 3D scenarios, on which the student is invited to reflect through quizzes whose positive answer determines the assignment of a score/badge and the updating of an overall ranking (*leaderboard*).

The definition of the game design and setting is up to the teachers who have to design according to their learning goals a specific scenario.

Figure 7 represent one frames from the AR scenario related to the behavioural expectation "I store my belongings". The scenario complete description is the following:

> Arpro (the name of the alien in the app) walks purposefully through the corridor towards the classroom together with his peer students. He sees the rack to store his backpack. Arpro takes a few steps and then turns 90 degrees, holding the backpack with his hands in front of his chest Arpro hangs his backpack on the peg. He takes out his lesson materials/books out of the backpack and walks to the classroom door.

After the play of this AR scenario, the student have to answer to the following question: "Why is this expected behaviour appropriate?"
S/he has three options:

1. Arpro avoids accidents by storing the backpack in the rack. (*correct*)
2. Apro is free to go to class by hanging up his backpack. (*uncorrect*)
3. Arpro does what his classmates do. (*uncorrect*)
4. Arpro does what the teacher tells him. (*uncorrect*)

This modality is similar for all the scenarios in the "Discovery mode", related to the others expected behaviors.

Fig. 7 A frame from the example scenario: 'I store my belongings'

5.3 *Practice Mode*

Another relevant feature designed for the PBIS–AR app is the 3D AR multi–user interaction, in Fig. 3 the "Practice mode". This feature is designed to give students the opportunity to practice behavioral activities through augmented reality, using 3D objects and characters, to interact with in a mixed environment. The *Orkestra* library will be used to support the practice. Using this library, the application will enable students to participate in a multi–user interactive behavioural activity. Students will interact through their mobile devices in an AR scenario to practice the PBIS lessons.

The idea was that collaboration with peers and interaction during practice can be essential factors in the achievement of behavioural skills and in reaching goals as a team. As proof of concept, three AR multi–user examples have been developed and included in the "Practice mode" section of the AR–PBIS app. In particular, these interactions was designed as activities to practice the following expected behaviors:

- 'Greet others';
- 'Stand up for others';
- 'Keep your workspace organised'.

For example, 'Greet others' scenario (Fig. 8) presents itself as a role–playing game to practice how to greet others through a quiz. Two students interact on the same AR scenario with their own devices: one plays the role of the 'teacher' and the other the 'student'. When they scan a marker, they see the AR characters and a question will be displayed on the screen asking them: "How would you greet the student/teacher?". When both have chosen their answer the characters show the greetings through

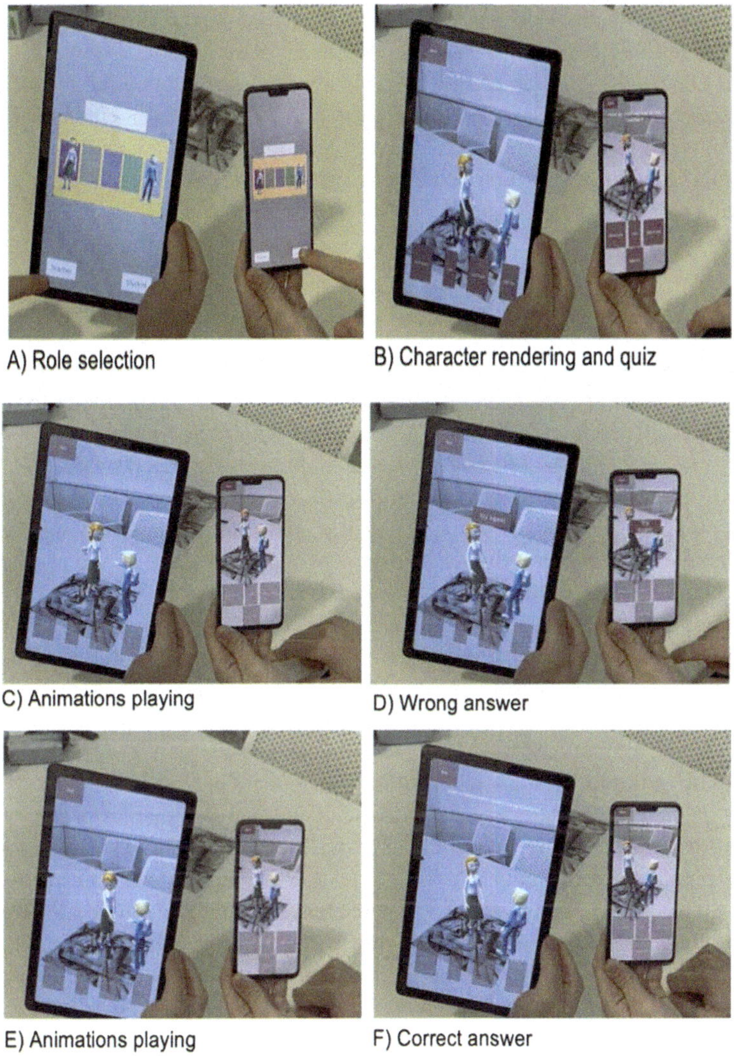

Fig. 8 Sequence at the 'Greet others' scenario

synchronised animations. If the animations are played and the answer is wrong, they are asked to try again. Otherwise, students are rewarded and the game is completed.

The PBIS-AR app integrates the *Orkestra library* for creating AR multi–user interaction behavioral activities during the "Practice mode". More details can be find in [33]. Briefly, the *Orkestra library* allows the creation of multi–device and multi–user applications and enables communication and data–sharing between different users. Common functionalities of the Web version of the library include sharing data between users, sending peer to peer or broadcast messages, providing timing and synchronization mechanisms, distributing content across multiple devices.

This concise description of the AR–PBIS app is aims at showing the strict relationship between the different phases of the behavioral lessons and the integration of different types of AR resources. All the AR technologies have been chosen after an a careful inspection of the their pedagogical affordances, avoiding to adopt solutions without a clear cut didactic value.

6 Conclusion

This chapter has illustrated an innovative application of AR technologies to behavior management. The idea was the integration of AR within the framework of the PBIS. PBIS (*Positive Behavioral Interventions and Supports*), or PBS (*Positive Behavioral Support*), is a systemic approach, including different strategies and procedures, aimed at improving the capacity of schools, families, and communities to design effective environments that improve the relationships between evidence–based practices and the places in which teaching and learning occur. The systemic vision of the PBIS framework is evident yet in its organization: PBIS schools provide a continuum of academic, behavioral, social and emotional support, across three tiers. The tier 1 support is for everyone across all school setting, the SWPBIS; the tier 2 support is for students with specific needs; finally, tier 3 is for a small number of student for whom the tier 1 and tier 2 supports have not been sufficient.

In the US, the Individuals with Disabilities Education Act (IDEA) establish the PBS as an intervention strategy to be used with children who display problem behavior. This framework, with suitable changes due to the cultural and social differences, is also spreading to other countries and contexts. PBIS aspires to transform the school climate in order to create positive, equitable and safe learning environment.

Technologies can play an important role in this transformation process. The introduction of new devices can contribute to changing the social landscape, opening new communication possibilities, modifying habits, and transforming traditional modalities and practices. The diffusion of smartphones among children at also very young is an example of these disruptive transformations. These processes are destined to continue shortly when new technologies and devices are ready to come into play.

AR is one of these innovative technologies destined to have an increasing impact on teaching and learning, together with Mixed and Virtual Reality, in the Extended Reality (XR) vision. The mixture of real and digital objects and characters in an environment in which the difference between the real and the artificial is more and more blur is becoming simple and accessible thanks to the spread of new devices, as head-mounted displays, smart glasses, etc.

In this work, AR was introduced to support the creation of an enriched learning space. Different technological solutions, designed as an interrelated ecosystem, permit to integrate of the AR solutions within the PBIS framework to foster students' behavioral learning experience.

Some drawbacks are currently inherent to this proposal: the AR technologies are not fully mature; the devices in use at school are not suitable for a gratifying experience; the authoring of the AR resources is not easy and without trouble. However, as in any ecosystem, evolution is not a possibility but a destiny. The solution proposed will be able to transform itself following the diffusion of new technologies and new learning experiences.

Acknowledgements This work was carried out thanks to funding from the *Augmented Reality Interactive Educational System*–ARETE project, from the European Union's Horizon 2020 research and innovation programme under grant agreement No 856533.

References

1. McEvoy, A., Welker, R.: Antisocial behavior, academic failure, and school climate: a critical review. J. Emot. Behav. Disord. **8**(3), 130–140 (2000). https://doi.org/10.1177/106342660000800301
2. Greenwood, C.R., Kratochwill, T.R., Clements, M.: Schoolwide prevention models: lessons learned in elementary schools. Guilford Press, New York, NY (2008)
3. Carr, E.G., Horner, R.H.: The expanding vision of positive behavior support: research perspectives on happiness, helpfulness, hopefulness. J. Posit. Behav. Interv. **9**(1), 3–14 (2007). https://doi.org/10.1177/10983007070090010201
4. Sugai, G., Horner, R.H.: Defining and describing schoolwide positive behavior support. In: Roberts, M.C., Sailor, W., Dunlap, G., Sugai, G., Horner, R. (eds.) Handbook of Positive Behavior Support, pp. 307–326. Springer, New York (2009)
5. Dunlap, G., Kincaid, D., Horner, R.H., Knoster, T., Bradshaw, C.P.: A comment on the term "positive behavior support". J. Posit. Behav. Interv. **16**(3), 133–136 (2014). https://doi.org/10.1177/1098300713497099
6. Sugai, G., Sprague, J.R., Horner, R.H., Walker, H.M.: Preventing school violence: the use of office discipline referrals to assess and monitor school-wide discipline interventions. J. Emot. Behav. Disord. **8**(2), 94–101 (2000). https://doi.org/10.1177/106342660000800205
7. Horner, R.H., Sugai, G., Smolkowski, K., Eber, L., Nakasato, J., Todd, A.W., Esperanza, J.: A randomized, wait-list controlled effectiveness trial assessing school-wide positive behavior support in elementary schools. J. Posit. Behav. Interv. **11**(3), 133–144 (2009). https://doi.org/10.1177/1098300709332067
8. Nelen, M.J.M., Willemse, T.M., van Oudheusden, M.A., Goei, S.L.: Cultural challenges in adapting SWPBIS to a Dutch context. J. Posit. Behav. Interv. **22**(2), 105–115 (2020). https://doi.org/10.1177/1098300719876096
9. Lynass, L., Tsai, S.-F., Richman, T.D., Cheney, D.: Social expectations and behavioral indicators in school-wide positive behavior supports: a national study of behavior matrices. J. Posit. Behav. Interv. **14**(3), 153–161 (2012). https://doi.org/10.1177/1098300711412076
10. Bradshaw, C.P., Waasdorp, T.E., Leaf, P.J.: Effects of school-wide positive behavioral interventions and supports on child behavior problems. Pediatrics **130**(5), e1136–e1145 (2012). https://doi.org/10.1542/peds.2012-0243
11. Horner, R.H., Sugai, G., Anderson, C.M.: Examining the evidence base for school-wide positive behavior support. Focus Except. Child. **42**(8), 1–14 (2010). https://doi.org/10.17161/foec.v42i8.6906
12. Bradshaw, C.P., Koth, C.W., Thornton, L., A., Leaf, P.J.: Altering school climate through school-wide positive behavioral interventions and supports: findings from a group-randomized effectiveness trial. Prev. Sci. **19**(2), 100 (2009). https://doi.org/10.1007/s11121-008-0114-9

13. Waasdorp, T.E., Bradshaw, C.P., Leaf, P.J.: The impact of schoolwide positive behavioral interventions and supports on bullying and peer rejection: a randomized controlled effectiveness trial. Arch. Pediatr. Adolesc. Med. **166**(2), 149–156 (2012). https://doi.org/10.1001/archpediatrics.2011.755
14. Bradshaw, C.P., Koth, C.W., Bevans, K.B., Ialongo, N., Leaf, P.J.: The impact of school-wide positive behavioral interventions and supports (PBIS) on the organizational health of elementary schools. Sch. Psychol. Q. **23**(4), 462–473 (2008). https://doi.org/10.1037/a0012883
15. Kelm, J.L., McIntosh, K.: Effects of school-wide positive behavior support on teacher self-efficacy. Psychol. Sch. **49**(2), 137–147 (2012). https://doi.org/10.1002/pits.20624
16. Ross, S.W., Romer, N., Horner, R.H.: Teacher well-being and the implementation of school-wide positive behavior interventions and supports. J. Posit. Behav. Interv. **14**(2), 118–128 (2012). https://doi.org/10.1177/1098300711413820
17. Horner, R.H., Sugai, G.: Future directions for positive behavior support: a commentary. J. Posit. Behav. Interv. **20**(1), 19–22 (2018). https://doi.org/10.1177/1098300717733977
18. Ozdemir, M., Sahin, C., Arcagok, S., Demir, M.K.: The effect of augmented reality applications in the learning process: a meta-analysis study. Eurasian J. Educ. Res. **18**(74), 165–186 (2018). https://doi.org/10.14689/ejer.2018.74.9
19. Garzón, J., Pavón, J., Baldiris, S.: Systematic review and meta-analysis of augmented reality in educational settings. Virtual Real. **23**(4), 447–459 (2019). https://doi.org/10.1007/s10055-019-00379-9
20. Radu, I.: Augmented reality in education: a meta-review and cross-media analysis. Pers. Ubiquitous Comput. **18**(6), 1533–1543 (2014). https://doi.org/10.1007/s00779-013-0747-y
21. Tosto, C., Matin, F., Seta, L., Chiazzese, G., Chifari, A., Arrigo, M., Taibi, D., Farella, M., Mangina, E.: The potential of AR solutions for behavioral learning: a scoping review. Computers **11**(6), 87 (2022). https://doi.org/10.3390/computers11060087
22. Chiazzese, G., Goei, S.L., Pronk, J., Tosto, C., Seta, L., Arrigo, M., Taibi, D., Farella, M., Mangina, E.: Teaching behavioural routines using augmented reality in the ARETE project. In: SCIFI–IT 2021, pp. 60–64. https://doi.org/10.5281/ZENODO.4916268
23. Ianì, F.: Embodied memories: reviewing the role of the body in memory processes. Psychon. Bull. Rev. **26**(6), 1747–1766 (2019). https://doi.org/10.3758/s13423-019-01674-x
24. Baragash, R.S., Al-Samarraie, H., Moody, L., Zaqout, F.: Augmented reality and functional skills acquisition among individuals with special needs: a meta-analysis of group design studies. J. Spec. Educ. Technol. **37**(1), 74–81 (2022). https://doi.org/10.1177/0162643420910413
25. Chen, Y., Zhou, Z., Cao, M., Liu, M., Lin, Z., Yang, W., Yang, X., Dhaidhai, D., Xiong, P.: Extended reality (XR) and telehealth interventions for children or adolescents with autism spectrum disorder: systematic review of qualitative and quantitative studies. Neurosci. Biobehav. Rev. **138**, 104683 (2022). https://doi.org/10.1016/j.neubiorev.2022.104683
26. Ocay, A.B., Rustia, R.A., Palaoag, T.D.: Utilizing augmented reality in improving the frustration tolerance of ADHD learners: an experimental study. In: Proceedings of the 2nd International Conference on Digital Technology in Education—ICDTE (2018), pp. 58–63. https://doi.org/10.1145/3284497.3284499
27. Lane, K.L., Oakes, W.P., Royer, D.J., Cantwell, E.D., Menzies, H.M., Jenkins, A.B., Hicks, T.: Using the schoolwide expectations survey for specific settings to build expectation matrices. Remed. Spec. Educ. **40**(1), 51–62 (2019). https://doi.org/10.1177/0741932518786787
28. Newcomer, L., Colvin, G., Lewis, T.J.: Behavior supports in nonclassroom settings. In: In Roberts, M.C., Sailor, W., Dunlap, G., Sugai, G., Horner, R. (eds.) Handbook of Positive Behavior Support, pp. 497–520. Springer, New York (2009)
29. Goei, S.L., Van Joolingen, W., Goettsch, F., Khaled, A., Coenen, T., In 't Veld, S., de Vries, S., Schipper, T.: Online lesson study: virtual teaming in a new normal. Int. J. Lesson Learn. Stud. **10**(2), 217–229 (2021). https://doi.org/10.1108/IJLLS-09-2020-0078
30. Carter, D.R., Pool, J.L.: Appropriate social behavior: teaching expectations to young children. Early Childhood Educ. J. **40**(5), 315–321 (2012). https://doi.org/10.1007/s10643-012-0516-y
31. Ennis, R.P., Hirsch, S.E., MacSuga-Gage, A.S., Kennedy, M.J.: Positive behavioral interventions and supports in pictures: using videos to support schoolwide implementation. Prev. Sch.

Fail. Altern. Educ. Child. Youth **62**(1), 1–12 (2018). https://doi.org/10.1080/1045988X.2017.1287048
32. Akkerman, S.F., Bakker, A.: Boundary crossing and boundary objects. Rev. Educ. Res. **81**(2), 132–169 (2011). https://doi.org/10.3102/0034654311404435
33. Dominguez, A., Cabrero, A., Simões, B., Chiazzese, G., Arrigo, M., Farella, M., Seta, L., Chifari, A., Tosto, C., Goei, S.L., Mangina, E. Masneri, S.: Collaborative augmented reality tools for behavioral lessons. In: Proceedings of ICL2022 (2022) (in press)

A New Framework for Learning Patterns and Social Presence in Virtual Reality for Learning

Thommy Eriksson and Maria Sunnerstam

Abstract In this chapter, we will combine existing concepts and frameworks from both research on virtual reality as a media and research on teaching methodology. We will combine our findings to present a new framework for planning and organizing learning activities in virtual reality (VR). There is a wide variety of pedagogical theories and frameworks available, but they are quite seldom thoroughly utilized in work on virtual reality for learning. VR has now been evolving technically and creatively for many decades. Even though it has been slow in widespread adoption, we believe that the area of VR for learning is now mature enough to be properly supported by contemporary pedagogical frameworks. After going through first the basics of VR, as well as Laurillard's learning patterns, we will present our suggested framework and use it on two case studies.

1 Introduction on ICT and Learning and VR Learning

Information and communication technology (ICT) has been developed to be used in teaching and learning for quite some time now. Learning management systems (LMS) have gradually been implemented since the late 1990s and are now commonly used in most schools and universities. Video technology for video meetings, streaming of lectures and recording of lectures has matured alongside LMS platforms and is now widely adopted as well. A wide range of different ICT tools have been developed and implemented in higher education practices, such interactive whiteboards, audience response systems, discussion forums etc. Similarly, pedagogical methods and delivery models—such as flipped classroom, active learning, distance learning, hybrid and Massive Open Online Courses (MOOCs)—have been developed and supported by the affordances of new infrastructures where software, hardware and

T. Eriksson (✉)
Chalmers University, Gothenburg, Sweden
e-mail: thommy@chalmers.se

M. Sunnerstam
Gothenburg University, Gothenburg, Sweden
e-mail: maria.sunnerstam@gu.se

Y. Cai et al. (eds.), *Mixed Reality for Education*, Gaming Media and Social Effects,
https://doi.org/10.1007/978-981-99-4958-8_15

digital platforms can be combined in various ways. A large amount of applied and theoretical research has been conducted on ICT and learning. Bates [1] has summarized the field in an excellent and comprehensive open digital textbook. Säljö [2] has thoroughly problematized and theorized the impact of new technology in society and how digital technology affect teaching and learning. Laurillard [3] has made a highly comprehensive and researched interdisciplinary summary of what we know about learning and arrived at a set of learning patterns, which we will use in the framework that will be presented later in this chapter.

This brief sketch of the field of ICT and learning is intended as a reminder that it is a well evolved and thoroughly researched area, even though not all questions are answered nor will probably ever be.

Virtual reality has been around for quite some time as well, despite the current impression that it is a "new" technology. The first prototype head-mounted display was developed in 1968 by Ivan Sutherland [4]. Sutherland also wrote a seminal vision for what he called an ultimate display [5], and visionary statements like this have been accompanying the development of VR. Fictional stories have also played a not negligible role in guiding anticipation and actual development. Science fiction author William Gibson coined the term cyberspace in his novel "Neuromancer" [6] and during the late 1980s VR had its first hype; many visionary and extravagant claims about the promise of this technology as well as the ontological status of VR as a new world, were proclaimed. The publications of Heim [7] and Benedikt [8] are two typical examples, and a general summary of this period can be found in the PhD thesis by one of the authors of this chapter [9]. Jaron Lanier, early VR entrepreneur, highly experimental and provocative VR researcher and technology skeptic, has chronicled the development of VR during this hype [10]. In the first half of the 1990s the first VR hype died out and the technology was confined to research laboratories. It very seldom found practical use, mostly because the hardware was very expensive and complex to use. In parallel, technology for augmented reality (AR) was continuously developing. Milgram et al. explored how VR and AR could be used in different settings, resulting in a framework organizing the different technologies along a continuum (1994). The nowadays well-established term mixed reality (MR) has sprung from this research.

The second hype cycle of VR started in the early 2010s when Oculus launched their first VR headset, Oculus Rift. This was, and still is, an affordable VR headset, raising the hope for widespread consumer adoption of VR. Facebook bought Oculus in 2014 and has since then developed the technology, emphasizing low-cost, ease of use and portability. The hope for mainstream adoption has not yet been fulfilled, most likely due to the lack of content or tools of interest for a wide set of users. As with the former introduction to ICT and learning, this recap is intended as a reminder that VR and its applications are fairly well researched.

2 The Need for a Framework for VR Learning

However, when exploring the research on VR for learning and combining this with pedagogical research, another pattern emerges. Much research is done on VR for learning, but it is often either laboratory experiments with limited connection to real-life applications or case studies of applied research that is largely disconnected from other similar case studies [11]. Even though this research is necessary and vital for further development of the field, there is also a need for frameworks combining pedagogical theory and best practice with research on VR for learning. Therefore, we intend to explore how Laurillard's learning patterns might be combined with a number of VR affordances.

The research question guiding this work has been:

How can existing frameworks from pedagogy and VR research be combined into a new framework that will give helpful guidance for teachers and course designers?

In order to clarify the scope, a couple of limitations have to be emphasized. First, only virtual reality is included in the proposed framework. This is not because we don't believe AR has a role to play in teaching and learning practices, it most certainly has, but rather to limit ourselves to a scope that is possible to grasp and to focus on technology where we have thorough experience and knowledge. We would be happy to see others develop a similar framework for AR and learning, or maybe even combine VR and AR for learning in one cohesive framework. The second limitation is that focus has been on the planning and organization of teaching and learning activities, and not on the teaching in itself, or on the learners' own actions. Maybe this framework can be useful for those perspectives as well, as is or in a further developed form.

To our knowledge, there are no established and frequently used frameworks combining pedagogy and VR. We have come across a few VR and learning frameworks; the frameworks suggested by [12–14]. However, these frameworks are generally focused more on the technical aspects of VR, or the properties of VR as a media. These frameworks include references to and discussions about pedagogical aspects, but they do not include established learning patterns which is why we argue that a VR learning framework needs to be more firmly based on pedagogical research.

3 What Is a Framework and Why Do We Need It?

Before presenting our framework, we will illustrate what a framework can be—and not—by discussing a few existing frameworks commonly used in teaching and learning. These are particularly useful to bring up since they have been to some degree misunderstood, and therefore highlight the intentions with our own framework.

In everyday discussions with teachers and others working with education, there is often some confusion regarding what the differences between a model, a theory and a framework really are. In our opinion, the result of this is sometimes a misguided critique against frameworks in general. The two examples to be discussed are the TPACK framework and the SAMR framework, which we claim to be precisely that: frameworks.

Concepts such as models, theories and frameworks are difficult to discuss in multidisciplinary contexts (such as in this chapter) since the usage and interpretation of terms vary between research disciplines. There are some distinctions however, that can be made on a very general level.

- A model is an abstract and simplified representation of *a section* of what exists and happens in reality. It can be a model of how the brain process visualizes stimuli, or the air flow in a hurricane, and so on [15].
- A theory is a theoretical explanation of an observed phenomenon, verified by further experiments or observations. The theory of natural selection explains the development of species, and the theory of cognitive load explains how the brain handles different amount of information [16].
- A framework is an analytical tool that is used to organize ideas and concepts into a structure, with the intention to guide and help planning and organization of different actions [17].

Note that the first two, models and theories, represent and explain parts of reality, and therefore need to be evidence-based. One characteristic of a theory is that a hypothesis becomes a theory only when it is supported by observations. A framework on the other hand, does not have the same requirement to be fully supported by evidence or observations. What determines its value is primarily if it is useful as a guide for planning. Therefore, frameworks cannot be evaluated on the same grounds as models or theories. This does not mean that a framework might as well be a fantasy product, generally a framework is more useful the closer it is connected to empirical observations. Frameworks can be built on interpretations, which might open up for conflicting frameworks—that nevertheless can co-exist. One example is the concept of the three-act structure in movies, and other forms of storytelling. Many, but not all, stories can be divided into a three-act-structure. Stories for instance, can also be divided into five- or seven-act-structures. Different frameworks for how a story is told can, despite their differences, all make sense at the same time. Frameworks describing different act structures are not based on evidence grounded in the actual artifact (i.e., the movie) but based on different viewpoints which separately can add insights into how a particular story is told, or how it should be told.

With this in mind, let's consider the SAMR model created by Ruben [18]. The SAMR model has been criticized quite harshly [19], one of the issues being the claim that the SAMR model is a model, i.e., evidence-based. The abbreviation SAMR stands for

- Substitution
- Augmentation
- Modification
- Redefinition.

These categories are explained as being different levels of technology implementation that might be applied in a learning situation. The SAMR model is intended to be used for selecting, using and evaluating technology, primarily in K-12 settings [20]. We do agree with the critique of SAMR not being evidence-based, since SAMR is presented as a model, and as such, should be an abstract subset of reality. However, since the SAMR model is primarily used as a guiding tool for teachers working with technology in the classroom, it makes more sense to regard the SAMR model as a framework. That would also defuse some of the critique about lack of evidence, simply because evidence is formally not needed in a framework. Another critique is that the four categories are described, and visually structured, giving the impression that the top level is preferable and what inherently always should be strived for [20]. The latter is a quite problematic issue since the case is not always that for example Redefinition is the best choice. On the contrary the implementation of new technology is very context dependent, and in many cases a simplified technology utilization is more efficient and appropriate than a more complex. An important take-away from this is not to build in explicit or implicit assumptions or preferences that are not supported by evidence (our logic is that if a framework starts to be prescriptive rather than helpful by structuring, then actual evidence becomes important).

Now to the TPACK framework, properly referred to as a *framework* by the creators Koehler and Mishra [21]. The TPACK framework consists of three components: *technology*, *pedagogy* and *content*, the K stands for *knowledge*, including the relationships among and between them. The TPACK framework suggests that the three components have roles to play individually and together, and the framework is usually visualized as a Venn diagram with these three components as partially overlapping circles. Usually, the TPACK framework is used when considering important skills and roles to include in different small scale or large scale ICT and learning projects in schools and universities [20]. One reflection is that in everyday discussions TPACK is occasionally referred to as a model, which is contrary to what Mishra and Koehler intended. Mishra and Koehler themselves were explicit and precise with their own expectations saying that "having a framework goes beyond merely identifying problems with current approaches, it offers new ways of looking at and perceiving phenomena and offers information on which to base sound, pragmatic decision making".

This is a balanced and sound view on what can be expected from a framework, and that is also the anticipations concerning the framework that will be presented next in this chapter; Laurillard's learning patterns [3]. These learning patterns, which will be explained in detail in the next section, have a very thorough contextualization. In her book, Laurillard goes through and makes extensive research references to a wide range of aspects on pedagogy, teaching, and learning. Even though these learning patterns appear very grounded, it doesn't mean that their existence can be proved with evidence. Since this is a framework, and not a model or a theory, that is not a problem. The value of Laurillard's learning patterns is rather based on whether they can help us understand and change reality.

What can a framework be used for? The SAMR framework, even though criticized, can be a platform for discussions on how much a new tool or technology can transform the practice of teaching. VR can be used for Substitution (for example watching a video in a virtual cinema theater), or it can be used for Redefinition (the possibility to choose an avatar with a gender opposite your own, and to experience how that might change other people's behavior towards you).The TPACK framework is a useful reminder of what it actually takes—in skills and responsibilities—to plan or implement for example an ICT and learning intervention. The learning patterns of Laurillard have been developed into the ABC method, a methodology for curriculum planning. By categorizing learning activities, the patterns can help to identify missed opportunities. Learning activities can be put into a broader perspective; one example, there is often much debate concerning the advantages and disadvantages with a live lecture compared to a recorded lecture, or a recorded lecture compared to a textbook. Granted, these are important choices, but when Laurillard categorizes all the three of them in the same category—learning through acquisition—it emphasizes that the most important difference (in the case of these three options) is a matter of delivery, and not so much a matter of the even more important aspects of learning; what do the students do and how do they learn by doing this?

Similar insights are ultimately what we hope for concerning our own proposed framework.

4 Laurillard's Learning Patterns

Diana Laurillard's learning patterns are described in her excellent book "Teaching as a Design Science: Building Pedagogical Patterns for Learning and Technology" [3]. The learning patterns Laurillard outlines categorize different learning activities into:

- Learning through *acquisition*. The teacher provides narrative explanations or descriptions of which the students take part. According to Laurillard, this should preferably be supplemented by one or more of the other patterns, and it is vital that the teacher verifies that the narrative has been understood.
- Learning through *inquiry*. The teacher gives the students a challenge to explore a question or a problem, with guidance from the teacher.
- Learning through *discussion*. The teacher creates a careful plan and set-up for a fruitful discussion about specific topics between students, often with the teacher acting as facilitator.
- Learning through *practice*. The teacher creates a goal-oriented task or project that through its design give students experiential learning and intrinsic feedback from the situation in itself.
- Learning through *collaboration*. According to Laurillard, collaboration combines all other forms of learning, especially discussion and practice.
- Learning through *production*. In Laurillard's description, production means primarily the creation and publication of essays, reports, videos, e-portfolios and such.

One of the benefits with these learning patterns is that they highlight learning patterns available for courses to be planned, learning patterns in existing courses, and what the actual similarities and differences actually are between different tools and methods that teachers commonly use. For example, Laurillard makes the difference between books, lectures, and videos less dramatic, since they are all learning through acquisition whilst it becomes obvious that a literature seminar (learning through discussion) is actually something very different from discussing a project problem (learning through collaboration).

A comment concerning learning through production is that Laurillard misses the opportunity to be specific concerning project work. In many situations, especially in design but also in engineering, project work is an absolute corner stone, but it is difficult to interpret where this is supposed to be categorized. Under learning through production, she mentions "designs" and "representations of designs", and under learning through collaboration she mentions "small group projects". But where would a big movie production or the creation of a finalized smartphone app fit in? It can barely be regarded a "small" project, and it is a very different artifact from for example a report.

Laurillard's book offer many more insights beyond the learning patterns framework, and some have a specific relevance to our own framework on learning in VR. Laurillard points out that any learning activity involves both what the teacher does and what the learners do: an obvious fact that is nevertheless easy to forget. We have incorporated this in our own framework, as well as Laurillard's suggestions that learning through acquisition shouldn't be used stand-alone but preferably be supplemented by other learning patterns, also that learning through inquiry must be completed with guidance for the learner. When using learning through practice, a repetitive cycle of goal-action-feedback-revision is beneficial.

Laurillard also discusses the use of a model environment—sometimes referred to as a micro world—and how this is well suited for learning through practice as well as several of the other learning patterns. Such a micro world should give feedback to the learners, helping them to refine their understanding or skill, and this feedback is either extrinsic (coming from the outside of the model environment, for example from a teacher) or intrinsic (coming from the model environment itself). The latter is often more efficient since it encourages the learners to understand by themselves what they have done wrong and how to adjust their actions based on this experience.

5 The Different Aspects of Virtual Reality

What follows is an overview of different aspects of VR that are of relevance for teaching and learning with and in VR.

Immersion and Presence

Immersion means the technical aspects of creating the virtual reality experience, while presence is the phenomenological experience. This (technical) immersion deals with hardware specifics such as Field of View angles, frame rate and image resolution for the VR headset, as well as audio quality, hand tracking, and so on.

Presence is the feeling of existing in an environment other than the actual place that the physical body is, of extending, projecting somewhere else. According to Witmer and Singer [22] presence is the "subjective experience of being in one place or environment, even when one is physically situated in another" (p. 225).

Many attempts have been made to determine the components of presence Lombard and Ditton [23] propose five aspects:

- Social richness (intimacy and immediacy through self-expression and the presence of others)
- Realism (the fidelity and accuracy of the representation)
- Social actors within medium (virtual, interactive agents)
- Transportation (the feeling of being somewhere else)
- Immersion (the perceptual and psychological immersion in a space).

Steuer [24] emphasizes two dimensions:

- vividness ("the ability of a technology to produce a sensorially rich mediated environment") (p. 10)
- interactivity ("the degree to which users of a medium can influence the form or content of the mediated environment") (p. 11).

Finally, Zeltzer [25] suggests three salient components:

- autonomy (a simulations ability to react to stimuli)
- interaction (the degree of control the user can have over the simulation)
- presence as a bath of sensations.

Clearly, there are many ways to conceptualize aspects that are important for creating presence. A few of these are repeated in different forms—most notably different aspects of richness and interaction—and there is common ground to find among these suggestions.

Social Presence

The concept of social presence has evolved during some time. Williams and Christie [26] considered social presence a single dimensional concept related to intimacy and immediacy when they introduced the term in the context of telecommunications. They claimed that the level of social presence was determined by attributes of the media and assumed that unmediated in-person communication had the highest social presence. Compared to that, the social presence would essentially be diminished when the mediation became less rich. According to this logic, a video call would have higher social presence than a telephone call, and both would have higher social presence than a text chat [27]. However, later research showed that the level of social presence was not solely determined by the affordances and richness of the mediating technology [28, 29]. Different users experienced quite diverging levels of social presence in the same computer mediated conference, and some even perceived higher social presence in situations with less media richness. Gradually social presence was redefined into a concept focusing on the phenomenological experience of the user [23]. Nowadays, social presence is described as when users in a co-habited virtual space project themselves socially and emotionally into this environment, so they can communicate directly with others [30, 31].

Another concept is co-presence, and this is the sense of being together, the impression of co-habiting an environment [32]. Yet another related concept is self-presence, the capability of a VR application to track the physical body movements of a user (primarily head and hands) and move the user's avatar in accordance [33, 34].

This can be summarized as follows, adapted from [35].

- Presence—the experience of being somewhere ("I can see a world")
- Self-presence—the experience of being embodied some-where (e.g., "I can see my hands")
- Co-presence—the experience of being somewhere together with others (e.g., "I can wave to that person over there")
- Social presence—the experience of having cognitive and emotional connection with others in the environment (e.g., "When I wave to that person, she waves back in a welcoming way")

Interaction

Interaction is a fundamental aspect of VR, and closely connected with presence and immersion. Interaction means agency, the ability to act, and there are three fundamental levels of agency in VR. The most basic is looking around. In a VR experience the user can turn her head and this matches what she can see, just as in an actual environment [36]. Secondly, the user can move through the environment, providing a spatial interaction between user and world, mimicking what we expect in an actual environment [36, 37]. Thirdly, the user can have agency in the world, being able to interact with objects in the world. Important for this agency is to what level the user can change aspects of the world, and get perceptual feedback confirming the change [37–40]?

Technically, the first two levels of interaction, and partially the third, are enabled by degrees of freedom (DoF), which is the number of spatial axis that can be used. In the simplest example of looking around, the VR headset's direction is tracked in all three spatial dimensions (x, y, z) enabling turning the head in three ways (up-down, left–right, and leaning the hand left or right). This is 3 DoF. When the headsets position is tracked in all three dimensions, then you can move through the environment in all directions (back–forth, left–right, up-down). This is almost always combined with directional tracking, so this will be 6 DoF. In a similar way the hand controllers can be tracked in 3 DoF or 6 DoF (there are other combinations as well, for example the 2 DoF tracking of the mouse pointer position on a conventional computer).

Space, Place and World

A virtual world in which we are represented as avatars is a world that we can inhabit. The presence of others gives the world even more feeling of being inhabitable, of worldliness, and once again interaction is a key word since it is the interaction with others that makes the world appear inhabited [41]. Inhabitation of virtual worlds creates a narrative. The participants in online virtual worlds live and create and tell a story together [42, 43]. Klastrup calls it "story-living rather than story-telling" (ibid., p. 104). For a virtual world to have worldliness and not just be an environment, it needs persistence. Persistence means that the world has a continuous, autonomous history. When a user is not present the world nevertheless continues to evolve and change; the other users are persistent as well. Klastrup describes the following types of virtual spaces.

- Virtual world (persistent). A full-fledged virtual world is persistent in time and extended in a three-dimensional spatial space. Usually, the participant interacts with both the world and the other inhabitants.
- Virtual environment (impermanent). A virtual environment is a spatial environment but it is not permanent in time. Usually, the participant only interacts with the environment itself, not other users.
- Virtual community (social, not spatial). Non-VR, non-3d forums, and therefore not relevant.

Thus, Klastrup defines a virtual world as an online VR experience that is persistent, inhabited, and provides interaction with the world and its users.

6 Embodiment and Avatars

What do you do with your body when you are in a VR experience? In the simplest VR experiences, where you only can look around and you don't have any hand controllers, you don't have a body. When you raise your hands or when you look down where your actual body is in actual space, then you don't see it. In a CAVE virtual reality, where you don't wear a VR headset but instead stand inside a small cube with the virtual environment projected on the walls, then the answer is simple; you bring your meat body with you into the virtual environment. But, in most VR environments you are provided with a virtual body, an avatar. The avatar is important for achieving presence, and especially your self-presence.

Avatar is originally a Sanskrit word meaning the incarnation of Hindu gods, as a material manifestation. Incarnation literally means embodied in flesh or taking on flesh [44, 45]. The modern use of avatar was popularized by Neal Stephenson in his 1992 novel Snow Crash [46]. An avatar is like a mask we take on, a role we act [37]. The avatar splits us into two, giving us a double nature [44]. The meat body is left behind, and we project our attention and our presence to the virtual world.

The avatar is personal; it is my body, but it also connects me to others since I can be seen. The affordance result in a unifying concept of same time and same place, a concept tied in closely with the concept of worldliness.

Avatar Realism

Generally, avatars on social VR platforms are stylized and come in many different designs. The simple style of many avatars has to some extent technical reasons—minimizing performance demands on rendering—but it is to a large extent also a design choice [47]. However, simple style (low texture quality, low 3d mesh polygon count) does not mean a lack of extravagant designs. The many different avatar designs available in social VR platforms both offer quite high levels of customizability, and fantastical designs such as robots, animals, and anime characters as is popular in Japanese social VR [48].

Since users can both choose existing avatars, one interesting question is whether they choose an avatar that resembles themselves, or not. According to an interview study performed by Kolesnichenko et al. [47] the designers of avatars in commercially available social VR platforms acknowledge that the appearance of avatars probably contributes to one's own identity and personal brand. According to McLeod et al. [49], "people high in extraversion tended to have high similarity to their avatars", as well as conscientious people. The explanation would be that extrovert people to an extent prefer their avatars to represent themselves, and that conscientious people want to give an impression of being truthful and honest. The latter is similar to results described by Hooi and Cho [50], claiming that people who choose avatars similar to themselves perceive that their online self is highly similar to their real-life self. Their research also suggests that "most people would prefer an avatar that is similar to them or which they can identify with".

However, there are strong advocates for non-realistic avatars as well. VR pioneer Jaron Lanier has often emphasized the opportunities for experimentation and joy afforded by non-human avatars such as octopuses [51]. Dylan Paré, one of the developers behind the VR experience Queer & Trans Narratives in Virtual Reality, claims that the highly abstract avatars (participants are represented by moving pile of rocks) open up for being more open for conversation and capable of empathy, especially concerning the sensitive gender topics of the VR experience [52]. Dylan suggests, similar to Lanier's argument, that it is important to be able to play with and explore identities with avatars. Interestingly, Vtubers (Youtuber that represent themselves using avatars) have similar arguments for using avatars and not their real faces, according to a study by Liudmila [53]. They emphasize the freedom and enjoyment of "being anyone and anything" and "performing in an ideal form". They also mention the safety aspect of avoiding harassment and potential stalking.

Zach Deocadiz [54] have suggested a framework for conceptualizing and evaluating social VR applications, and even if this is a blog post and not a research paper or article, it do have some reasonable guidelines. It lists several design issues for social VR platforms.

- Guided to Self-Taught (how does the user learn the interaction with the VR application)
- Public to Private (are the social spaces open for everyone or closed for a smaller, controlled group?)

- Prescribed to User-Generated (to what extent can users control the look of their avatars, and change the environment?)
- Anonymous to Identified (can the user be anonymous?)
- Reactive to Preemptive (management of safe social interactions)
- Simple to Complex Interactions (can the users interact in complex ways?)
- Persistent to Temporary (is the environment and changes to it persistent from one session to next)
- Shareable to Real-Time (can artefacts such as notes be saved and shared outside the VR application?)

Kolesnichenko et al. [47] have interviewed industry experts working with several of the larger social VR platforms, and summarized the following aspects for avatars and social VR, suggested by their interviewees.

- Embodied locomotion (usually teleporting)
- Avatar aesthetics
- Personal Space Mechanics (e.g., blocking, muting, flagging and reporting)
- Social mechanics (interactions such as handshake, dance animations etc.)
- Create Custom Worlds (allowing the user to build their own environments)
- Import Avatar (allowing the user to build their own avatars)
- Humanoid or non-humanoid avatar (the design of avatars to choose from).

7 Social Aspects of Virtual Reality and Learning

Learning can take place, and can be successful and effective, both in solitude and in groups with others. This is true for conventional classroom teaching, for online teaching, and most likely for teaching with VR as well. In our framework we will include both, but we give specific attention to social VR. The reason for this is the wide-spread focus in general on learning in groups, and the general assumption that learning in groups are good pedagogy [55, 56]. There are many frameworks, models and theories in pedagogy that discuss learning in groups. Two of Laurillard's learning patterns—learning through discussion and learning through collaboration—require interaction with other learners. In sociocultural theory it is suggested that individuals learn through the interaction with culture as well as others [2, 57]. Another concept is Communities of Practice, a theory of learning in groups based on Foucault's claim that knowledge is embedded in social relations and social interactions in communities of experts or at least of shared interest [58]. Finally, another example is Communities of Inquiry, a concept first introduced by philosophers Peirce and John Dewey, emphasizing the social dimension of knowledge and scientific research, and later being adapted into a pedagogical approach by Matthew Lipman [59]. The idea is that a group of learners should form a community of inquiry that together ask questions, explore, and come to conclusions. During the 2010s there has been an increasing focus on the use of social VR for learning activities [60], to a large extent facilitated by inexpensive and easy to use VR headsets that can easily connect with other users via Internet connections.

Finally, it should not be forgotten that not all learning needs to take place in groups. There are situations where deep focus of solitary work is highly beneficial for learners, especially those with an introvert personality [61].

8 Suggestion for a Framework for Learning Patterns and Social Presence in Virtual Reality for Learning

The framework, depicted in Table 1, is our suggestion for how to select and organize all different aspects of VR that is relevant for learning and how to combine it with Laurillard's learning patterns. Many aspects, such as interactivity, have been brought up in several of the discussed frameworks, and they have then been combined into a single aspect. We arranged the framework as a matrix since two perspectives are combined, this topology emphasizes that it is the combination of learning patterns and different affordances that is the main purpose and advantage.

We assume that any given learning activity can be represented by more than one combination (more than one cell in the matrix).

Table 1 The learning patterns for VR framework

			LEARNING PATTERNS					
			acquisition	inquiry	discussion	practice	collaboration	production
VIRTUAL REALITY AFFORDANCES		**Levels of interaction**						
		look around (3 DoF headset)						
		move around (6 DoF headset)						
		agency to modify world (6 DoF headset, 6DoF hand controllers						
		Worldliness						
		Persistent virtual world						
		Impermanent virtual environment						
		Levels of presence						
		presence						
		self-presence						
		co-presence						
		social presence						
		Realism, flexibility and plausibility						
	avatar	plausible design (humanoid)						
		fantasy design (non-humanoid)						
		user-generated						
	world	plausible design (repreesnt actual places)						
		fantasy design (fictional places)						
		user-generated						
Learning activities What do the teacher do? What do the learner do? **If a micro world is used, then:** What are the extrinsic fedeback (from the outside of the model, e.g. a teacher)? What are the intrinsic fedeback (from the model itself)? **How is VR's unique affordances utilized?** How is spatiality utilized? How is embodiment utilized?			If this is used by its own, can it be supplemented with another learning patterns before, during, or after the acquisition?	If learning through inquiry is used, then what guidance is offered for the learner?		When using learning through practice, a repetitive cycle of goal-action-feedback-revision is beneficial.		

9 Discussing the Usage of the Framework

For what purpose can we use the learning patterns for VR? As mentioned earlier, it is our belief that a framework like this can primarily be used to guide the planning and organization of learning, either a full course, or smaller learning activities. It can point out what activities a teacher is planning to do with the students, and what other options or useful supplements there are. What we want to achieve with the proposed framework is both a set of hands-on guidelines for successful and efficient learning activities in VR, and a mapping of what we do with students in the VR environment. It can help us find unexplored and promising blind spots, but also highlight learning activities that might not be appropriate for virtual reality.

As a demonstration, and a stress test of how the framework corresponds to real-life teaching and learning, we will now show how two of our very recent case studies fit into the framework. User feedback was collected in both case studies. These user studies do not have any direct relevance to the application of the framework, but for the supervision in Virtual Reality case study, the results can be found in Eriksson [62]. For the CityAirSim project, user feedback is being analyzed and is planned to be published during 2022 and 2023.

Case Study: Supervision in Virtual Reality

During 2020 and 2021 one of the authors, Thommy Eriksson, had extensive supervision with students in the social VR application Mozilla Hubs [62]. In the 7, 5 credit university course Digital Movie Making as well as in five master's thesis projects, all at Chalmers University of Technology, I had about 50 supervision meetings of a duration of approximately half an hour. I was usually in immersive VR (using an Oculus Quest or an Oculus Rift), while the students were in desktop 2d mode. Table 2 shows these supervisions reflected in the learning patterns for VR framework. Clearly this set-up afforded a quite social activity. One drawback with the framework is obvious; it does not reflect that the meetings in VR were asymmetrical (I had a 3 DoF headset, while the students at-tended via 2d mode). The framework could be developed to reflect this, but would then become even more complex.

Case Study: CityAirSim

The second case study is a research project including development and field test of a VR application visualizing how city planning affect air quality on the street level. The application will be used as a guided inquiry physics lab with high school students. Each learner is alone in the VR experience, but collaborate with peers outside VR, and the learning activity is scaffolded by a debriefing discussion with the teachers. Table 3 shows these lab exercises reflected in the learning patterns for VR framework. One observation is that the learning activity has potential to include socially oriented learning patterns, even if the VR application itself does not offer a social VR feature.

Table 2 Supervision in Mozilla Hubs according to the learning patterns for VR framework

			LEARNING PATTERNS					
			acquisition	inquiry	discussion	practice	collaboration	production
VIRTUAL REALITY AFFORDANCES		**Levels of interaction**						
		look around (3 DoF headset)						
		move around (6 DoF headset)						
		agency to modify world (6 DoF headset, 6DoF hand controllers						
		Worldliness						
		Persistent virtual world						
		Impermanent virtual environment						
		Levels of presence						
		presence						
		self-presence						
		co-presence						
		social presence						
		Realism, flexibility and plausibility						
	avatar	plausible design (humanoid)						
		fantasy design (non-humanoid)						
		user-generated						
	world	plausible design (repreesnt actual places)						
		fantasy design (fictional places)						
		user-generated						
Learning activities What do the teacher do? What do the learner do? **If a micro world is used, then:** What are the extrinsic fedeback (from the outside of the model, e.g. a teacher)? What are the intrinsic fedeback (from the model itself)? **How is VR's unique affordances utilized?** How is spatiality utilized? How is embodiment utilized?			If this is used by its own, can it be supplemented with another learning patterns before, during, or after the acquisition?	If learning through inquiry is used, then what guidance is offered for the learner?		When using learning through practice, a repetitive cycle of goal-action-feedback-revision is beneficial.		

Table 3 Lab exercise in the VR application CityAirSim, according to the learning patterns for VR framework

			LEARNING PATTERNS					
			acquisition	inquiry	discussion	practice	collaboration	production
VIRTUAL REALITY AFFORDANCES		**Levels of interaction**						
		look around (3 DoF headset)						
		move around (6 DoF headset)						
		agency to modify world (6 DoF headset, 6DoF hand controllers						
		Worldliness						
		Persistent virtual world						
		Impermanent virtual environment						
		Levels of presence						
		presence						
		self-presence						
		co-presence						
		social presence						
		Realism, flexibility and plausibility						
	avatar	plausible design (humanoid)						
		fantasy design (non-humanoid)						
		user-generated						
	world	plausible design (repreesnt actual places)						
		fantasy design (fictional places)						
		user-generated						
Learning activities What do the teacher do? What do the learner do? **If a micro world is used, then:** What are the extrinsic fedeback (from the outside of the model, e.g. a teacher)? What are the intrinsic fedeback (from the model itself)? **How is VR's unique affordances utilized?** How is spatiality utilized? How is embodiment utilized?			If this is used by its own, can it be supplemented with another learning patterns before, during, or after the acquisition?	If learning through inquiry is used, then what guidance is offered for the learner?		When using learning through practice, a repetitive cycle of goal-action-feedback-revision is beneficial.		

10 Conclusion

Pragmatically, the conclusion is that we have suggested a framework for combining established learning patterns with established VR characteristics, based on previous research on VR, social VR, and pedagogy. We have explored whether this framework maps in an interesting way with two case study VR learning projects, and the framework seems to be useful.

There are other frameworks for VR learning, such as the frameworks suggested by [12–14]. However, these frameworks do not include established learning patterns and we argue that they are not firmly based on pedagogical research. The contribution of our framework is such a connection to pedagogical research.

An important limitation of the suggested framework is that it has not been properly applied and evaluated by others. The next step will be for us, and hopefully others, to continue testing this framework in different settings where VR is developed for learning activities. Thus, we will use it to plan future learning activities in VR, as well as analyzing our own and other teachers' teaching activities in VR. Again, we want to emphasize that this testing is not intended to verify whether the framework corresponds to observations, but to evaluate to which extent the framework is useful and helpful in the development of learning activities and courses.

References

1. Bates, T.: Teaching in a digital age: guidelines for designing teaching and learning. Tony Bates Associates, Vancouver BC (2015)
2. Säljö, R.: Digital tools and challenges to institutional traditions of learning: technologies, social memory and the performative nature of learning. J. Comput. Assist. Learn. **26**, 53–64 (2010)
3. Laurillard, D.: Teaching as a Design Science: Building Pedagogical Patterns for Learning and Technology. Taylor and Francis Ltd, London, United Kingdom (2012)
4. Sutherland, I.: A head-mounted three dimensional display. AFIPS, San Francisco, California (1968)
5. Sutherland, I.: The Ultimate Display, Information Processing Techniques Office, ARPA (1965)
6. Gibson, W.: Neuromancer, Ace (1984)
7. Heim, M.: The Metaphysics of Virtual Reality. Oxford University Press, New York (1993)
8. Benedikt, M.: Cyberspace: First Steps. MIT Press, Cambridge MA (1991)
9. Eriksson, T.: A Poetics of Virtuality. Chalmers University of Technology, PhD (2016)
10. Lanier, J.: Dawn of the New Everything: Encounters with Reality and Virtual Reality. Henry Holt and Co, San Francisco (2017)
11. Radianti, J., Majchrzak, T.A., Fromm, J., Wohlgenannt, I.: A systematic review of immersive virtual reality applications for higher education: design elements, lessons learned, and research agenda. Comput. Educ. **147**, 1–29 (2020)
12. Cochrane, T., Cook, S., Aiello, S., Christie, D., Sinfield, D., Steagall, M.: A DBR framework for designing mobile virtual reality learning environments. Australas. J. Educ. Technol. **33**(6), 54–68 (2017)
13. Kommetter, C., Ebner, M.: A Pedagogical Framework for Mixed Reality in Classrooms based on a Literature Review. EdMe-dia + Innovate Learning. Amsterdam, Netherlands (2019)
14. Christopoulos, A., Pellas, N., Laakso, M.J.: A learning analytics theoretical framework for STEM education virtual reality applications. Educ. Sci. **10**(317), 317 (2020)

15. Winter, S., Johansson, P.: Digitalis filosofi Människor, modeller och maskiner. .SE:s Internet-guide. Stiftelsen för Internetinfrastruktur. Report #13 (2009)
16. Sutton, R., Staw, B.: What theory is not. Adm. Sci. Q. **40**, 371–384 (1995)
17. Ravitch, S.M., Riggan, M.: Reason and Rigor: How Conceptual Frameworks Guide Research. CA, SAGE Publications, Thousand Oaks (2017)
18. Puentedura, R.: Transformation, technology, and education. http://hippasus.com/resources/tte/ (2006)
19. Linder, M.: En betraktelse utifrån Jonas Linderoths "Open letter to Dr. Ruben Puentedura". I'm All For It! https://marielinder.wordpress.com/2013/10/20/en-betraktelse-utifran-jonas-linderoths-open-letter-to-dr-ruben-puentedura-2/ (2013)
20. Hamilton, E.R., Rosenberg, J.M., Akcaoglu, M.: The substitution augmentation modification redefinition (SAMR) model: a critical review and suggestions for its use. TechTrends **60**, 433–441 (2016)
21. Mishra, P., Koehler, M.: Technological pedagogical content knowledge: a framework for teacher knowledge. Teach. Coll. Rec. **108**(6), 1017–1054 (2006)
22. Witmer, B., Singer, M.: Measuring presence in virtual environments: a presence questionnaire. Presence **7**(3), 225–240 (1998)
23. Lombard, M., Ditton, T.: At the heart of it all: the concept of telepresence. J. Comput. Mediat. Commun. **3**(2) (1997)
24. Steuer, J.: Defining virtual reality: dimensions determining telepresence. In: Biocca, F., Levy, M. (eds.) Communication in the Age of Virtual Reality. Lawrence Erlbaum Associates, Hillsdale (1994)
25. Zeltzer, D.: Autonomy, interaction and presence. Presence **1**, 127–132 (1992)
26. Short, J., Williams, E., Christie, B.: The Social Psychology of Telecommunication. Wiley, London (1976)
27. Kim, J.: Developing an instrument to measure social presence in distance higher education. Br. J. Edu. Technol. **42**(5), 763–777 (2011)
28. Gunawardena, C.N.: Social presence theory and implication for interaction and collaborative learning in computer conferences. Int. J. Educ. Telecommun. **1**(2/3), 147–166 (1995)
29. Perse, E.I., Burton, P., Kovner, E., Lears, M.E., Sen, R.J.: Prediction computer-mediated communication in a college class. Commun. Res. Rep. **9**(2), 161–170 (1992)
30. Garrison, D.R., Anderson, T.: E–Learning in 21st Century: A Framework for Research and Practice. Routledge Falmer, London (2003)
31. Akyol, Z., Garrison, D.R., Ozden, M.Y.: Online blended communities of inquiry: exploring the developmental perceptional differences. Int. Rev. Res. Open Dist. Learn. **10**(6), 65–83 (2009)
32. Youngblut, C.: Experience of Presence in Virtual Environments. Alexandria, VA (2003)
33. Braun, N., Debener, S., Spychala, N., Bongartz, E., Sörös, P., Müller, H.H.O.: The senses of agency and ownership: a review. Front. Psychol. **9**(535), 1–17 (2018)
34. Kilteni, K., Groten, R., Slater, M.: The sense of embodiment in virtual reality. Presence Teleop. Virt. **21**, 373–387 (2012)
35. Biocca, F.: The Cyborg's dilemma: progressive embodiment within virtual environments. J. Comput. Mediat. Commun. **3**(2) (1997)
36. Robinett, W.: Synthetic experience: a proposed taxonomy. Presence **1**(2), 229-247(1992)
37. Murray, J.: Hamlet on the Holodeck: the future of narrative in cyberspace. The Free Press (1997)
38. Dreyfus, H.: Telepistemology: Descartes's last stand. In: Goldberg, K. (eds.) The Robot in the Garden: Telerobotics and Telepistemology in the Age of the Internet. The MIT Press (2001)
39. Goldberg, K.: Introduction: the unique phenomenon of a distance. In: Goldberg, K. (eds.) The Robot in the Garden: Telerobotics and Telepistemology in the Age of the Internet. The MIT Press (2001)
40. Lovén, S.: Even better than the real thing—counterfeit realities and twentieth century dystopian fiction. Hum. IT **5**(2–3), 233–289 (2001)
41. Gunnarsson, M.: En plats vid sidan av. master, Göteborgs Universitet (2002)

42. Klastrup, L.: Towards A Poetics of Virtual Worlds—Multiuser Textuality and the Emergence of Story. Ph D. (2003)
43. Turkle, S.: Life on the Screen Identity in the Age of the Internet. Simon and Schuster, New York (1995)
44. Doyle, D.: Embodied Presence: The Imaginary in Virtual Worlds. Digital Arts and Culture. Irvine, CA (2009)
45. Nusselder, A.: Interface Fantasy—A Lacanian Cyborg Ontology. MIT Press, London (2009)
46. Lovén, S.: Also Make the Heavens, Avdelningen för litteratur-sociologi vid Litteraturvetenskapliga institutionen i Uppsala (2010)
47. Kolesnichenko, A., McVeigh-Schultz, J., Isbister, K.: Under-Standing Emerging Design Practices for Avatar Systems in the Commercial Social VR Ecology. DIS. San Diego, CA (2019)
48. Bredikhina, L., Kameoka, T., Shimbo, S., Shirai, A.: Avatar driven VR society trends in Japan. In: IEEE Conference on Virtual Reality and 3D User Interfaces Abstracts and Workshops (2020)
49. McLeod, P.L., Liu, Y.C., Axline, J.E.: When your second life comes knocking: effects of personality on changes to real life from virtual world experiences. Comput. Hum. Behav. **39**, 59–70 (2014)
50. Hooi, R., Cho, H.: Avatar-driven self-disclosure: the virtual me is the actual me. Comput. Hum. Behav. **39**, 20–28 (2014)
51. Lanier, J.: You Are not a Gadget: A Manifesto. Allen Lane, London (2010)
52. Hoguet, B.: Body, Avatar and Gender in Virtual Reality. https://cmf-fmc.ca/now-next/articles/body-avatar-and-gender-in-virtual-reality/ (2020)
53. Liudmila, B.: Designing identity in VTuber Era. ConVRgence Virtual Reality International Conference Proceedings (2020)
54. Deocadiz, Z.: How to Design Social VR Spaces: A Framework for Conceptualizing and Evaluating Social VR Applications. Virtual Reality Pop. https://virtualrealitypop.com/how-to-design-social-vr-spaces-fc06f532ef4a(2019)
55. Laffey, J., Lin, G.Y., Lin, Y.: Assessing social ability in online learning environments. J. Interact. Learn. Res. **17**(2), 163–177 (2006)
56. Sung, E., Mayer, R.E.: Five facets of social presence in online distance education. Comput. Hum. Behav. **28**(5), 1738–1747 (2012)
57. John-Steiner, V., Mahn, H.: Sociocultural approaches to learning and development: a Vygotskian framework. Educ. Psychol. **31**(3/4), 191–206 (1996)
58. Foucault, M.: Discipline and Punish: The Birth of the Prison. Tavistock, London (1975)
59. Wang, M.J., Chen, H.C.: Social presence for different tasks and perceived learning in online hospitality culture exchange. Australas. J. Educ. Technol. **29**(5), 667–684 (2013)
60. Schultze, U., Brooks, J.A.M.: An interactional view of social presence: making the virtual other "real." Info Syst. J. **29**, 707–737 (2019)
61. Cain, S.: Quiet: The Power of Introverts in a World That Can't Stop Talking. Crown Publishing Group, New York (2012)
62. Eriksson, T.: Failure and Success in Using Mozilla Hubs for Online Teaching in a Movie Production Course. Immersive Learning Research Network. Online (2021)

Security, Ethics and Privacy Issues in the Remote Extended Reality for Education

Muhammad Zahid Iqbal, Xuanhui Xu, Vivek Nallur, Mark Scanlon, and Abraham G. Campbell

Abstract The adoption of Extended Reality (XR), an umbrella term for augmented, virtual, and mixed reality, has grown over the last few years. But this adoption has been accelerated with the impact of the pandemic, which has demonstrated the value of this technology as future of learning technology. As a result, XR is becoming a popular solution to facilitate remote learning, remote conferences, and remote working. To mitigate the problems of remote learning, a trend is emerging among authorities to emphasize the potential for new technologies such as artificial intelligence and virtual, augmented, or mixed reality to create engagement, providing a kinesthetic aspect of learning and addressing students' attention problems. XR with wearable devices can make the learning process more productive and even more interesting. Despite the fact that remote learning with XR offers several interesting educational advantages as compared to in-person classroom environments, it has its own downsides that have not been addressed in previous research and considerable research gaps remain in this area. When these devices are used in public places, they can infringe on other people's rights as well. As for security concerns, the more we live our lives online and virtually, the more vulnerable we can become to hackers. These privacy and security risks include input and output data, user interaction data, and identification of both the user and devices. This chapter addresses these ethics, security, and privacy-related issues in line with XR in education. In the broader view, this chapter will focus on a better understanding of the human value of XR for learning purposes in the remote setting with responsible design.

Keywords eXtended reality · Immersive learning · Privacy issues in immersive learning · Ethical issues · Mixed reality · Augmented reality · Virtual reality

M. Z. Iqbal (✉) · X. Xu · V. Nallur · M. Scanlon · A. G. Campbell
School of Computer Science, University College Dublin, Dublin, Ireland
e-mail: Muhammad-zahid.iqbal@ucdconnect.ie

Y. Cai et al. (eds.), *Mixed Reality for Education*, Gaming Media and Social Effects,
https://doi.org/10.1007/978-981-99-4958-8_16

1 Introduction

Extended Reality (XR) technology which includes virtual, augmented and mixed reality, is maturing very rapidly and becoming more accessible to a broader audience.

- ***Augmented Reality (AR)***: Interactive experience by superimposing virtual objects on the real world
- ***Mixed Reality (MR)***: Merging of real and virtual environments, where physical and digital objects co-exist allowing user real-time interaction
- ***Virtual Reality (VR)***: Completely virtual environment providing simulated experience

This promising technology has the capability to entertain and educate, to connect and enhance our real environments with immersiveness. Figure 1 has shown the difference between real and immersive environments, which tends from superimposing virtual objects on a real environment to a completely virtual setting.

The revolutionary progress of internet connectivity like 5G and its accessibility to the world has added more attention towards adopting these technologies for creating completely immersive experiences [59]. According to statistics, mixed reality market will increase to USD 3.7 billion by 2025 [45].

The adoption of new technologies changes human values in society. It is always a question about understanding "technological value dynamism" and dealing in a responsible way [39]. XR technology provides new innovative possibilities for learners for an engaging experience [50], immersive environment and interactive contents [2, 55]. These immersive experiences help the learner to learn deeply in an interactive way using the immersive and engaging power of Mixed Reality (Fig. 2).

Research has shown that XR technologies increase overall success and improve learning outcomes when learning with 3D interactive technologies [24, 47].

Rapidly evolving progress in the XR enabled devices are more accessible, creating opportunities for learning in remote settings [37, 79]. Considering the multi-user approach, Collaborative Mixed-Reality (CMR) applications are gaining popularity in gaming, social interaction, design, and health-care [43]. A vast majority of published research work focuses on display technology advancements, benefits in visualization, software, collaboration architectures, and applications. However, the potential privacy, ethical and security concerns that affect collaborative and remote platforms have not received enough research attention.

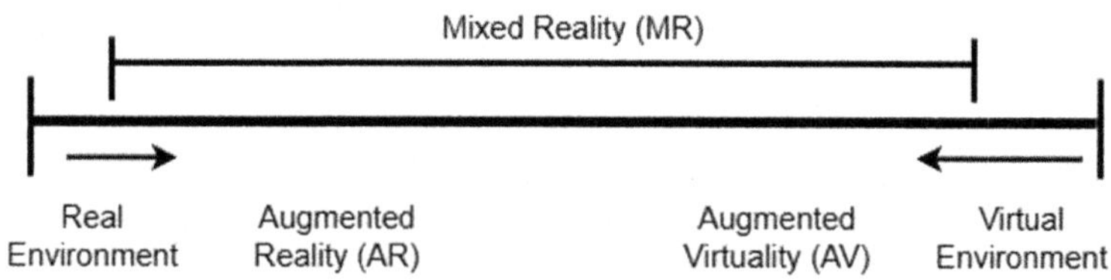

Fig. 1 Virtuality continuum: from real environment to virtual environment [51]

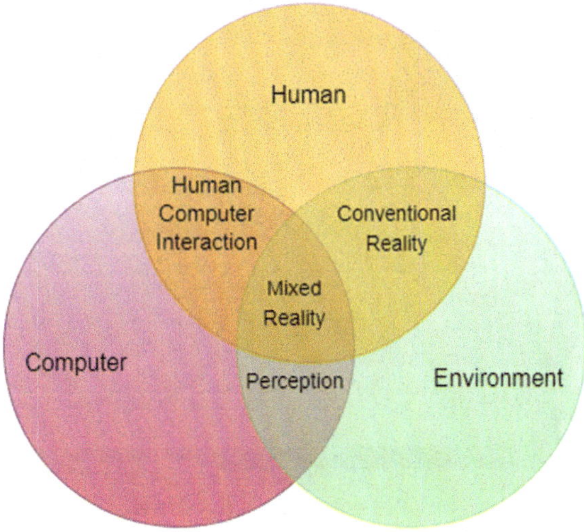

Fig. 2 Interactions among computers, humans, and environments

This chapter will outline the background of XR in education and existing general security, privacy, and ethical issues in Sect. 2. Section 3 will discuss the issues using remote XR in terms of teaching and learning while Sect. 4 will discuss assessment issues with remote XR. Section 5 will discuss the potential future of XR remote education, and finally, the conclusions will summarise the chapter.

2 Background

2.1 Existing Case Studies of XR in Education

The adoption of XR (including augmented, virtual and mixed reality) in education has increased during recent years due to its capability to immerse the users into an artificial world. XR technology has been pitched as a next-generation computing platform. These technologies have the potential to revolutionize the learning process by making learners' experiences more "engaging". The potential benefits of XR as educational technology include increased understanding of the spatial structure, functionality, long-term memory retention [41], improved kinesthetic learning, and increasing motivation [3].

The journey to adopt XR in the classroom though still in the developing and designing phase of the curriculum, there is a need for consideration of social constructivism. Social constructivism drives a connection between societal factors and the adoption of new technology such as XR [72].

XR within remote learning requires these devices to be used in personal settings such as at home or outside if location-based. For these devices to operate within such environments, new definitions of social spaces and legal issues around privacy must

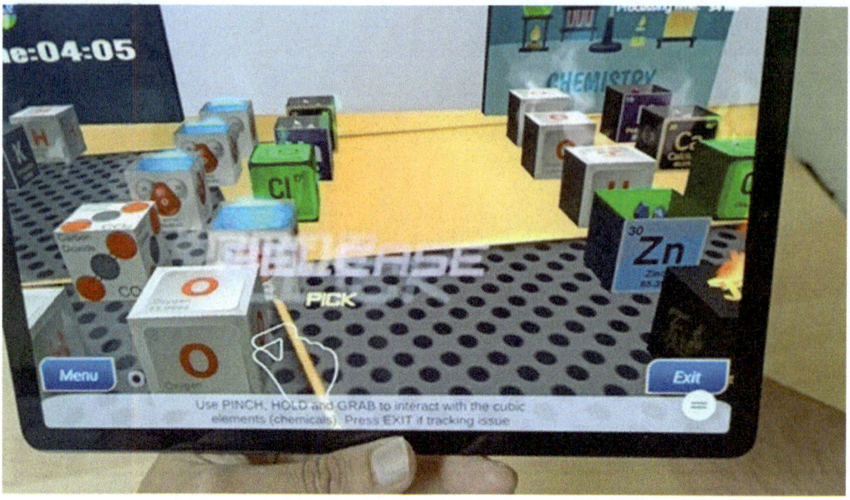

Fig. 3 Use of real-time hand interaction in the augmented reality for learning chemistry

be considered [76]. These devices allow better-visualized learning and for concepts such as remote laboratories. However, these devices have nascent risks to our privacy and security due to the personal data collection, data sharing, and capturing the surroundings data [14].

From basic alphabet learning to anatomy learning in medical science [75] and skill-based technical learning in STEM subjects, XR is supporting exploration and simulation activities using different learning pedagogies [18, 29]. Studies have reported exploration lessons with better visualization, kinesthetic learning and simulators with positive perceptions and outcomes [28]. The ability of XR technology to support and promote collaboration and shared experiences is powerful when users are not physically collocated. The use of tracking devices like Kinect, leapmotion, Azure Kinect and APIs like Manomotion (hand interaction in smartphones as shown in Fig. 3) and Google Mediapipe [80] are adding value to the state-of-the-art hands-on learning experiences. The recent development of Oculus hand tracking and Microsoft Hololens with hand interaction ability opens up new options for experiences of immersive kinesthetic approach. Use of Head Mounted Display(HMD) [23] and PHANToM with tangible models and haptic feedback provides a natural way of understanding and learning closer to realism [65].

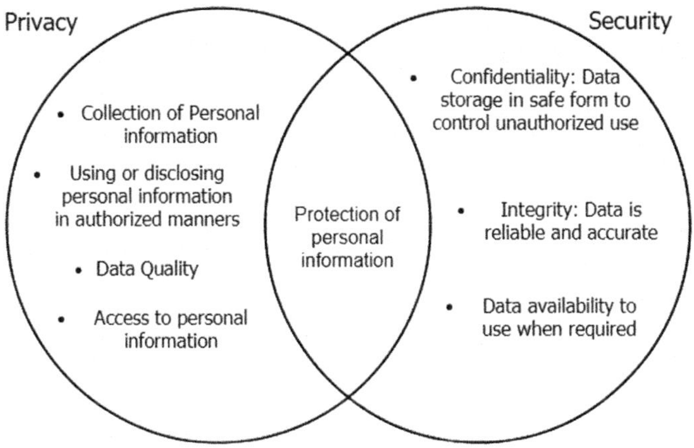

Fig. 4 Differentiating privacy and security

2.2 *Existing XR Privacy Issues*

The concerns about privacy are not exclusive to extended reality but have long been associated with management information systems, human-computer interaction, robotics and drones technology.

As XR allows truly powerful utilities and games in the world mapped in 3D space, for a user this could mean allowing a lot of personal information sharing over the network. These concerns are slightly different when comparing virtual, augmented and mixed reality (Fig. 4).

XR devices allow interaction with the real world using cameras, microphones, and sensors, which means that information about our environment can also be collected as explained in Fig. 6. Adams et al. interviewed 20 VR users and found that the majority of users were concerned about their devices' microphone and IR sensor data and information that can be gleaned about their physical and virtual environments [1]. A potential case of privacy-preserving with cloth try-on use of augmented reality can be used as a privacy steal tool because it can collect enough data with a full-body scan [71].

There are models of XR headsets in development that would not only allow you to record your everyday life by taking pictures and videos using the headsets, but also to live stream what you see, and even to perform live analytics such as facial recognition on the footage you capture. And if the headsets integrate with third-party apps, this could even add new ways to violate the rights of bystanders. XR browsers facilitate the augmentation process in the virtual space but the contents are delivered by third-parties which raises questions about the unreliability. There are various cyberthreats like data manipulation and sniffing which can create unreliability in the contents even coming from authentic sources.

2.3 *Existing XR Security Issues*

As there is already an exorbitant number of privacy and personal data theft issues with the use of social networks [68], yet these issues would be minor compared with the level of personal data generated within a mixed reality device that could be exploited. As always, these risks are heavily dependent on the use cases [27].

It might be surprising that in the XR landscape, the biggest security concerns are not about virtual intrusion, they are about physical compromise. XR technology requires access to multiple sensors, data and camera (data flow explained in the Fig. 5). In fact, it has been proven that even the digital video cables themselves, such as HDMI, are vulnerable to remote video stream eavesdropping [69]. This poses a huge security risk when AR and VR devices/applications are hacked or breached. When AR and VR devices and applications use authentication from social media accounts, it adds another layer of vulnerability for the data. Moreover, VR social networks themselves record a multitude of data about the user and their virtual interactions [77]. There are potential risks and threats to the data exchange with the use of mixed reality for educational purposes. Security risks and vulnerabilities in devices, mobile apps, and web browsers used by XR will give attackers the opportunity to compromise information, steal highly valuable and sensitive intellectual property, send false information to XR headsets, and even prevent access to the XR systems themselves.

The privacy of a user can be easily compromised if hackers gain access to the device and it can not be understated how much more critical a breach would be than a similar attack on other computing devices. When a constant stream of data will flow between the digital and physical worlds, with attacks on the digital world directly it creates dire consequences for privacy and personal safety. Casey et al. [11] explored the possibility of compromising a VR session and performing several attacks including chaperone, disorientation, human joystick, overlay, and camera stream and tracking. While some VR attacks may initially seem minor in their implications,

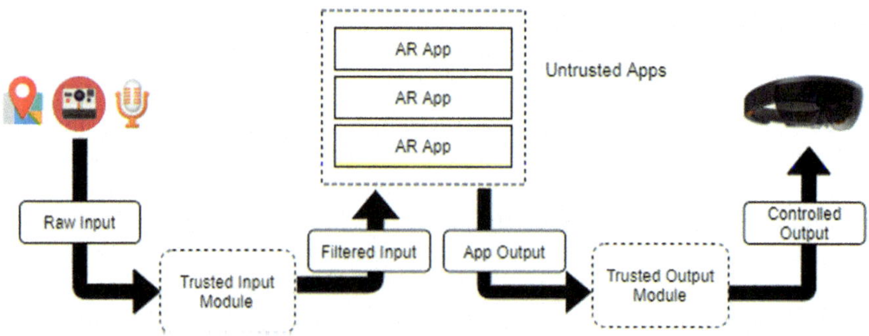

Fig. 5 Flow of information through an XR platforms

XR systems are being employed in professional settings, e.g., industrial control systems [46].

Research practices in human-computer interaction (HCI) have changed due to innovative modern interactive technologies. The new methodologies are actually shifting the whole diversity of contexts in which novel digital technology is being used. In terms of research ethics and responsible research and innovation, new ethical challenges emerge for the HCI community.

Privacy, security, and ethical concerns are emerging as these technologies are becoming more popular and accessible.

2.4 Existing XR Ethics Issues

In a scientific context (such as universities and academic labs), the use of XR technology is governed by ethics guidelines and laws. These may vary across countries but tend to abide by some general principles.

The designers and developers of XR technology often consider ethics-in-practice as potential including the cognitive, physiological and behavioural impact. The ethics-in-practice has a role in the field of XR design and instructional curriculum development.

XR technologies are considered as future computing platforms; these headsets are the potential successors of ubiquitous technologies such as smartphones and desktops but with very high capability to get into our lives. These systems, with higher ergonomics and better design, can offer personal, private, increased availability of sensing and augmentation of reality, allowing a near-constant flow of digital content delivery [20].

XR technologies are rapidly being adopted by consumers, and increasingly being used for work and for education. It is essential that guidelines are set to ensure privacy and safety while business models are being established. As AR smart glasses automatically screen and process a user's environment, the privacy of users, and those around them can also be affected [33]. This is very unique to AR, since most existing technologies only collect information about the user. In decision making, people often consider how other people perceive their behavior, so called'social norms'. The term 'glassholes' [17] became a prototypical term for insulting Google Glass users.

For developers and designers of XR, there are many concerns regarding accountability for ethical design of XR devices and applications in virtual remote learning environments [9]. Ethical design consideration and privacy issues are things to consider for design research when integrating interactive media into learning content and placing the user in a virtual environment.

3 Issues Using Remote XR in Terms of Teaching and Learning

Though XR technology is a strong tool to bridge geographic distances between the learners and instructors with realism, it has several complex issues.

3.1 Inconsistent Quality of Learning—Equipment/Resources

To facilitate learning in the resource constrained environment, XR technology is providing virtual learning material and enables innovative interactive techniques like gestures, hand interaction, and haptic feedback. The quality of learning with such approaches fully depends on the devices used, application and strength of the network connection. In a remote XR education unlike an XR lesson in a classroom, most student will be using their own devices, with the huge differences in the processing power of devices this could create inconsistency in the frame processing which leads to a different learning experience between users.

3.2 Privacy Issues

Fundamentally this involves three forms of privacy concerns; keeping a user's behavior private when recording their actions in the real world, keeping private users' actions and data in the virtual world, and finally other people's privacy who are around when using their device. Many people when interacting with digital devices are aware about the privacy risks but paradoxically do not behave in a manner that acknowledges this fact [4]. In the context of AR, the findings are similar.

3.2.1 Privacy of Environment

The use of XR remote for education creates an additional burden on maintaining privacy within their environment. In an immersive Virtual reality use case, a user may be unaware of users entering the room in which they are conducting a lesson through virtual reality. Due to the use of headphones and the natural privacy of the VR HMD, the external party would only be aware of the user's physical actions and speech with others in the virtual world. There have been innovations in improving existing boundary systems like the Oculus Guardian system, such as from O'Hagan and Williamson [58], which allowed a user to be made aware of a bystander through multiple techniques such as an avatar represented, text notification, silhouette, and even a sonar radar approach.

Without such systems in place, users have been documented to reveal personal information. One experiment conducted by Campbell et al. [10] found that during the testing of a Tele-presence VR system, users would openly discuss personal issues, as they were so immersive in their communications that they forgot that in reality, they were both standing in a bank lobby with dozens of people around them. The experimenter needed to briefly stop the experiment to remind the participants that they were not alone.

3.2.2 Third-Party Privacy

Third party privacy means the privacy of the people nearby who are being captured in your virtual environment setting as explained in Fig. 6. Remote XR education by its very definition is remote from traditional classroom environments and thus third-party privacy needs to be considered in any remote XR education application.

3.2.3 Forcing Students to Sign EULAs

One area of deep concern is the requirement for users to login to a service to use a HMD unnecessarily. It's understandable that a user may need to login to a service such as an online world but it seems unnecessary when the user is simply using an education service. This always connected and login approach has become common in many devices in the mobile market and its introduction into XR space is logical but comes with even more challenges. This login inevitably comes with the user

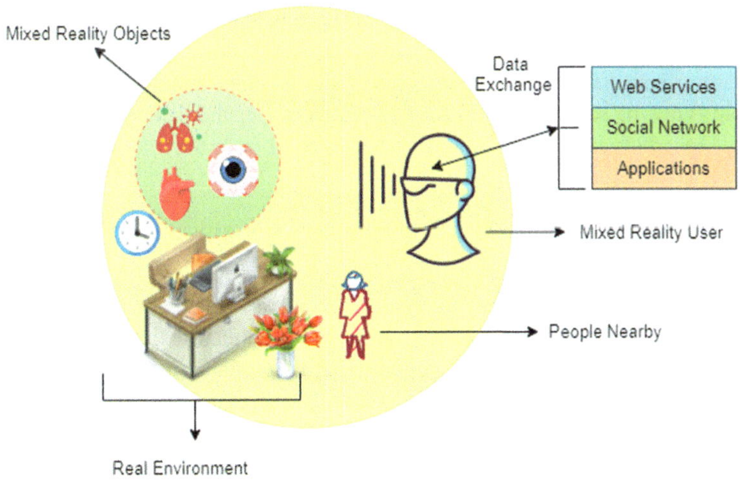

Fig. 6 Extended reality environment: explaining XR user, people nearby, real environment and exchange of data

needing to agree to an End-user license agreement (EULA). The major example of this is the Facebook EULA that needs to be signed by a user if they use the consumer grade version of their Quest series. This EULA is not required if the user purchases the more expensive business edition. Some have accused these requirements and the EULA [16] of creating a "Wild West" where users can have "missing rights". For instance, users' basic property rights are in danger if their access to applications bought on the Quest is not accessible, if they violate Facebook's EULA while posting something on their social media account. This lack of access to a device based on social media activity is completely unacceptable from an educational perspective. It can result in the suppression of freedom of speech of a user, knowing that the provider can, at any given time, could remove access to their device. This constraint on education could infringe on a student's basic fundamental rights. In classroom XR allows the school to potentially bypass this requirement if the institution sets up the devices themselves, but in remote XR education where users are using their own personal devices in general, this issue is a major issue going forward.

3.3 Security Issues

As an XR user, we generate data about our behaviour and movements in virtual environments and often share it over the network, which could be used to steal our virtual assets or virtual identities [21].

3.3.1 Data Collected by Sensors

An immersive experience is based on high level sensor accuracy and precise data processing. For high performance, XR headsets require extremely sensitive and responsive motion sensors. These devices collect lots of data about the user with these multiple types of sensors. With the new advances in the sensors used in the XR applications, there is a large amount of data needed from the real world.

3.3.2 Location Data

There is a significant amount of augmented and mixed reality applications in education that are designed for localized learning. These applications are using the location of the users with GPS sensors and collecting location data [60]. Adams et al. found that the majority of VR users they interviewed were concerned about the data collected from cameras and sensors commonly found on XR devices, where it is being send and what it could be used for [1].

3.3.3 Biometric Data

Biometric data can be generated in multiple ways through the interaction with an XR device. Eye tracking has a long history to use it for assessing user behavior. This technology is becoming efficient, cheap, and compact with increasing use in gaming, marketing, education, military, and healthcare. This rapidly expanding technology coming up with serious privacy issues. Research has evidences that for possibilities of infer personality traits automatically from eye tracking data [26]. For educational purposes in XR, eye tracking technology has been used for emotional learning, accessing reading patterns and student behaviours [63]. Devices which have capability of eye tracking, can potentially capture more information than the user expects to reveal. Some of this personal data protection is prescribed in European General Data Protection Regulation (Article 9 GDPR) [38]. According to Bloomberg's Privacy and Security Law Report, eye tracking is not just a tool for interaction, it is possible to approximate the identity by utilizing the attributes from gaze patterns [42].

XR devices provide hand tracking abilities which are collecting gestures data as shown in a case study in the Fig. 3, real-time hand interaction in augmented reality. Hand tracking technology has gained more attention recently as touchless interaction technology in the post-COVID world [31] provided with leap motion or other devices for hands-on (kinesthetic) learning in education [32, 34].

Speech data includes very sensitive and personal characteristics which can be used for a plethora of diverse range of applications, from medical profiling to digital biometric recognition [54]. Speech command data can be a security threat for users as speech commands are also used as identification data in numerous applications. Voice recognition technology marketing is surging as touchless technology is emerging in the world with the "avoid touching" concept. The privacy and security risks associated with voice recognition technology cannot be overlooked.

3.3.4 Software/Hardware Vulnerabilities

The security and privacy issues associated with XR are strongly linked with the system architecture and flow of the data. Some applications are designed to render, process and store data in the device while others are designed with cloud resources integrated. On the other hand, the provision of software applications by the platform and third party creates a difference in cyberthreats in the virtual world. One of the solutions to prevent inconsistency in the quality of learning is to make sure that all students use the same device or a small subset of approved devices. This can aid in the development of software as testing becomes far easier on a narrow range of devices or simply one device. Security-wise though this can result in a situation where a homogeneous environment, where a single exploit could allow a hacker access to the system. This situation also results in a lack of choice for the student on what devices they can use to access their education.

For remote XR education to function, students are sent software to access their classes and to aid in their learning. In main cases due to rapid prototypes in this space, there can be a lack of content signing leading to students bringing used to ignoring warnings of unsigned content, even going as far as asking students to allow content onto their devices, by turning off existing security settings. This allows for multiple ways for hackers to embed malicious content into XR applications to trap the users from collecting data. There needs to be more tailored content signing mechanisms in place, where once-off permissions are given without having to set the entire device into developer mode. Also, XR HMD manufacturers create walled gardens that force developers into requiring end-users to use side-loading applications so it's not simply a case where developers ignore the best practices.

3.4 Ethical Issues with XR in Teaching and Learning

There are many important ethical questions linked with the use of XR in remote learning. These ethics-related issues will increase with the development of learning systems with more collaboration capabilities in remote learning for the users as the data involved in a collaboration can also create new more sensitive problems. There are issues of consent when children use XR for distance learning and issues of *depersonalization* and *derealization* when they operate such devices in an unsupervised environment. Designing applications from an ethical perspective where they give maximum benefit to the learners while attempting to minimize any potential harms [73]. Acknowledgment of the potential ethical issues of using XR within remote learning applications is critical.

3.4.1 Bullying

Bullying has been identified as a group process, where children subconsciously take on various roles for social acceptance [64]. It has also been shown that peer pressure can also affect behaviour in XR situations [57]. This indicates the potential for "behaviour without consequences" to be perpetuated in the virtual world, and even transferred to the physical world. That is, there are not only likely to be situations where bullying takes place in remote XR environments, but also that since there is no physical harm done, little (or less) is done to control it. The remote XR education will require new policies for schools, similar to the development of cyberbullying [8] or it could become the perfect tool for a bully. This has two implications, the first on the mental health of the bullied, and second on all the students co-present in that situation who learn that bullying is without consequences.

3.4.2 Erosion of Physical World Feedback

Linked to the lack of behavioural restraints that operate in the real world, is the lack of physical consequences that also serve an important role in the development of

Fig. 7 Case of explosion in chemistry in augmented reality

children. As every parent knows, children from a very young age learn a lot about their environment through spatial exploration and construct highly useful models of causation, and qualitative physics, that enable them to navigate the world. While XR offers the opportunity to scale up certain kinds of learning (e.g., larger class sizes can experiment with chemistry kits without having to fear breakage), it also disconnects the students from the real-world consequences of their actions. For example, a mistake or carelessness in handling chemicals in the virtual world may lead to a predictable virtual explosion or burns (a case presented in Fig. 7), but given the absence of a physical counterpart, it may not result in the correct lessons being learned by students.

The decoupling of sensory experience may lead to incomplete (and even dangerously wrong) physical and spatial models of the world.

3.4.3 Lack of Common Environments and Social Isolation

Sociologists tell us that social interaction due to a situated environment acts as a grounding force in our conception of the world, and interaction norms. XR creates a situation where people can start to prefer a virtual world, and not learn how to interact with unpleasant people/events in the physical world. An extreme case would be an inability of children to interact with peers and other social groups [62]. **Note**: We do not claim that extreme versions of this condition (such as *Hikikomori* [74]) will inevitably result, however even mild versions in large classroom cohorts could have a significant impact on society. Furthermore in remote XR education environments, the common start point and end point of the classroom is removed which could further erode the connection between their peers.

3.4.4 Re-entry and Disembodiment

Immersive experiences, whether positive or negative, inside an artificial world, can cause problems with adjusting to the real, physical world. The same cues, environments, and effects are not present, and behaviour that is valid or learned within the augmented world, may not be appropriate in the physical world. These are labeled as re-entry problems. Not only has it been shown that children are particularly susceptible to not distinguishing the virtual from the real, but also that the effective experience lasts longer than the actual immersion experience. That is, any emotions that are generated during the immersive phase are carried over into the re-entry process [5]. This is not necessarily a bad thing; students that are excited about experimentation will carry over their curiosity into the real world. However, students that experience any intense emotions (e.g., during depictions of extreme events in geography, or frightening events in history/ literature) may carry over these into the real world. Therefore, the teacher must be aware of the potential emotional impact of their educational immersive experience, this could prove to be more difficult in a remote XR environment and thus this issue may be more pronounced.

3.5 Ethical Issues with Remote XR in Public

The impact of Corporate Social Responsibility (CSR) on society has gained attention in the last two decades with the emergence of digital technology [36].

Just similar to other innovative technologies, XR comes with significant challenges which are related to personal data. The collected data can be very personal and sensitive as well regarding identification, also involving our behaviors. Just like eye-tracking technology which brings many advantages for graphics and responsiveness quality but it also enables the collection of highly personal data. Data of patterns during eye movements focusing on at any given time provides insight into preferences and thoughts of users.

4 Issues Using Remote XR for Assessment

There are several challenges associated with using XR for remote assessment that can question the integrity of results. XR provides a perfect tool to allow for anonymous marking that has been shown to potentially remove bias in grading [6]. XR use could allow for realistic training scenarios instead of common role plays that can sometimes lack immersion for the student knowing they are interacting in front of an assessor role-playing.

There exists the possibility to reduce the need for assessment materials allowing for the possibility of remote assessment where none existed before. However, there are numerous issues that could prove difficult to overcome and in some cases lead to ethical dilemmas.

These include headset discomfort [53], lack of knowledge about navigating in the virtual world, and content alignment gaps. The student may wish to opt-out of using the devices for the assessment and ethics surrounding that decision need to be explored. The addition of any device into an assessment adds additional degrees of freedom for something to go wrong. The student may attempt to cheat the system or could accidentally break the system for their benefit or detriment.

4.1 Students Opting Out

There are multiple reasons why a student may wish to opt-out of using XR as an assessment tool. The first issue is for medical reasons, should they be allowed to if the tool has been shown to be more effective than other assessment tools. Medical issues such as dizziness are still commonplace [53]. As a result, until such time as XR matures, this is purely academic for now. Alternative forms of assessment should be provided even when the XR is considered to be the best practice. This is commonplace right now for how current disabilities are treated in education right now, if someone is unable to take a written exam then they can take an oral exam. XR could be that form of alternative assessment for a student in which case they would be opting in. XR does have the potential though to be seen as a lightning rod issue, as it's a novel assessment tool, which may cause fear, uncertainty, and doubt (FUD) for new users if its adoption is forced.

4.2 Cheating, Accidental Use and Detection of User

Cheating in an exam using technology is a very well-known and documented issue [13]. Every educational institution will have policies against students cheating. Any new assessment tool has the potential to be exploited but also it must be understood that sometimes with new technology the tool could accidentally break for a student benefit or detriment. The use of XR for remote assessments suffers from the same issues as any current remote assignment such as an open book exam where a student may enlist others to help them. In an XR context, this is further exacerbated as code could be injected to allow the co-conspirator to be virtually present within the virtual world as an avatar only visible to the student.

Other exploits could be as simple as the student, partially removing the XR devices to look at notes, there even exist patents to detect a user's ear to avoid such an exploit [66].

4.3 Re-grading Queries and Ability to Record

For many institutions around the world, the right for a student to appeal their grade is fundamental [22]. This can prove difficult for assessments such as presentation, in which case they tend to be marked by multiple examiners as best practice and the marking scheme and examiners notes can be reexamined if requested by the student. For XR assessments this basic right would need to be maintained, however, how would this be achieved. Logically given the digital nature of an XR environment, a digital record could be generated for actions the student took in the virtual world, and potentially a video recording of their actions with the physical world too. This approach could be onerous in practice and similar policies for other performance-based exams could be used. A compromise approach could be taken to allow a recording of just the interaction within the world, while not recording all of the student's movements. Any recording be it full or partial of course leads us to data protection issues with the use of XR in assessment.

4.4 Data Protection

The final issue with the assessment being conducted in XR is that of Data Protection. The user has a right to privacy in a digital world under laws such as GDPR within Europe. Any unnecessary data should not be recorded, and the data recorded should be with the consent of the user. Any assessment records from an XR tool should be only kept until graded, and any potential appeal or review period is over. In this scenario, it appears that XR use does not differ from traditional assessment data protection, but XR data reveals potentially far more personal data than, for example, a simple exam script.

As even if you were anonymously accessing a virtual world, someone's gait and movements could possibly identify them [44], thus the ability to conduct an anonymous assessment in XR could be problematic if steps are not taken to remove such data from the final record. Just to emphasize how problematic anonymity could be, even just throwing a ball in a virtual environment can be used to identify an individual [52]. XR devices could, through body movements alone, be used to diagnose someone with an illness [12]. Ethically, we may need to ensure that these tools do not perform that function.

5 Potential Future of XR Remote Education

Immersive XR will get more adoption in the near future in higher education due to the growing accessibility of 5G technology and edge computing. As XR use in remote education and education, in general, is only in its infancy, this is the time that

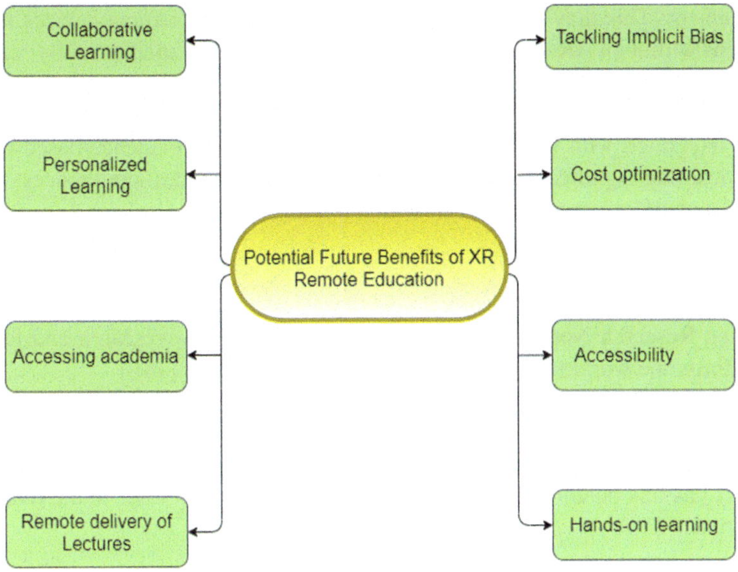

Fig. 8 Potential future benefits of XR technology in remote learning

these issues need to be addressed so we can help protect users so they can enjoy the rich potential of using this learning technology in the remote environment.

5.1 *Potential Future Benefits of XR Remote Education*

XR due to its compelling features for learning, getting attention from the researchers and policymakers across the world in different disciplines from formal to informal learning explained in Fig. 8. New technologies come up with their own complexities but over time through multiple versions, evaluation of the technology acceptance approach brings a change in usability. XR in education can help with two approaches; technology-driven or learner-driven. In the technology-driven approach, the capabilities of the technology are used for the development of learning material while the learner-driven approach is focusing on pedagogical frameworks to improve them with technology integration.

Collaborative Learning

The broader use of XR devices and growing adoption can lead towards the creation of new potential for mass collaboration in content authoring [15]. It can progress towards more sustainable open-source communities for collaborations to develop virtual world platform development [81]. This collaboration naturally feeds in the potential of XR remote education, where both teacher and peer can exist in a shared environment. This may take the form of a VR space but example devices like the Hololens within an AR space are more suitable for longer experiences.

Personalized Learning

When integrated with intelligent agents, XR can enhance evidence-based personalized learning significantly which reflects individual choices, preferences, performance, and effectiveness of the instructional design.

Accessing Academia

The rise of online conferences in VR, and the success of conferences like Immersive Learning Research Network (iLRN Virtual Campus) allow research and teachers from across the world to meet and discuss education online [48].

Remote Delivery of Lectures

Converting remote learning into a virtual classroom concept as in Fig. 9, immersive VR has this potential which can move virtual learning closer to the real world environment [78].

Tackling Implicit Bias

One of the key tasks of educationists is to help nurture a student cohort that is both aware of implicit biases as well as taking steps to avoid it. With a virtual world where students can embody avatars of different races, ethnicities, and gender to help combat these implicit biases.

Cost Optimization

With the provision of virtual learning material, XR technologies are drastically lowering the cost of learning as compared to physical laboratories and costly scientific equipment. It also saves the maintenance and operational costs linked with physical laboratories which can make the use of XR much more economical in the long run.

Fig. 9 Immersive virtual reality classroom for remote learning [19]

Accessibility

XR technology as being a nascent technology can offer exciting equity and inclusion opportunities with a more diverse user experience. These opportunities include potential as an empathy tool, adapting to meet the accessibility needs of people with disabilities, and trying to minimize the barriers arising due to physical distance to enhance person-to-person interaction.

Hands-on Learning in Resource-Constrained Environment

In resource-constrained environments, immersive technology provides virtual learning material to learn skill-based learning or technical topics with the interaction ability of hand tracking [23].

5.2 Potential Future Downsides of XR Remote Education

XR is consistently evolving with productive accomplishments, though it has its own downsides explained in Fig. 10.

Data Protection

Through the use of XR devices, there is a potential as outlined previously to reveal personal information. Steps need to be taken to protect the personal data of any student using XR for remote education, as they are now using XR in own their

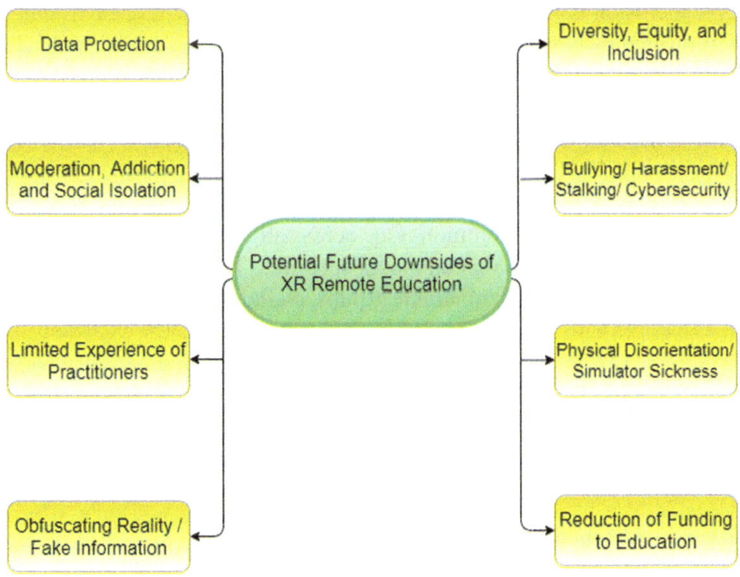

Fig. 10 Future downsides of XR remote education

personal space or potentially a public space with third parties involved and not in safety of a classroom.

Collaborative Spaces: Moderation, Addiction and Social Isolation

The use of new virtual Collaborative spaces to allow for remote education, creates a difficult balancing act. These new spaces need to be engaging and truly collaborative to avoid an isolating effect. This may require a moderator, and some design considerations so that they avoid the mistakes of social media platforms that have been identified as causing/exacerbating potential mental health problems among adolescents [40].

Limited Experience of Practitioners

With any new educational technology, there is a lack of competency of the instructors to generate the new content. This can cause a divide between the learners [30] and limited instructional design [35] for XR learning can hurt progress. This is only compounded by remote XR scenarios where the practitioners are not present with the students, leading to further training requirements on top of learning how to generate content for this new medium.

Diversity, Equity, and Inclusion

Accessibility of highly immersive innovative technology to a small population is already creating a gap of knowledge and especially the quality of learning within the same age group. This gap could grow with the introduction of XR in remote education. XR education within a classroom allows the potential to standardize devices thus promoting equity. Inclusion in the physical world of educators and peers can help solve inclusion issues if they emerge within the XR teaching environment. The inclusion problems that could exist in XR remote education could hinder learning indirectly without the educators being aware of it. In a remote XR education scenario issues such as gender diversity may not be as obvious and across the world gender can be a major factor in students' ability to access technology.

Remote Physical Bullying/Harassment/Stalking/Cybersecurity

Harassment and bullying is common issue across all communications technologies, from private chat platforms to social media networks. These risks have higher possibilities for users with already vulnerable groups, including women, racial or religious minorities, people with disabilities, and LGBTQ community. The extremely adaptive nature of immersive technologies, including collaborative approaches, can grab the attention of developers and organizations in practice to put effective safety measures in place with a comprehensive understanding of the potential misuse. This misuse follows similar issues of Cybersecurity in the present use of remote learning using the internet.

Physical Disorientation/Simulator Sickness

Current AR and VR HMD still can cause physical disorientation and simulator sickness, this effect is not going away any time soon. Numerous tests have been created to document this issue [25]. However in many cases "content significantly influences" the effect of simulator sickness, but even with the best intentions, XR devices will always suffer these issues in some form or another [67]. The best example of this is car sickness, in cases where someone is severely affected, other forms of transport are used [70]. Alternative educational arrangements would need to be accommodated, but the very nature of using remote XR suggests this will disadvantage students who for whatever reasons, and no matter how perfectly the virtual worlds sync up with reality, may always suffer from physical disorientation of some kind. Remote XR education compounds this issue when compared with in-class XR education as if a student experiences this, they are not in a physical classroom environment and could be aided by an instructor.

Obfuscating Reality/Fake Information

XR benefits are clear in remote education, however, with the remote nature of these tools, a bad actor could influence a learners opinion though obfuscating reality around them. As realism in virtual and augmented reality continues to improves, the veracity of these worlds can be controlled, and false information could abound [73]. Without physical classmates in a remote XR scenario this issue takes on a more sinister risk than in XR education scenarios that is conducted in a classroom as potential for nefarious actions by the developer without any third parties present to mitigate any harm.

Reduction of Overall Funding to Education at all Level/Race to Bottom

Presently, the future appears bright for digital education technology; in general, funding is increasing across all technologies [7]. However, as with remote work, there will be a temptation for funding bodies, be they governments or private enterprises, to reduce costs using remote education, XR offers impressive potential, but its adoption could reduce overall funding to education at all levels and theoretical race to the bottom could manifest itself. Where XR maintains current education standards, its adoption is used to cut in-person instruction. Outcomes in terms of grades may appear similar, but as we have seen during the COVID crisis, where grades and standards have been broadly maintained [61], remote education may not create the intellectual culture [56] of peers that we seek to create.

6 Conclusion

Due to advances in computer vision, sensor fusion, and realistic displays, XR technology is gaining momentum in every field of life. The potential for XR in the remote learning context is very substantial and will improve over time. Using HMDs with high-fidelity graphics and immersive content can allow students to explore and learn complex technical topics in a way impossible with traditional teaching. Planning in a rapidly evolving technology environment is a constant challenge. Making XR technology more accessible and safe is the right way to accommodate this pace of change. It is the right time for organizations, policymakers, and curriculum designers to consider XR for remote education.

The privacy, ethics, and security concerns attached to XR technology are real and have increased over time. Therefore, there is a need to stress notions like ethics, responsibility, safety, and trust in XR technologies, mainly when they are used in a remote environment. Continued discussion in this area has stressed the need for a code of conduct [49] for using XR technologies that is similar to the codes of conduct emerging in artificial intelligence, such as the IEEE Global Initiative on Ethics of Autonomous and Intelligent Systems in 2019. The IEEE Global Initiative on Ethics of Extended Reality is the first worldwide effort to achieve this goal. This process is ongoing, and it is crucially important that the use of XR in remote education is highlighted within this process as its needs differ from in-person XR education.

There is a need to ensure that any XR technology designed for educational purposes should be as affordable as possible. In addition, users must be informed about the type of data the system is collecting and allow them to opt-out. For the collection of personal data, data security measures must be considered to be a top priority. To protect XR users, most corporate solutions like *Responsible Innovation Principles* by Facebook will need to collaborate with international regulatory authorities, government policies, and the users themselves.

References

1. Adams, D., Bah, A., Barwulor, C., Musaby, N., Pitkin, K., Redmiles, E.M.: Ethics emerging: the story of privacy and security perceptions in virtual reality. In: Fourteenth Symposium on Usable Privacy and Security (SOUPS 2018), USENIX Association, Baltimore, MD, pp. 427–442 (2018). https://www.usenix.org/conference/soups2018/presentation/adams
2. Akers, J., Zimmermann, J., Trutoiu, L., Schowengerdt, B., Kemelmacher-Shlizerman, I.: Mixed reality spatial computing in a remote learning classroom. In: Symposium on Spatial User Interaction, pp. 1–3 (2020)
3. Bacca Acosta, J.L., Baldiris Navarro, S.M., Fabregat Gesa, R., Graf, S., et al.: Augmented reality trends in education: a systematic review of research and applications. J. Educ. Technol. Soc. **17**(4), 133–149 (2014)
4. Barth, S., de Jong, M.D., Junger, M., Hartel, P.H., Roppelt, J.C.: Putting the privacy paradox to the test: online privacy and security behaviors among users with technical knowledge, privacy awareness, and financial resources. Telemat. Inform. **41**, 55–69 (2019)
5. Behr, K.M., Nosper, A., Klimmt, C., Hartmann, T.: Some practical considerations of ethical issues in VR research. Presence **14**(6), 668–676 (2005)
6. Brennan, D.J.: University student anonymity in the summative assessment of written work. High. Educ. Res. Dev. **27**(1), 43–54 (2008)
7. Brown, A., Green, T.: Issues and trends in instructional technology: access to mobile technologies, digital content, and online learning opportunities continues as spending on it remains steady. In: Educational Media and Technology Yearbook, pp. 3–12. Springer (2019)
8. Brown, K., Jackson, M., Cassidy, W.: Cyber-bullying: developing policy to direct responses that are equitable and effective in addressing this special form of bullying. Can. J. Educ. Adm. Policy (57) (2006)
9. Bye, K., Hosfelt, D., Chase, S., Miesnieks, M., Beck, T.: The ethical and privacy implications of mixed reality. In: ACM SIGGRAPH 2019 Panels, pp. 1–2 (2019)

10. Campbell, A.G., Holz, T., Cosgrove, J., Harlick, M., O'Sullivan, T.: Uses of virtual reality for communication in financial services: a case study on comparing different telepresence interfaces: virtual reality compared to video conferencing. In: Future of Information and Communication Conference, pp. 463–481. Springer (2019)
11. Casey, P., Baggili, I., Yarramreddy, A.: Immersive virtual reality attacks and the human joystick. IEEE Trans. Dependable Secure Comput. **18**(2), 550–562 (2021). https://doi.org/10.1109/TDSC.2019.2907942
12. Cherniack, E.P.: Not just fun and games: applications of virtual reality in the identification and rehabilitation of cognitive disorders of the elderly. Disabil. Rehabil. Assist. Technol. **6**(4), 283–289 (2011)
13. Curran, K., Middleton, G., Doherty, C.: Cheating in exams with technology. Int. J. Cyber Ethics Educ. (IJCEE) **1**(2), 54–62 (2011)
14. De Guzman, J.A., Thilakarathna, K., Seneviratne, A.: Security and privacy approaches in mixed reality: a literature survey. ACM Comput. Surv. (CSUR) **52**(6), 1–37 (2019)
15. Dengel, A., Iqbal, M., Grafe, S., Mangina, E.: A review on augmented reality authoring toolkits for education. Front. Virtual Real **3**, 798032 (2022). https://doi.org/10.3389/frvir
16. Devereaux, A.: The digital wild west: on social entrepreneurship in extended reality. J. Entrepr. Publ. Policy (2020)
17. Due, B.L.: The social construction of a glasshole: Google glass and multiactivity in social interaction. PsychNol. J. **13**(2) (2015)
18. Estrada, J.G., Prasolova-Førland, E.: Running an XR lab in the context of covid-19 pandemic: lessons learned from a Norwegian university. Educ. Inf. Technol. 1–17 (2021)
19. Gao, H., Bozkir, E., Hasenbein, L., Hahn, J.U., Göllner, R., Kasneci, E.: Digital transformations of classrooms in virtual reality. In: Proceedings of the 2021 CHI Conference on Human Factors in Computing Systems, pp. 1–10 (2021)
20. Gugenheimer, J., McGill, M., Huron, S., Mai, C., Williamson, J., Nebeling, M.: Exploring potentially abusive ethical, social and political implications of mixed reality research in HCI. In: Extended Abstracts of the 2020 CHI Conference on Human Factors in Computing Systems, pp. 1–8 (2020)
21. Gulhane, A., Vyas, A., Mitra, R., Oruche, R., Hoefer, G., Valluripally, S., Calyam, P., Hoque, K.A.: Security, privacy and safety risk assessment for virtual reality learning environment applications. In: 2019 16th IEEE Annual Consumer Communications Networking Conference (CCNC), pp. 1–9 (2019). https://doi.org/10.1109/CCNC.2019.8651847
22. Gynnild, V.: Student appeals of grades: a comparative study of university policies and practices. Assess. Educ. Princ. Policy Pract. **18**(1), 41–57 (2011)
23. Hahn, N., Fuchs, B., Fortna, M., Cobb, E., Iqbal, M.Z.: Learning sustainable locust control methods in virtual reality. In: ACM International Conference on Interactive Media Experiences, pp. 271–274 (2022)
24. Harley, J.M., Poitras, E.G., Jarrell, A., Duffy, M.C., Lajoie, S.P.: Comparing virtual and location-based augmented reality mobile learning: emotions and learning outcomes. Educ. Tech. Res. Dev. **64**(3), 359–388 (2016)
25. Hirzle, T., Cordts, M., Rukzio, E., Gugenheimer, J., Bulling, A.: A critical assessment of the use of SSQ as a measure of general discomfort in VR head-mounted displays. In: Proceedings of the 2021 CHI Conference on Human Factors in Computing Systems, pp. 1–14 (2021)
26. Hoppe, S., Loetscher, T., Morey, S.A., Bulling, A.: Eye movements during everyday behavior predict personality traits. Front. Hum. Neurosci. **12**, 105 (2018)
27. Hosfelt, D.: Making ethical decisions for the immersive web. In: 2019 USENIX Conference on Privacy Engineering Practice and Respect (PEPR 19), USENIX Association, Santa Clara, CA (2019). https://www.usenix.org/node/238153
28. Hsu, Y.S., Lin, Y.H., Yang, B.: Impact of augmented reality lessons on students' stem interest. Res. Pract. Technol. Enhanc. Learn. **12**(1), 1–14 (2017)
29. Ibáñez, M.B., Delgado-Kloos, C.: Augmented reality for stem learning: a systematic review. Comput. Educ. **123**, 109–123 (2018)

30. Iqbal, M.Z., Campbell, A.: Covid-19, education, and challenges for learning-technology adoption in Pakistan. Interactions **29** (2021)
31. Iqbal, M.Z., Campbell, A.G.: From luxury to necessity: progress of touchless interaction technology. Technol. Soc. **67**, 101796 (2021)
32. Iqbal, M.Z., Campbell, A.G.: Touchless hand interaction and machine learning agents to support kinesthetic learning in augmented reality. In: Proceedings of the 52nd ACM Technical Symposium on Computer Science Education, pp. 1322–1322 (2021)
33. Iqbal, M.Z., Campbell, A.G.: Adopting smart glasses responsibly: potential benefits, ethical, and privacy concerns with Ray-Ban stories. AI Ethics 1–3 (2022)
34. Iqbal, M.Z., Mangina, E., Campbell, A.G.: Exploring the real-time touchless hand interaction and intelligent agents in augmented reality learning applications. In: 2021 7th International Conference of the Immersive Learning Research Network (iLRN), pp. 1–8. IEEE (2021)
35. Jaradat, M.I.R.M., Imlawi, J.M.: Virtual reality applications in education: a developing country perspective. Int. J. Mobile Commun. **19**(4), 492–519 (2021)
36. Jiménez, D.L., Dittmar, E.C., Portillo, J.P.V.: New directions in corporate social responsibility and ethics: codes of conduct in the digital environment. J. Bus. Ethics 1–11 (2021)
37. Kaur, A., Bhatia, M., Stea, G.: A survey of smart classroom literature. Educ. Sci. **12**(2), 86 (2022)
38. Kröger, J.L., Lutz, O.H.M., Müller, F.: What does your gaze reveal about you? On the privacy implications of eye tracking. In: IFIP International Summer School on Privacy and Identity Management, pp. 226–241. Springer (2019)
39. Kudina, O., Verbeek, P.P.: Ethics from within: Google glass, the Collingridge dilemma, and the mediated value of privacy. Sci. Technol. Human Values **44**(2), 291–314 (2019)
40. Kuss, D.J., Griffiths, M.D.: Online social networking and addiction-a review of the psychological literature. Int. J. Environ. Res. Publ. Health **8**(9), 3528–3552 (2011). https://www.mdpi.com/1660-4601/8/9/3528
41. Lam, M.C., Sadik, M.J., Elias, N.F.: The effect of paper-based manual and stereoscopic-based mobile augmented reality systems on knowledge retention. Virtual Real. **25**, 217–232 (2021)
42. Law, B.: Privacy Issues in Virtual Reality: Eye Tracking Technology (2017)
43. Lebeck, K., Ruth, K., Kohno, T., Roesner, F.: Towards security and privacy for multi-user augmented reality: foundations with end users. In: 2018 IEEE Symposium on Security and Privacy (SP), pp. 392–408. IEEE (2018)
44. Liebers, J., Abdelaziz, M., Mecke, L., Saad, A., Auda, J., Gruenefeld, U., Alt, F., Schneegass, S.: Understanding user identification in virtual reality through behavioral biometrics and the effect of body normalization. In: Proceedings of the 2021 CHI Conference on Human Factors in Computing Systems, pp. 1–11 (2021)
45. Liu, S.: Mixed reality (MR) market size worldwide in 2017, 2018 and 2025 (2017). https://www.statista.com/statistics/897595/world-mixed-reality-market-value/
46. Lv, Z., Chen, D., Lou, R., Song, H.: Industrial security solution for virtual reality. IEEE Internet Things J. **8**(8), 6273–6281 (2021). https://doi.org/10.1109/JIOT.2020.3004469
47. MacCallum, K.: Supporting steam learning through student-developed mixed reality (MR) experiences. Pac. J. Technol. Enhanc. Learn. **3**(1), 6–7 (2021)
48. MacIntyre, B.: Remote conference participation in social virtual worlds. Semant. Sch. **5** (2020)
49. Madary, M., Metzinger, T.K.: Real virtuality: a code of ethical conduct recommendations for good scientific practice and the consumers of VR-technology. Front. Robot. AI **3**, 3 (2016)
50. Hernandez-de Menendez, M., Escobar Díaz, C., Morales-Menendez, R.: Technologies for the future of learning: state of the art. Int. J. Interact. Des. Manuf. (IJIDeM) **14**(2), 683–695 (2020)
51. Milgram, P., Kishino, F.: A taxonomy of mixed reality visual displays. IEICE Trans. Inf. Syst. **77**(12), 1321–1329 (1994)
52. Miller, R., Banerjee, N.K., Banerjee, S.: Using Siamese neural networks to perform cross-system behavioral authentication in virtual reality. In: 2021 IEEE Virtual Reality and 3D User Interfaces (VR), pp 140–149. IEEE (2021)
53. Moro, C., Štromberga, Z., Raikos, A., Stirling, A.: The effectiveness of virtual and augmented reality in health sciences and medical anatomy. Anat. Sci. Educ. **10**(6), 549–559 (2017)

54. Nautsch, A., Jiménez, A., Treiber, A., Kolberg, J., Jasserand, C., Kindt, E., Delgado, H., Todisco, M., Hmani, M.A., Mtibaa, A., et al.: Preserving privacy in speaker and speech characterisation. Comput. Speech Lang. **58**, 441–480 (2019)
55. Nesenbergs, K., Abolins, V., Ormanis, J., Mednis, A.: Use of augmented and virtual reality in remote higher education: a systematic umbrella review. Educ. Sci. **11**(1), 8 (2021)
56. Newman, J.H.: The Idea of a University Defined and Illustrated: I. In Nine Discourses Delivered to the Catholics of Dublin. Longmans, Green, and Company (1888)
57. Neyret, S., Navarro, X., Beacco, A., Oliva, R., Bourdin, P., Valenzuela, J., Barberia, I., Slater, M.: An embodied perspective as a victim of sexual harassment in virtual reality reduces action conformity in a later milgram obedience scenario. Sci. Rep. **10**(1), 6207 (2020). https://doi.org/10.1038/s41598-020-62932-w
58. O'Hagan, J., Williamson, J.R.: Reality aware VR headsets. In: Proceedings of the 9TH ACM International Symposium on Pervasive Displays, pp. 9–17 (2020)
59. Orlosky, J., Kiyokawa, K., Takemura, H.: Virtual and augmented reality on the 5g highway. J. Inf. Process. **25**, 133–141 (2017)
60. Paucher, R., Turk, M.: Location-based augmented reality on mobile phones. In: 2010 IEEE Computer Society Conference on Computer Vision and Pattern Recognition-Workshops, pp. 9–16. IEEE (2010)
61. Pócsová, J., Mojžišová, A., Takáč, M., Klein, D.: The impact of the covid-19 pandemic on teaching mathematics and students' knowledge, skills, and grades. Educ. Sci. **11**(5), 225 (2021)
62. Popper, K.R., Popper, K.R.: The Open Society and Its Enemies: The Spell of Plato. Routledge & Kegan Paul (1957)
63. Porta, M., Ricotti, S., Perez, C.J.: Emotional e-learning through eye tracking. In: Proceedings of the 2012 IEEE Global Engineering Education Conference (EDUCON), pp. 1–6. IEEE (2012)
64. Salmivalli, C., Lagerspetz, K., Björkqvist, K., Österman, K., Kaukiainen, A.: Bullying as a group process: participant roles and their relations to social status within the group. Aggress. Behav. **22**(1), 1–15 (1996). https://doi.org/10.1002/(SICI)1098-2337(1996)22:1<1::AID-AB1>3.0.CO;2-T
65. Sankaranarayanan, G., Weghorst, S., Sanner, M., Gillet, A., Olson, A.: Role of haptics in teaching structural molecular biology. In: Proceedings of the 11th Symposium on Haptic Interfaces for Virtual Environment and Teleoperator Systems, 2003, HAPTICS 2003, pp. 363–366. IEEE (2003)
66. Sapozhnykov, V., Harvey, T.I., Erfaniansaeedi, N., Luke, R.: Headset on ear state detection. U.S. Patent US10812889B2 (2019)
67. Saredakis, D., Szpak, A., Birckhead, B., Keage, H.A., Rizzo, A., Loetscher, T.: Factors associated with virtual reality sickness in head-mounted displays: a systematic review and meta-analysis. Front. Hum. Neurosci. **14**, 96 (2020)
68. Sarikakis, K., Winter, L.: Social Media Users' Legal Consciousness About Privacy, vol. 3, no. 1. Social Media+ Society. 2056305117695325 (2017)
69. Sayakkara, A., Le-Khac, N.A., Scanlon, M.: Accuracy enhancement of electromagnetic side-channel attacks on computer monitors. In: The Second International Workshop on Criminal Use of Information Hiding (CUING), Part of the 13th International Conference on Availability, Reliability and Security (ARES). ACM, Hamburg, Germany (2018)
70. Schmidt, E.A., Kuiper, O.X., Wolter, S., Diels, C., Bos, J.E.: An international survey on the incidence and modulating factors of carsickness. Transport. Res. F: Traffic Psychol. Behav. **71**, 76–87 (2020)
71. Sekhavat, Y.A.: Privacy preserving cloth try-on using mobile augmented reality. IEEE Trans. Multimed. **19**(5), 1041–1049 (2016)
72. Sharma, H.N., Toups, Z.O., Jain, A., Kerne, A.: Designing to split attention in a mixed reality game. In: Proceedings of the 2015 Annual Symposium on Computer-Human Interaction in Play, pp. 691–696 (2015)
73. Slater, M., Gonzalez-Liencres, C., Haggard, P., Vinkers, C., Gregory-Clarke, R., Jelley, S., Watson, Z., Breen, G., Schwarz, R., Steptoe, W., et al.: The ethics of realism in virtual and augmented reality. Front. Virtual Real. **1**, 1 (2020)

74. Stip, E., Thibault, A., Beauchamp-Chatel, A., Kisely, S.: Internet Addiction, Hikikomori Syndrome, and the prodromal phase of psychosis. Front. Psych. **7**, 6 (2016). https://doi.org/10.3389/fpsyt.2016.00006
75. Stretton, T., Cochrane, T., Narayan, V.: Exploring mobile mixed reality in healthcare higher education: a systematic review. Res. Learn. Technol. **26**, 2131–2131 (2018)
76. Wassom, B.: Augmented reality law, privacy, and ethics: law, society, and emerging AR technologies. Syngress (2014)
77. Yarramreddy, A., Gromkowski, P., Baggili, I.: Forensic analysis of immersive virtual reality social applications: a primary account. In: 2018 IEEE Security and Privacy Workshops (SPW), pp. 186–196 (2018). https://doi.org/10.1109/SPW.2018.00034
78. Yoshimura, A., Borst, C.W.: Evaluation and comparison of desktop viewing and headset viewing of remote lectures in VR with Mozilla hubs. In: International Conference on Artificial Reality and Telexistence, and Eurographics Symposium on Virtual Environments (ICAT-EGVE) (2020)
79. Yoshimura, A., Borst, C.W.: Remote instruction in virtual reality: a study of students attending class remotely from home with VR headsets. Mensch Comput. (92020-Workshopband) (2020)
80. Zhang, F., Bazarevsky, V., Vakunov, A., Tkachenka, A., Sung, G., Chang, C.L., Grundmann, M.: Mediapipe hands: on-device real-time hand tracking (2020). arXiv:2006.10214
81. Ziker, C., Truman, B., Dodds, H.: Cross reality (XR): challenges and opportunities across the spectrum. In: Innovative Learning Environments in STEM Higher Education, p. 55 (2021)

An Immersive Learning Environment to Improve User Experience in Science Museums

Peidi Gu, Wenjing Li, Xinyi Ye, Jing Wang, and Yanlin Luo

Abstract Science museums are not only places for scientific research but also places for the public to access various aspects of science fields. With the increasing inquiries and interest of the public to science, traditional exhibitions cannot fulfill the needs of their audiences to acquire science knowledge nowadays. Diversities and various inquiries of visitors should be paid attention to for science museum administrators. Thus, new formats of learning are necessary to be implemented into the construction of science museums. Among various types of learning means, the immersive learning supported by Virtual Reality (VR) technologies can be a good candidate to build immersive and interactive virtual learning environments so that visitor's experience can be enhanced. Guided by the sensory perceptual system introduced by James Gibson, this study proposes a construction framework for science museums to systematically build immersive learning environments for visitors. We also embedded force feedback technology for enhancing the haptic perception of system users to increase their level of immersion. A pilot study of thirty senior year high school students from Beijing Experimental Middle School was conducted to evaluate the usability and user experience of the developed virtual immersive learning environment. Results showed that students reported comparatively higher feeling of presence and engagement than those in the control group. But overall, students reported satisfying user experience through the developed learning system.

Keywords Virtual reality · Immersive learning · Science museum

P. Gu
College of Future Education, Beijing Normal University, Zhuhai, China

W. Li · X. Ye · J. Wang · Y. Luo (✉)
School of Artificial Intelligence, Beijing Normal University, Beijing, China
e-mail: luoyl@bnu.edu.cn

Y. Cai et al. (eds.), *Mixed Reality for Education*, Gaming Media and Social Effects,
https://doi.org/10.1007/978-981-99-4958-8_17

1 Introduction

Science museums are one of the essential public spaces and informal learning environments that aim at promoting citizen's science literacy and interest [1, 2]. Traditionally, exhibitions are the major presentation formats for science museums to offer visitors opportunities with close observation of the exhibits [3]. Such visiting means is constrained by time, space, and more importantly, presents a single learning mode so that visitors' knowledge acquisition and enjoyment are limited [4]. With the advantages of modern computer technology, museums start to be aware of diverse features and inquiries of visitors [5–7]. Thus, new formats of learning are to be explored by museum designers and experts in order to improve people's visiting experience and knowledge acquisition.

Immersive learning is a situative, cognitive and associative learning method that aims to provide learners a realistic and well controlled learning environment, which is supported by modern technology such as Virtual Reality (VR), Augmented Reality (AR) and Mixed Reality (MR), and to promote learning experience through natural or embodied interaction with the learning environment [8]. An immersive learning environment with natural and enjoyable interaction enables learners to focus on the learning tasks and has been proven to be effective in enhancing people's learning motivation [9]. According to the constructivism, learners usually take proactive roles to collect diverse information and participate learning activities in an immersive learning environment, to construct knowledge and conceptual learning based on their prior knowledge with motivation and creativity [10]. Immersive learning methodology has great potential in educational research field.

With the development of VR technology and Human–Computer Interaction (HCI) technology, VR based immersive learning environment is able to provide people with a high immersive and intensified interactive virtual environment, which has been previously adopted in science museum exhibitions to improve people's visiting experience and further enhance their knowledge acquisition [1, 11–13]. The British Museum provides visitors with the opportunities to browse collections through Head-Mounted Devices (HMD) in an online virtual immersive environment [14]. HMD assisted onsite learning experience is also available in the British Museum designed with some theme topics related to different field of studies such as history and biology. In China, the naked-eye 3D digital movie *Forbidden City: Palace of the Son of Heaven* uses the VR technology to reproduce the archaeological excavation scene, where the audience can watch all the attractions of the Forbidden City [15]. The China Science and Technology Museum launched the first virtual reality expo in June 2016, applying VR technology to many fields such as aerospace, history, culture, scientific research, etcetera, displaying more than 30 interactive exhibition items such as *Gene Exploration, Chang'e Flying to the Moon, Aircraft Carrier Virtual Tour*, and *Fighting for the Luding Bridge* on a naked-eye 3D screen [14]. These applications of VR technology in science museum immersive learning have been proved to be effective in promoting visitors' browsing interest and knowledge gain.

Even though virtual immersive learning has been applied to several science museums to diversify and improve visitors' experience, how to design a systematic, highly immersive and intensified interactive learning environment is seldom discussed in previous research. Current implementation of VR based immersive learning environment in science museums is usually limited by several obstacles such as high cost, extended construction period, lagged-behind content updates and lack of explanation [4]. Thus, a systematic and effective immersive learning environment design is significant for science museums who plan to adopt VR technology to improve the efficiency of scientific knowledge dissemination. Thus, this paper proposes a construction framework of an immersive learning environment that aims to enhance sensory immersion of visitors in science museums. The purpose of this study is to embed force feedback technology into the VR based learning environment to foster intensified interaction between users and the virtual items, and furthermore improve people's immersive experience with natural interaction in science museums. Thus, our research question for this study is: whether the suggested construction framework and force feedback technology application together can enhance people's experience in virtual immersive learning?

2 Method

This study aims to propose a construction framework for immersive learning in science museums. To better provide users a fully immersive learning environment based on VR, sensory immersions are considered one of the major factors that affect users' feeling of immersion [16]. Hence, this study embeds force feedback systems into the immersive learning design to enhance users' haptic experience, which were less discussed in previous virtual immersive learning environment studies [17]. A pilot study of users' satisfaction survey about using the developed system was conducted to evaluate the effectiveness of the immersive learning design.

2.1 Sensory Immersion

Sensory immersion is 'a variant of immersion within an environment … in which all the person's senses are immersed into that environment, with each one, as much as possible, reporting back data from that environment' [18]. According to Gibson [19], the human perceptual system is consisted with visual system, auditory system, haptic system and smell/taste system.

In constructing a virtual immersive environment, each perception system can be enhanced by modern computer technology to achieve better user sense of presence as being there [20]. As one of the most important human perceptions, vision captures 70–80% of information received by humans [21]. In the virtual environments, immersion and interactivity can be enhanced with improved visual perception generated by the real-time rendering of realistic graphics technology [22]. Auditory sense, the second important human perception, can supplement received information outside of the field of vision, augmenting the sense of space and realism in virtual worlds [23]. The rendering of the audio characteristics and spatial information of the sound source in the virtual environment is the key to achieving immersive auditory perception [24]. The third perceptual system, haptic system, is a process of perception through tactile and kinaesthetic activities. The simulation of haptic perception brings together the research of biomechanics, psychology, neurology, engineering and computers in human tactile and force feedback [25], which allows users to perceive three-dimensional virtual objects and obtain information about their shape, weight, surface texture, temperature and other attributes [26]. Enhanced haptic perception in the virtual environment can make users confuse the boundary between the real world and the virtual environment, thereby enhancing their sense of reality and immersion [27]. Lastly, taste/smell perception requires chemical stimulation, which is rarely discussed in building of virtual environments. However, understanding taste and olfactory characteristics and applying them appropriately to immersive experiences could be a novel consideration for future research.

As a result, the human perception systems may provide virtual environment designers a theoretical background to achieve better user immersion and interaction. This study tries to construct a fully immersive and interactive science museum virtual learning system that aims to enhance learners' visual, auditory and haptic perceptions. Based on previous research about visual and auditory perception for immersion, our study focuses more on improving learners' haptic perceptions through force feedback technology to achieve better sense of immersion and interaction.

2.2 Immersive Learning Environment

This article proposes a technical framework for constructing an immersive learning environment for science museums, as shown in Fig. 1, which includes the device level, data level, processing level, and application level.

The device level is the hardware foundation for building a virtual environment, which includes servers, display terminals and three-dimensional interactive devices. High-performance servers are used for data storage for high-quality virtual environment rendering. Learners join in the immersive learning environment through manipulating three-dimensional interactive devices (such as Touch force-feedback devices, HTC Vive handles, and etc.) and watching the terminal screens (such as LED displays or holographic screens) by wearing HTC helmets or stereo glasses.

The data level contains all the data required by the system and those collected during the running process, which includes model and scene data, multimedia data and user data collected by device sensors. Model data refers to the three-dimensional model built by 3DsMax, Maya or other modeling software, similarly, scene data includes the position, shape, physical parameters and related logic scripts of virtual objects; multimedia data is collected science knowledge materials stored in the form

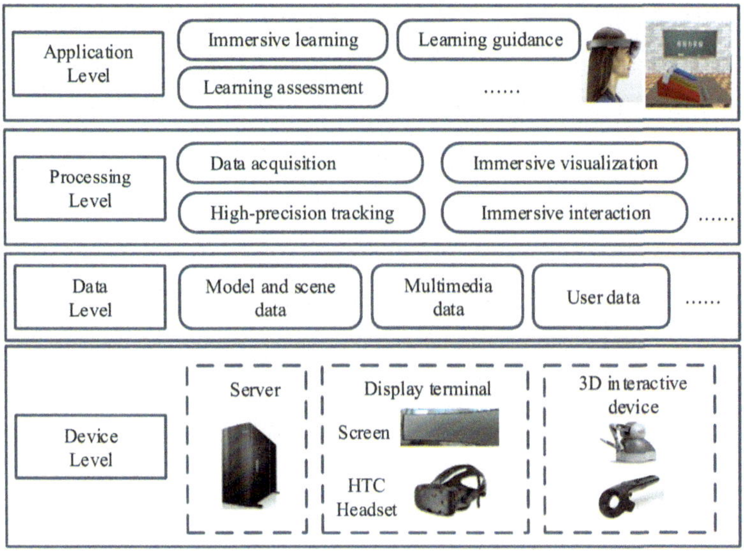

Fig. 1 The overall technical framework for constructing an immersive learning environment for science museums

of pictures, audio or text; user data is the learning trajectory information generated during the learning process. It is stored in the learner's file as a generative resource and stored in the user database for subsequent analysis.

The processing level is responsible for the acquisition of sensor data based on high-precision tracking. It involves the core technologies of this study as immersive visualization and immersive interaction. Through data acquisition, high-precision tracking and other means, the processing level deals with information from users to support immersive interaction with characteristics such as mobility and natural interaction, expanding user interaction from 2 to 3D.

The application level provides the main functional modules of immersive learning, including a series of instructional applications such as immersive learning, learning guidance, and learning assessment. Learners participate in immersive learning through the learning guidance system. And the system analyzes and evaluates the effectiveness of their learning behaviors. The intelligent agent 'virtual tutor' supports and guides interactive learning.

2.3 Design of Immersive Learning Environment

The theme topics of immersive learning in this study are selected from two of the most popular science fields as physics and biology. In order to construct a physical and biological immersive learning environment based on 'law of inertia', 'action force

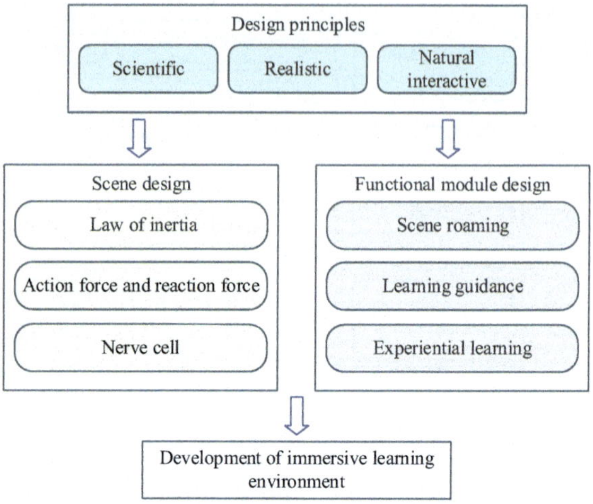

Fig. 2 Design flow of the immersive learning environment

and reaction force' and 'nerve cell' of popular science venues, this paper proposes an overall design flow as shown in Fig. 2. Based on the design principles, there are three design processes: scene design, functional module design and development.

2.3.1 Design Principles

The design of immersive learning environment for science museums should follow the three design principles: scientific principle, realistic principle and natural interactive principle, to create realistic and vivid learning environment to improve visitors' learning experience. First of all, the scientific principle requires consistency between the virtual space and the real world [28]. That is to say, the results of immersive visualization of science exhibitions must follow the scientific laws of the professional field. Secondly, the realism of the immersive learning environment is an important factor influencing the user's immersion and experience [29]. If the virtual scene lacks a sense of reality (inconsistent proportions of virtual objects, unreasonable materials, too simple models, etc.), the user will subjectively feel too unreal to produce a sense of immersion, and consequently decrease their learning motivation [30]. Thus, realistic principle requires high realism of scene simulation and rationality of interaction. Lastly, the natural interactive principle not only requires the interaction design to be authentic, but also it requires the interaction to be more real and convenient so that users can interact with the virtual environment easily [31]. Previous research has also proved the positive relationship between naturalism of interaction and learning motivation [32].

Table 1 Script of science topics' logic

	Contents of scripts
Law of inertia	Two buttons (slow or fast) control the train to move at a constant speed
	The balls carried by the train pop up when the train passes the arch bridge
Action force and reaction force	Three buttons (small, medium or big force) control the force to launch the projectile
Nerve cell	Relative information can be presented if the learner selects one of the structures of the nerve cell

2.3.2 Scene Design

This study uses 3DsMax modeling software to build three-dimensional models, and uses Unity3D engine to build virtual scenes. The simulation of the operation mode of science exhibition items is realized by adding physical components and control scripts to the related virtual objects in the scene in Unity3D. Unity3D has built-in NVIDIA's powerful Physx physics engine, which enables objects to respond to collisions and subject to forces such as gravity. Therefore, in Unity3D, we added a Rigidbody component and a Collider component to the object so that it can move under the control of the laws of physics. In order to realize the operation logic of each exhibition item, it is also necessary to add the C# script shown in Table 1 to the object to make it respond correctly to a specific event. As a result, it vividly simulates the operation logic of the exhibition item and exhibits the corresponding scientific laws.

2.3.3 Functional Module Design

The functional module design includes three parts: scene roaming, learning guidance and experiential learning. Considering the large virtual space of science museums, this study uses a combination of mobile roaming and static roaming technology. Visitors can not only experience the virtual scene through their own movement, but also control their movement in the virtual space by inputting motion instructions through the HTC Vive handle. With regard to the learning guidance part, it includes explanation audio, prompt text, sound and light effects when the exhibition item is running. These contents are designed to help the visitors to better experience and explore the virtual environment independently. Lastly, experiential learning can be realized by using head-mounted displays, gamepads, and force feedback devices to enter the virtual world. Visitors have a multi-sensory immersive experience so that they can interact with science exhibitions immersively, and learn knowledge during their personal visiting experience.

2.4 Interactive Design

According to the human perception system [19], this study aims to enhance the user interaction experience through visual system, auditory system and haptic system. We use Unity3D to build an interactive immersive learning environment. Visual immersion is supported by computer display devices. And auditory immersion is achieved by audio playback devices to present audio interpretations to specific knowledge when learner approached the marked spots. And lastly, HTC Vive is used to realize handle interaction and Geomagic Touch to realize force feedback to complete an immersive haptic interaction, with the aid of additional devices such as Leap Motion to collect gesture data. The overall setting of immersive learning environment is shown in Fig. 3.

HTC Vive can convert the captured motion into data through laser positioning, and then reproduce the motion by data analysis to achieve interaction. It mainly includes a head-mounted display (helmet) (HMD), two single-handed controllers (handles), and a positioning system (locator) that can simultaneously track the display and the controller in the effective space. In the design of the exhibition items, the HTC Vive handle is used to provide gesture information. The user observes the virtual environment through the HMD of HTC Vive, and can realize interactive operations such as forward, backward, and turn in the virtual scene by walking around. Figure 4 is an example of a user trying to manipulate virtual biology object through HTC Vive.

Geomagic Touch is a mid-range professional tactile force feedback device. As shown in Fig. 5 it can apply force feedback to the user's hands, allowing the user to feel the virtual object and produce a 'touch' when manipulating a 3D object on the

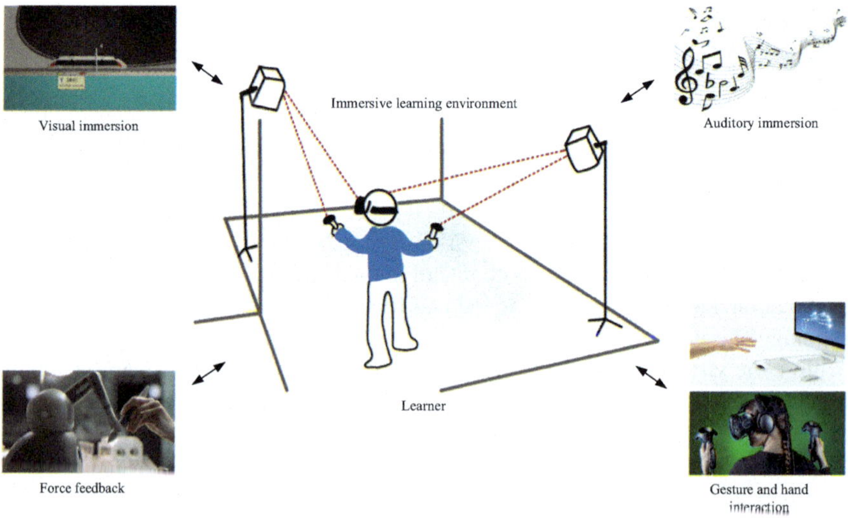

Fig. 3 Setting of the immersive learning environment

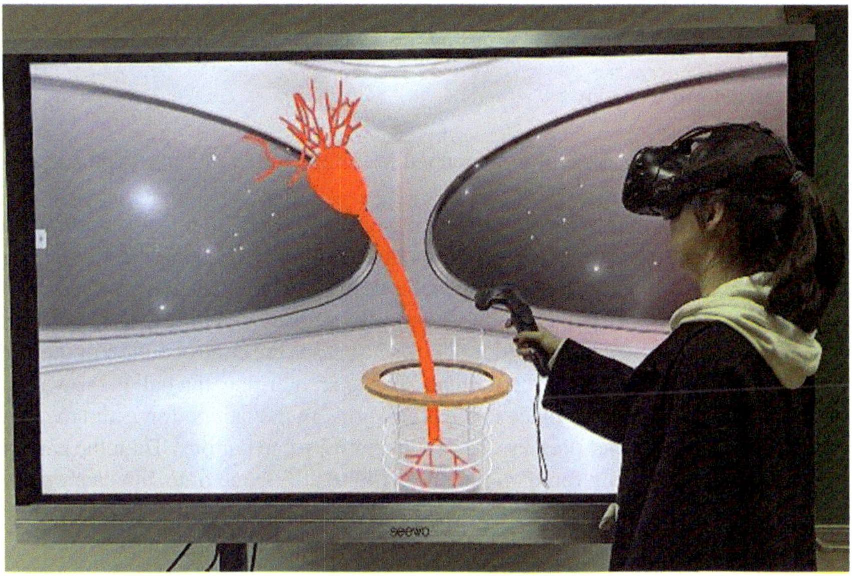

Fig. 4 HTC Vive manipulation in the virtual learning environment

screen. Touch combined with the application of the OpenHaptics toolkit can create tactile examples. The user interacts with the force bar, while the force bar returns force information to the user, making the user feel like touching the virtual object in real so as to change the control strategy accordingly. Through the two twisting axis points, the position sensing of 6 degrees of freedom is realized by the movement of the pen tip, which is connected to the mechanical arm with two twisting axis points.

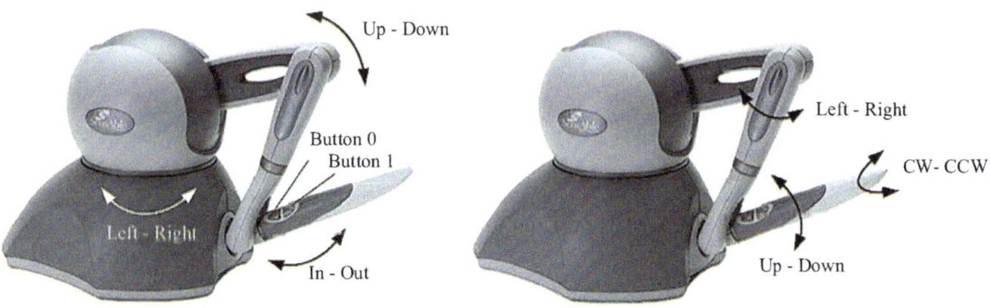

Fig. 5 Operations of 6 degrees of freedom by touch

2.5 Experiment Design

2.5.1 Learning Content

Two topics of the science exhibition content are simulated in this study: physics and biology.

This study builds the virtual scenes of the physical exhibition items as shown in Fig. 6a, b. The two exhibition items are respectively simulation of the 'Law of Inertia' and 'Action Force and Reaction Force' in the 'Law of Motion' section of the science museum. The 'Law of Inertia' exhibit is mainly composed of a little train that carries two balls, the track and an arch bridge. The user presses the control button to start the train at an initial speed. When the train passes the arch bridge, the ball is thrown out and gets back to the original place in the afterwards. In the 'action force and reaction force' exhibition, the user presses one of the control force buttons. Then the cannon will exert a forward firing force on the projectile. At the same time, the cannon will also retreat due to the opposite reaction force with the same magnitude, pushing the hanging pointer to rotate accordingly.

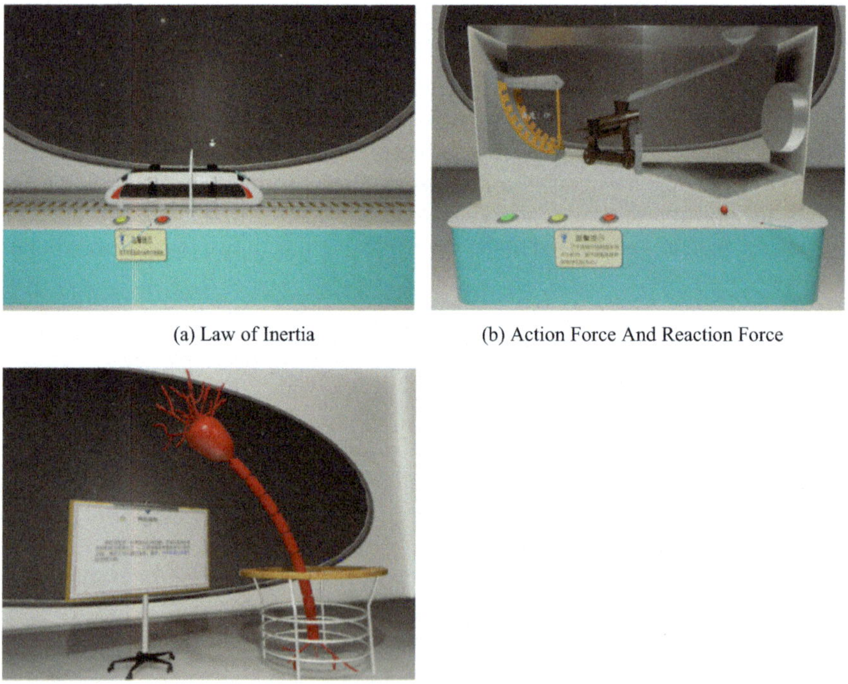

(a) Law of Inertia

(b) Action Force And Reaction Force

(c) Nerve Cell

Fig. 6 Virtual scenes of science exhibitions

The virtual scene of the biological exhibition in this study is as shown in Fig. 6c. It was designed based on the 'Nerve Cell' exhibition in the 'Secret of Life' exhibition session of the Science Museum. During the modeling process of this project, the structure and materials of axon, dendrites, cell body, myelin sheath and axon terminals of nerve cell were restored 1:1 according to the original model of the Science Museum. In this exhibition, when the user selects the specific part of the nerve cell, relevant introduction information will be displayed on the virtual information board as well as on the side of the target part. At the same time, a playback of the interpretive audio will be triggered and be played automatically.

2.5.2 System Application and Assessment

A pilot study of user satisfaction survey was conducted in this study to assess the interaction experience of using the developed virtual learning environment in the purpose of finding problems in the system in time and making improvements before it is officially put into use.

The equipment setting of the experiment is shown in Table 2. We used computer systems equipped with CPU as Intel Core™ i7-10875H, 16 GB RAM, NVIDIA GeForce RTX™ 2080 Graphics card and 1 TB SSD hard disc. The VR device used in this study are HTC Vive HMD and the force feedback device are Geomagic Touch.

In total 30 senior high school students from Beijing Experimental Middle School were enrolled in this study. They were randomly divided into 6 groups with 5–6 students in one group. Right before the experiment, each group of students were demonstrated the operation of HTC handles and Touch force feedback devices. Then, they were required to participate in immersive learning about all three virtual exhibitions for about 8 min. By the end of the experiment for each group, students were asked to fill in a user satisfaction survey about their VR learning experience by using the developed learning system. Demographic information was also collected at the beginning of the survey in aspects of students' age, gender and prior experience of using VR devices. Experiment was completed until all 6 groups of students finished the survey.

Table 2 Experimental equipment settings

CPU	Intel Core™ i7-10875H (8 cores, 16 threads)
RAM	16G
Graphics card	NVIDIA GeForce RTX™ 2080 (8G video memory)
VR device	HTC Vive HMD (Resolution: single eye 1080 × 1200 pixels, refresh rate: 90 Hz, field angle: 110°)
Force feedback device	Geomagic Touch

2.6 Outcome Measures

The user experience questionnaire for immersive virtual environment developed by Tcha-Tokey and colleagues [33] was used as the instrument to evaluate students' experience and level of satisfaction towards the developed science museum exhibitions. This questionnaire for immersive virtual environment approaches users' experience from seven dimensions, namely, engagement, presence, immersion, emotion, usability, experience consequence, and technology adoption. Engagement refers to the relationship between people and activities in the learning process, which usually includes three dimensions: behavioral participation, emotional participation, and cognitive participation [34]; while presence refers to the feeling of being in the created environment [35]. In addition, immersion refers to the illusion that the sensory stimulation of the virtual environment can produce the same effect as the sensory stimulation of the actual physical world [36]. Emotion refers to the subjective feelings of happiness, satisfaction, disappointment, tension, etcetera, generated by the user in the virtual environment [37]. Usability is defined as ease of learning and ease of use of the system [38]. Among them, experience consequence refers to the symptoms that the user may experience in the virtual environment, such as cyber sickness, nervousness, dizziness, headache, etcetera [39]. Lastly, technology adoption refers to actions and decisions taken by users in the future using or intending to use an immersive virtual environment [33]. All seven dimensions together have been developed to gather information about the user's opinion, beliefs and preferences towards the virtual learning environment. The questionnaire is composed of 10 subscales and 87 items with 75 items having a 10-point Likert scale, 12 items having a differential scale items and 3 items as open questions. The questionnaire has been proved to be valid and reliable with Cronbach's alpha from 0.71–0.91.

In order to fulfill the expectation of our experiment design, we adapted the questionnaire according to the features of our developed virtual environment. Twenty-one items were selected from the original questionnaire and a 5-point Likert scale was used to evaluate students' experience using the immersive learning environment from 1 as 'strongly disagree' to 5 as 'strongly agree'. For translation reliability, questionnaire items were translated into Chinese by a faculty member of Beijing Normal University. The Chinese copy of the questionnaire was then translated from Chinese to English by another faculty member from the same university for accuracy of the inquiries. Corrections were made when there were discrepancies in understanding the questions between the original questionnaire and the reverse translated version of the questionnaire until agreement on all questions were achieved. The adapted questionnaire can be found in Appendix 1.

2.7 *Data Analysis*

As this experiment aims to assess the usability and interactivity of the developed immersive learning environment, descriptive statistics and mean scores of the survey questions were used to evaluate and explain students' experience of using the immersive learning system. In addition, a t-test analysis was also conducted to discuss the differences about the experience of virtual immersive learning between students who had prior experience with VR devices and those without such experience.

3 Results

In total 30 senior high school students from Beijing Experimental Middle School participated in this experiment and filled out the survey, among whom there were 20 males (67%) and 10 females (33%). All students aged from 16–17 years old. With regard to their prior experience with VR devices, 11 students (37%) reported they have never used VR devices before the experiment, while the remaining 19 students (63%) reported to have some prior VR experience. A complete summary of the demographic information is shown in Table 3.

The survey questionnaire items were coded according to the Likert scale scores, including questions 15, 17, 18 and 19 as reversely coded. Mean scores of each item can be calculated based on the Likert scale scores and were used for data analysis.

T-test was used to find the differences between students with or without prior VR experience about their attitude towards the developed immersive learning system. Mean score of students with no prior VR experience is 4.12 and that of students with prior VR experience is 3.96. Significance of difference was not found from the T-test results ($p = 0.266$) (Table 4).

Mean scores for each survey question were also calculated as shown in Fig. 7, with the highest mean score as 4.63 in question 5 for presence and the lowest score as 3.20 in question 15 for usability.

In order to find out how the immersive learning environment design affects users' VR experience, questions for each dimension in the survey were grouped together

Table 3 Demographic Information of the senior high school students

Variable	Options	Frequency	Percentage (%)
Gender	Male	20	67
	Female	10	33
Age	16	17	57
	17	13	43
Prior VR experience	Never	11	37
	Occasionally	19	63

Table 4 T-test results of experience between students with or without prior VR experience

GroupID	Mean (SD)	T	Sig. (two-taled)
Without prior VR experience	4.12(0.42)	1.134	0.266
With prior VR experience	3.96(0.33)		

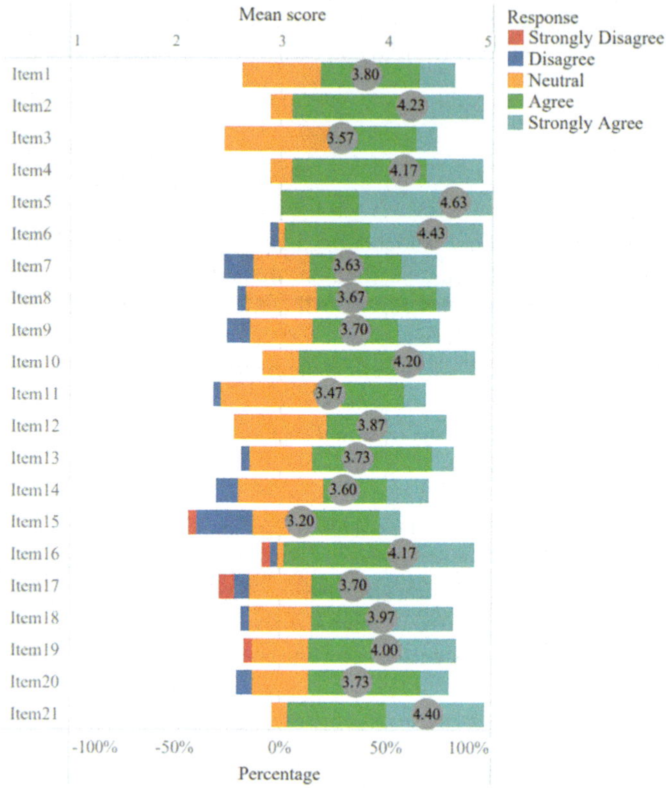

Fig. 7 Summary of the survey questions

and corresponding mean scores for each dimension were shown in Fig. 8, where technology adoption had the highest score as 4.07 and immersion had the lowest score as 3.69.

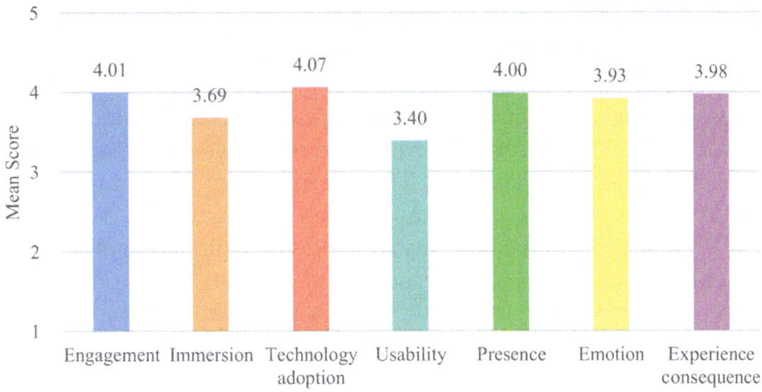

Fig. 8 Summary of each subscale of the questionnaire

4 Discussion

This article proposes a construction framework for immersive learning environment design for science museum exhibitions with force feedback technology application to improve user's haptic perceptual immersion. A pilot study about user experience evaluation for the immersive learning system was conducted with the participation of 30 senior year high school students.

Results have shown that there were no statistically significant differences about user experience between students with or without prior VR experience. The results can be explained by the dimensions of the questionnaire accordingly. Research by Saignier et al. [40] showed that prior knowledge of VR has an effect on self-assessment of performance, pragmatic quality and hedonic quality stimulation, but with insignificant effect on VR related components such as cyber sickness. First, Engagement mostly stands for peoples' feeling of being involved in the environment or an activity [41], whereas immersion means the level of consciousness of being aware of the simulated environment [42]. The use of HTC Vive and vivid virtual scene design in the developed virtual immersive learning system effectively attracted students' attention and provided sufficient physical autonomy for students to move around in the virtual environment, and hence enhanced students' feeling of engagement. On the other hand, students reported satisfactory score for immersion, which is comparatively less satisfied to engagement. The main reason for less immersive experience from students can be explained by the lower score for the usability of interaction devices in the survey. Previous research has shown that the ease of using VR technology and natural interaction format are positively related to users' feeling of immersion [43]. HTC Vive and Geomagic Touch are relatively less commonly used VR devices by public, prior VR experience from participated students in this study may play little effect on their experience using the developed virtual environment in terms of device familiarity. Thus the effect of prior VR experience on user experience of using virtual immersive learning can be decreased by reasonable designed VR equipment settings and natural interaction provided by haptic devices.

The developed virtual immersive learning system with force feedback technology can effectively enhance students' learning experience according to the survey results. The average of all items has a positive average score (greater than 3 points), of which 7 items (33.3%) are higher than 4 points. The more prominent items are item 5 and item 6, which have achieved average scores of 4.63 and 4.43 respectively, which shows that in this system, students could easily realize multi-angle and detailed observation of virtual exhibition items; meanwhile, item 2 and item 4 also achieved average scores of 4.23 and 4.17 respectively, which reflects that the system is attractive to students in terms of visual and roaming methods; And another, items 10, 16, and 21 achieved highs of 4.20, 4.17 and 4.40, respectively. It shows that the 3D audio of this system is effective in providing accurate signals to students for them to correctly identify the sound source. Overall, students reported that the interactive method of this project is interesting and emotionally enjoyed the experience in the virtual environment.

However, the average scores for items 3, 11, and 15 are low, and students reported that the interaction method is a bit cumbersome and unnatural, and they were not fully immersed in the learning process. During the experiment, it was observed that some students were unable to smoothly fit the projectile into the bullet hole when using the Touch force feedback device to experience the 'action force and reaction force' exhibition. Some students also reported that the resolution of the HTC helmet was not high enough. It may be the reason for the low scores of these three items. Nevertheless, items 18 and 19 to evaluate students' experience consequence were reported relatively higher than those of 3, 11 and 15 items. The experience consequence is closely related to users' comfort of using the VR devices, in which cyber sickness is one of the major factors that affect users' feeling of comfort [44]. The discrepancy between students' reported difficulties of using the interaction device and their reported experience consequence might because of the short period of time for students experiencing the virtual environment through HMD. Research reported that cyber sickness usually appears when using HMD by usual people with the time of more than 15 or 20 min [45]. We argue that by extending the learning time of students in this study, the negative effects of cyber sickness would unfold accordingly. At the same time, we suggest when designing science museum exhibition for general public, the length of time for learning should be considered to decrease the risk of discomfort from users caused by cyber sickness. Also, other types of immersive display and interaction methods should be researched to provide public better accepted and easier manipulated virtual learning environment.

There are still several limitations to this study. First of all, the generalization and dissemination of the VR environment to other science museums are still limited. Considering the cost of VR environment setting and the use of HTC Vive and Touch devices, it is hard to extend this research to some common science museums, especially for those with budget difficulties. Public education of how to use the interactive devices is also a problem for generalization. Future research will be suggested to search for more affordable and commonly familiar devices for the immersive learning environment design. Secondly, limited number of students participating in the study results in comparatively inaccurate predictions about how each dimension of user experience are approached by our proposed VR learning environment construction framework. Larger sample studies will be necessary in future research. Thirdly, the purpose of this study is to construct a virtual immersive learning environment for science museums, where learning outcome and motivation are two important factors to evaluate the effectiveness of immersive learning environment design. As this is a pilot study of the usability and feasibility of our developed system, learning outcome and motivation were not assessed in our current study. In the future, experiments will be conducted to test the effectiveness of the proposed immersive learning environment design on people's knowledge acquisition and learning motivation.

5 Conclusion

This study proposed a construction framework for science museum's virtual immersive learning environment. Sensory perception systems [19] was used as a guidance for immersive and interactive design. First, this article proposes the overall technical framework of the immersive learning environment; then, the specific content of the scene design and functional module design based on the design principles is given, and the virtual learning environment and interactive experience are realized; finally, the virtual learning environment is analyzed through the design questionnaire collected from the user experience evaluation experiments. The experimental results showed that the immersive virtual learning environment designed in this paper has a good overall performance. In the future, it can be equipped with higher-resolution HTC Vive headsets and servers, as well as a simpler and easier-to-operate interactive interface designed for better immersion and usability.

Highly immersive and interactive immersive learning environment, as a new mode of science popularization, allows audiences to experience the content of science museum exhibitions anytime and anywhere. And at the same time, it is also conducive to stimulating the visitor's spirit of exploration and innovation, which is useful for promoting public's science literacy. It is imperative to expand this research to other fields of science, such as mathematics, astronomy, and chemistry. At present, the project team has initially added a mathematics module to the system, and will gradually improve the guidance module to further accelerate the system to be put into practical application. In the future, we will combine immersive popular science projects with 5G technology to build popular science exhibition items with cloud experience, and provide effective solutions for science museum exhibitions, and even solving education fairness and education resource balance.

Acknowledgements This work is supported by the National Natural Science Foundation of China (No.61977063, No.12126508).

Appendix 1: User Experience Questionnaire for Virtual Immersive Environment

Items	Subscale
1. My interactions with the virtual environment seemed natural	Presence
2. The visual aspects of the virtual environment involved me	Engagement
3. The devices (controller handle or stylus) which controlled my movement in the virtual environment seemed natural	Presence
4. The sense of moving around inside the virtual environment was compelling	Engagement
5. I was able to examine objects closely	Presence
6. I could examine objects from multiple viewpoints	Presence
7. I was involved in the virtual environment experience	Engagement
8. I felt proficient in moving and interacting with the virtual environment at the end of the experience	Presence
9. I could concentrate on the assigned tasks rather than on the devices (controller handle or stylus)	Presence
10. I correctly identified sounds produced by the virtual environment	Presence
11. I become so involved in the virtual environment that I was not aware of things happening around me	Immersion
12. I become so involved in the virtual environment that it is if I was inside the game rather than manipulating a gamepad and watching a screen	Immersion
13. I felt physically fit in the virtual environment	Immersion
14. I thought the interaction devices (HTC headset, controller handle and/or stylus) was easy to use	Usability
15. I found the interaction devices (HTC headset, controller handle and/or stylus) very cumbersome to use	Usability
16. I enjoyed being in this virtual environment	Emotion
17. I got tense in the virtual environment	Emotion
18. I suffered from fatigue during my interaction with the virtual environment	Experience consequence
19. I suffered from vertigo during my interaction with the virtual environment	Experience consequence
20. Learning to operate the virtual environment would be easy for me	Technology adoption
21. The interaction devices (HTC headset, controller handle and/or stylus) would make work more interesting	Technology adoption

References

1. Pan, Z., Chen, W., Zhang, M., Liu, J., Wu, G.: Virtual reality in the digital Olympic museum. IEEE Comput. Graph. Appl. **29**(5), 91–95 (2009)
2. Yoon, S.A, Elinich, K., Wang, J., Van Schooneveld, J.G.: Augmented reality in the science museum: lessons learned in scaffolding for conceptual and cognitive learning. Int. Assoc. Dev. Inf. Soc. (2012)
3. Bedini, S.A.: The evolution of science museums. Technol. Cult. **6**(1), 1–29 (1965)
4. Guo, H.: Research and Application of Virtual Reality Technology in Popular Science Education [D]. Hubei University of Technology, Wu han (2016). (in Chinese)
5. Yoon, S.A., Wang, J.: Making the invisible visible in science museums through augmented reality devices. TechTrends **58**(1), 49–55 (2014)
6. Carrozzino, M., Bergamasco, M.: Beyond virtual museums: experiencing immersive virtual reality in real museums. J. Cult. Herit. **11**(4), 452–458 (2010)
7. Li, L., Zhou, J.: Virtual reality technology based developmental designs of multiplayer-interaction-supporting exhibits of science museums: taking the exhibit of" virtual experience on an aircraft carrier" in China science and technology museum as an example. In: Proceedings of the 15th ACM SIGGRAPH Conference on Virtual-Reality Continuum and Its Applications in Industry, vol. 1, pp. 409–412. (2016)
8. De Freitas, S., Neumann, T.: The use of 'exploratory learning' for supporting immersive learning in virtual environments. Comput. Educ. **52**(2), 343–352 (2009)
9. Ummihusna, A., Zairul, M.: Investigating immersive learning technology intervention in architecture education: a systematic literature review. J. Appl. Res. High. Educ. (2021)
10. Dengel, A., Mägdefrau, J.: Immersive learning predicted: presence, prior knowledge, and school performance influence learning outcomes in immersive educational virtual environments. In: 2020 6th International Conference of the Immersive Learning Research Network (iLRN), pp. 163–170. IEEE (2020)
11. Hirose, M.: Virtual reality technology and museum exhibit. In: International Conference on Virtual Storytelling, pp. 3–11. Springer, Berlin, Heidelberg (2005)
12. Jung, T., Tom Dieck, M.C., Lee, H., Chung, N.: Effects of virtual reality and augmented reality on visitor experiences in museum. In: Information and Communication Technologies in Tourism, pp. 621–635. Springer, Cham (2016)
13. Podgorny, J.: Studying visitor engagement in Virtual Reality based children's science museum exhibits. Unpublished master dissertation (2004). University Of Chicago, Chicago, IL
14. Zhou, R., Huang, Y., Ding, X.: Construction and counter measures of the science communication model of science and technology museum based on the immersive media [J]. J. Appl. Res. High. Educ. **1**(03), 22–27 (2016). (in Chinese)
15. Zhou, J., Xu, L., Liu, N., Gu, J., Wang, Y.: The application of Virtual Reality in science education: a case study of the insect VR video series in Shanghai natural history museum [J]. Sci. Educ. Mus. **5**(03), 208–213 (2019). (in Chinese)
16. Limbasiya, H.: Sense Simulation in Virtual Reality to Increase: Immersion, Presence, and Interactions. University of Dublin (2018)
17. Beck, D.: Augmented and virtual reality in education: immersive learning research. J. Educ. Comput. Res. **57**(7), 1619–1625 (2019)
18. VWN Virtual Dictionary: Sensory Immersion (2020)
19. Gibson, J.J., Carmichael, L.: The Senses Considered as Perceptual Systems, vol. 2, No. 1, pp. 44–73. Houghton Mifflin Boston (1966)
20. Slater, M., Usoh, M.: Representations systems, perceptual position, and presence in immersive virtual environments. Presence: Teleoperators Virtual Environ. **2**(3), 221–233 (1993)
21. Gibson, J.J.: A theory of direct visual perception. Vision and Mind: Selected Readings in the Philosophy of Perception, pp. 77–90 (2002)
22. Seng, D., Wang, H.: Realistic real-time rendering of 3D terrain scenes based on OpenGL. In: 2009 First International Conference on Information Science and Engineering, pp. 2121–2124. IEEE (2009)

23. Västfjäll, D.: The subjective sense of presence, emotion recognition, and experienced emotions in auditory virtual environments. Cyberpsychol. Behav. **6**(2), 181–188 (2003)
24. Hendrix, C., Barfield, W.: The sense of presence within auditory virtual environments. Presence: Teleoperators Virtual Environ. **5**(3), 290–301 (1996)
25. Lederman, S.J., Klatzky, R.L.: Haptic perception: a tutorial. Atten. Percept. Psychophys. **71**(7), 1439–1459 (2009)
26. Sreelakshmi, M., Subash, T.D.: Haptic technology: a comprehensive review on its applications and future prospects [J]. Mater Today: Proc. **4**(2), 4182–4187 (2017)
27. Kim, M., Jeon, C., Kim, J.: A study on immersion and presence of a portable hand haptic system for immersive virtual reality. Sensors **17**(5), 1141 (2017)
28. Loftin, R.B., Brooks, Jr. F.P., Dede, C.: Virtual reality in education: promise and reality. In: Virtual Reality Annual International Symposium, pp. 208–208. IEEE Computer Society (1998)
29. Daassi, M., Debbabi, S.: Intention to reuse AR-based apps: the combined role of the sense of immersion, product presence and perceived realism. Inf. Manag. **58**(4), 103453 (2021)
30. Yaqi, X., Lau, Y., Cheng, L.J., Lau, S.T.: Learning experiences of game-based educational intervention in nursing students: a systematic mixed-studies review. Nurse Educ. Today 105139 (2021)
31. Pasch, M., Bianchi-Berthouze, N., Van Dijk, B., Nijholt, A.: Immersion in movement-based interaction. In: International Conference on Intelligent Technologies for Interactive Entertainment, pp. 169–180. Springer, Berlin, Heidelberg (2009)
32. İbili, E.: Effect of augmented reality environments on cognitive load: pedagogical effect, instructional design, motivation and interaction interfaces. Int. J. Progress. Educ. **15**(5), 42–57 (2019)
33. Tcha-Tokey, K., Christmann, O., Loup-Escande, E., et al.: Proposition and validation of a questionnaire to measure the user experience in immersive virtual environments [J]. Int. J. Virtual Rity. **16**, 33–48 (2016)
34. Bouta, H., Retalis, S., Paraskeva, F.: Utilizing a collaborative macro-script to enhance student engagement: a mixed method study in a 3D virtual environment [J]. Comput. Educ. **58**(1), 501–517 (2012)
35. Witmer, B.G., Singer, M.J.: Measuring presence in virtual environments: a presence questionarire [J]. Presence **7**(3), 225–240 (1998)
36. Cadet, L.B., Chainay, H.: Memory of virtual experiences: role of immersion, emotion and sense of presence. Int. J. Hum. Comput. Stud. **144**, 102506 (2020)
37. Allcoat, D., von Mühlenen, A.: Learning in virtual reality: effects on performance, emotion and engagement. Res. Learn. Technol. 26 (2018)
38. Bangor, A., Kortum, P.T., Miller, J.T.: An empirical evaluation of the system usability scale. Int. J. Hum. Comput. Interact. **24**(6), 574–594 (2008)
39. Weech, S., Kenny, S., Barnett-Cowan, M.: Presence and cybersickness in virtual reality are negatively related: a review. Front. Psychol. **10**, 158 (2019)
40. Sagnier, C., Loup-Escande, E., Valléry, G.: Effects of gender and prior experience in immersive user experience with virtual reality. In: International Conference on Applied Human Factors and Ergonomics, pp. 305–314. Springer, Cham (2019)
41. Schaufeli, W.B.: What is engagement? In: Employee Engagement in Theory and Practice, pp. 29–49. Routledge (2013)
42. Cummings, J.J., Bailenson, J.N.: How immersive is enough? A meta-analysis of the effect of immersive technology on user presence. Media Psychol. **19**, 272–309 (2016)
43. Servotte, J.C., Goosse, M., Campbell, S.H., Dardenne, N., Pilote, B., Simoneau, I.L., Ghuysen, A.: Virtual reality experience: immersion, sense of presence, and cybersickness. Clin. Simul. Nurs. **38**, 35–43 (2020)
44. Rebenitsch, L., Owen, C.: Review on cybersickness in applications and visual displays. Virtual. Rity. **20**(2), 101–125 (2016)
45. Melo, M., Vasconcelos-Raposo, J., Bessa, M.: Presence and cybersickness in immersive content: effects of content type, exposure time and gender. Comput. Graph. **71**, 159–165 (2018)

The manufacturer's authorised representative in the EU is Springer Nature Customer Service Centre GmbH, Europaplatz 3, 69115 Heidelberg, Germany. If you have any concerns regarding our products, please contact ProductSafety@springernature.com

Printed and bound by CPI Group (UK) Ltd, Croydon, CR0 4YY
01/10/2026
02225868-0013